The Ecology of
Fossils

The Ecology of Fossils

an illustrated guide

edited by W. S. McKerrow

The MIT Press

Cambridge, Massachusetts

First MIT Press paperback edition, 1981

First MIT Press hardcover edition, 1978

First published in England in 1978 by
Duckworth & Co. Ltd.

This is an Alphabook,
designed and produced by Alphabet and Image
Sherborne, Dorset, England.

Community reconstructions drawn by Elizabeth Winson.
Cartographic reconstructions drawn by Peter Deussen.

Library of Congress catalog card number: 78–52250

ISBN 0–262–13144–7 (hard)
     0–262–63086–9 (paper)

Printed by Hillman (Printers) Ltd, Frome, England.

Contents

Contents

Permian 184

W. H. C. Ramsbottom

57 Barrier Reef Top Community
58 Back-reef Community
59 Hypersaline Landlocked Basin Community
60 Hypersaline Sea Community

Triassic 194

B. W. Sellwood B.Sc., D.Phil.
Department of Geology, Reading University

61 Triassic Lagoon Scene
62 Late Triassic Tidal-flats

Jurassic 204

B. W. Sellwood

63 Bituminous Mud Community
64 Restricted Clay Community
65 Silty Clay Community
66 Muddy Sand Community
67 Sand Community
68 Condensed Limestone Community
69 Ironstone Community
70 Calcarenite Community
71 Hardground Community
72 Carbonate Mud Community
73 Freshwater Communities
74 Restricted Marine Clay Community
75 Communities in Ironstones and Marine Sands
76 Oolitic Limestone Communities
77 Low Diversity Temporarily Stable Calcarenite Community
78 Diverse Calcarenite Community
79 Shelly Lime Mud Community
80 Muddy Lime Sand Community
81 Clear Water, Firm Substrate Communities
82 Lime Mud Coral Community
83 Muddy Sea Floor Communities
84 Marine Sand and Muddy Sand Communities
85 Calcarenite Community
86 Oyster-algal Bioherms Community
87 Coral-algal Patch-reef Community
    Oyster Lumachelle Community
88 & 89 Intertidal and Subtidal Algal Mat Communities
90 Freshwater Lagoon Communities

Cretaceous 280

W. J. Kennedy M.A., Ph.D.
Department of Geology and Mineralogy, Oxford University

91 Lower Cretaceous Terrestrial Communities
92 Lower Cretaceous Lake Communities

Contents

Cenozoic   323

J. Taylor B.Sc., Ph.D.
Department of Zoology, British Museum (Natural History)

Present day   352

J. Taylor

Introduction

Palaeontology is the study of fossils, the remains of past life preserved in old sediments. Throughout geological time, there have been marked changes in the dominant animal and plant groups, while the total range of physical environments has changed very little.

In some instances, the changes in the fauna and flora are the result of progressive evolutionary changes in particular fossil lineages, but much more commonly the fossil record shows that many organisms become widespread soon after they first appear; they may then flourish for a while before they become extinct and are replaced by another form. This changing pattern of forms of life in similar habitats is the basis for this book.

In a small village in western Newfoundland, I used to play cards with the schoolmaster's father in the evenings. I am sure he was the most intelligent man in the village. After I had been there for some time, he plucked up enough courage to ask me why I was not looking at the rocks up on the ridge where the old gold mines were. I told him I was not looking for gold, but (reaching into my bag and pulling out a slab of rock containing brachiopods) "I was looking for these". He studied the slab carefully and said, "Are these sea shells?" He had never seen a fossil before, nor even heard of fossils. This question reminded me of the best of the early eighteenth century palaeontologists. But my friend went on: "Does that mean the sea was once up here?" When I had replied "Yes", he then asked: "How long ago would that be then?" He clearly had all the makings of a first-class scientist. I replied, with a perfectly straight face, "About 450 million years ago", but I was not at all prepared for his final remark, "You'll be an atheist then".

This conversation demonstrates the type of questions asked by contemporary palaeontologists: Are these fossils marine? When and how did these animals live? What factors controlled where they lived?, and, though not really within the scope of this book, Why do they each have characteristic structures and shapes?

Plenty of books have been written on fossils, but in this book fossil assemblages, rather than individual genera or species, are the units of study. Only a selection of these is illustrated. Many fossils are not mentioned, for this book is not a treatise. Its main purpose is to educate and stimulate, rather than to describe every known fossil. We hope it will make the reader think about fossils in new ways, and lead to a better understanding of past life. Nowadays, most major fossil groups have been described (see, for example, the *Treatise on Invertebrate Paleontology*, (Ed.) R.C. Moore and C. Teichert, and *Vertebrate Paleontology* by A. S. Romer) and the most exciting and significant advances in palaeontology are being made by considering fossils as animals, and not just as formed stones. The study of palaeoecology includes attempts at reconstructing the relationships between organisms and sediments and the interactions between organisms, including those with no close relatives now living. This has direct implications for palaeogeography, stratigraphy, and certain aspects of structural geology (like the development and evolution of sedimentary basins, and the recognition of old oceans). In addition, the knowledge of past environments and their inhabitants has important contributions to make to evolutionary biology and to the whole history of the Earth.

The selection of communities illustrated in this book has been largely based on those found in the British Isles, but the range is such that most marine fossil communities throughout the world have some close parallel with a community illustrated here.

COMMUNITIES

A community is a group of organisms living together. The term is sometimes applied to organisms which depend on each other in some way (such as for food, or for protection), but in palaeontology it is normally only applied to organisms living in the same habitat.

At any one time there are different communities of animals living in different environments. The study of animals in relation to each other and to their habitats is known as ecology. Modern communities can be examined to determine not only where animals and plants live, but how they interact, especially in providing food and protection for each other. In looking at fossil associations, the palaeoecologist can describe those forms which occur together, but he can only speculate as to their interactions. Biologists argue as to whether a community should be defined according to the animals' habitat or according to their interactions. There is no doubt that, in the study of fossils, only the habitat definition can be used. The definition of "Community" used throughout this book is thus: a group of animals living in the same habitat.

The communities in this book are all known to be recurring

assemblages; they are not just random collections of fossils. Some of the faunas illustrated have been analysed statistically in great detail, but many others have only been qualitatively assessed. To some extent the diagrams must be considered as cartoons in that they emphasize those elements which are most easily preserved as fossils (by having hard parts or recognizable burrows). The soft parts of extinct organisms are, of course, conjectural. The reconstructions presented here are based on the best information available, but they are deduced largely from similar organisms alive today, so there is a considerable element of informed guesswork in many of the illustrations.

It may well be that further work will indicate the presence of other animals and plants in these communities. On the other hand, it may eventually be seen that some of the communities described in this book in fact have members which occupied slightly different habitats. In the geological record, it is not always easy to separate fossils which occur together in a deposit, but which lived at slightly different times. Most fossil collections include those animals which have lived in one area over a period of several years, or possibly several centuries. So, even if the shells have not been transported after death, there may be a greater variety of animals on a square metre of a bedding plane than could have existed at any one time while that bedding plane was exposed on the sea floor.

It should also be stressed that the communities illustrated have been selected either because they are the most commonly occurring or because they are of particular interest. The vast majority of ancient communities still await description, and other palaeontologists would certainly make a different selection.

Modern aquatic organisms can be classified according to where they live (Figure a). Those which float or drift are known as plankton and are dominantly pelagic, that is, they live mostly on or near the surface of the sea. Actively swimming animals, like many fish and cephalopods, are called nekton. Bottom-dwelling organisms (benthos) can be placed in one of two categories: epifauna which live on the substrate, which may be soft sediment, rock or vegetation, and infauna which live within the substrate (burrowers or borers).

Plankton can be subdivided according to size (microplankton are those organisms which can only be studied under a microscope); they can also be classified according to their methods of obtaining energy: whether they manufacture their foodstuffs by photosynthesis (like most plants) or whether they feed on organisms or organic debris (like most animals). In many simple small organisms, the difference between the animal and plant kingdoms becomes blurred. In groups of apparently closely related organisms, some may employ photosynthesis and some not, and

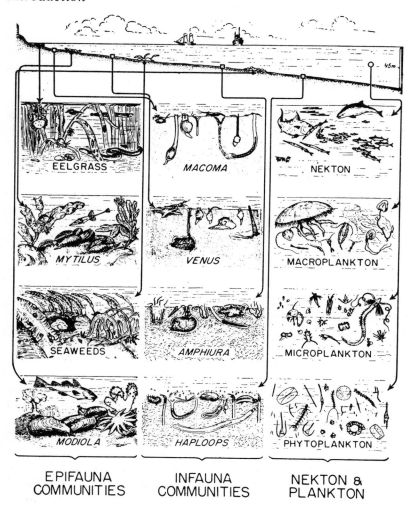

EELGRASS

MACOMA

NEKTON

MYTILUS

VENUS

MACROPLANKTON

SEAWEEDS

AMPHIURA

MICROPLANKTON

MODIOLA

HAPLOOPS

PHYTOPLANKTON

EPIFAUNA
COMMUNITIES

INFAUNA
COMMUNITIES

NEKTON &
PLANKTON

Fig. a. *Modern animal communities between Denmark and Sweden (from Hedgpeth, 1957, p. 31 with permission).*

in common with other recent publications (e.g. *Treatise on Invertebrate Paleontology*) we refer these simple organisms to a third kingdom: Protista.

There are some problems in our definition of a community. For example, are the animals living above (or below) the sea floor to be included in the same habitat as those on the sediment surface? It is quite reasonable to argue that they are not, but for the purpose of this book, we frequently show free-swimming animals in the same diagram as the benthos on the floor below. However, in Figure a (above) the epifaunal and infaunal organisms are shown as separate communities.

Within a single habitat there can be many niches. For example, on a particular patch of sea floor there may exist, side by side, epifaunal suspension-feeding bivalves attached by byssal threads to shell fragments on the sea floor, infaunal burrowing deposit-feeding crustaceans, epifaunal suspension-feeding serpulid worms encrusting some of the bivalves, and an epifaunal carnivorous gastropod boring the bivalves with its rasp-like radula. Each of these animals is in a distinct niche within the same habitat.

In many fossil assemblages, the niche occupied by a particular organism can often be determined by a comparison with present day animal and plant communities, but further back in time we see fewer and fewer close relatives of living organisms, and other criteria (like the type of sediments or geographical distributions) have to be used to determine the ecology of these ancient assemblages. In some cases geologists have produced very convincing evidence for accurate interpretations of ancient environments, but in other instances there is still much uncertainty.

Many modern marine communities have sharp boundaries. At the margin of a rock outcrop in the sea, the mussels, limpets and winkles (*Mytilus, Patella* and *Littorina*) living on the rock will suddenly give way to sand-dwelling cockles and heart urchins (*Cardium* and *Echinocardium*). There is no difficulty in recognizing a different bottom-dwelling community in a case like this. A change from clean sand to muddy sand will allow a great increase in deposit feeders and a marked reduction in sessile epifauna. Here again, though the change may be more gradual, there is a reasonable boundary between the two distinct habitat communities.

Many fossil collections from Palaeozoic rocks consist largely of suspension-feeding brachiopods, and thus contain very few fossils which were not suspension feeders. Some beds show burrows (perhaps made by annelid worms or arthropods) and there must have been carnivores or scavengers around, but their fossil record is very poor by comparison with that of the brachiopods. Five Silurian communities have been described (Ziegler, 1965) living on the sea floor between the coast and the deep sea. A single sample from each community is very distinctive, but when a large number of samples are made gradations between most of the communities become evident, and some arbitrary decisions, depending on the proportions present of certain selected species, have to be made to define the community boundaries. We still do not yet know in many cases which fossil communities are gradational and which are not, but it must be borne in mind that those communities described here (especially those with a high proportion of suspension feeders) may be arbitrary points on an ecological gradient.

ECOLOGICAL CONTROLS

In the sea, the factors controlling the distribution of organisms can be classified as:

1. Biological: food supply, competition from other organisms, and sometimes protection by or dependence on other organisms.

2. Physical: the sedimentary environment, turbulence, currents, temperature, light intensity.

3. Chemical: salinity, water and substrate chemistry.

Some of these controls are related to depth (light, turbulence, temperature fluctuations and suspended organic matter all show a general decrease with depth), but hydrostatic pressure only has marked effects in the deepest environments, so depth alone is not a significant control.

Latitude is closely related to temperature, but temperature can vary according to the distribution of oceanic currents. Many marine organisms are restricted by temperature and other factors which may change with latitude, but latitude by itself is not a primary control.

Every animal needs to feed, and different methods of feeding can restrict animals to certain habitats. Marine animals (and it is these we are chiefly concerned with in this book) can be classified into four types of feeders (Raup and Stanley, 1971):

1. Grazers, which remove algal and other encrusting organic material from rocks and other hard surfaces.

2. Deposit feeders, which ingest sediment and feed on the organic matter coating the grains or mixed with the grains.

3. Suspension feeders, which select organic matter suspended in water.

4. Carnivores and scavengers, which eat other animals alive or dead.

Grazers normally inhabit rocky areas which are coated by organic films, they are thus generally confined to shallow water areas where bare rock is exposed, but they can occur in deeper water if suitable substrates are present.

Deposit feeders are often burrowers in muddy sediments, especially if the mud is mixed with a little sand or silt so that stable burrow systems can be developed. Organic matter in sediments normally occurs as a thin coating on individual grains of sediment. With decreasing grain size, the proportion of surface area in a given volume will increase. Thus fine grained sediments contain more organic matter and can support greater densities of deposit feeders. Fine grained sediments can occur in all depths of water.

The infauna are protected by their burrows; many of them have no external hard parts, and those that do often have thinner shells (or other external covering) than the epifauna exposed above the substrate. Many burrowing animals (especially worms and some decapod crustaceans) are thus rare as fossils although their burrow systems may be quite common. Deposit feeders often have characteristic complex burrow systems, which are developed as they search for food (these are in marked contrast to the simple burrows produced by infaunal suspension feeders).

Suspension feeders may be epifaunal, infaunal or pelagic. They include sedentary animals like corals, bryozoans, brachiopods, many bivalves and the crinoids, all of which have well-developed hard parts and are thus common as fossils. Their main food supply is probably diatoms and other protists which rely on photosynthesis, and which can thus only develop in abundance very near the surface of the sea (Ryther, 1963). The pelagic larval stages of many marine invertebrates also feed on these protists and live with them near the surface of the oceans. These larvae are also an important part of the marine food chain. Much more of this (protist and larval) food supply is thus available on the sea floor in shallow water areas; these are the areas, both today and in the Palaeozoic, where the majority of suspension feeders live.

Although well endowed with food supplies, the shallow water faunas have to contend with high environmental stresses, including much variation in sedimentary deposition rates, the unsettling effects of storms, and (except in low latitudes) fluctuations in temperature. So the only places for a quiet life (in a stable environment) are those where food is scarce. In the deep sea, there are sparse, but diverse, populations of specialized benthos. It is possible that some epifauna (e.g. spiriferide brachiopods) developed especially efficient feeding systems to cope with a limited food supply in these deep bottom environments.

Suspension-feeding epifauna usually require a stable anchorage, and are thus more common (at least since the Mesozoic) on stable sands and silts rather than on muds. Moreover, a large mud supply would tend to choke the filter-feeding structures of many bivalves. It has recently been suggested (Steele-Petrović, 1975) that some brachiopods can tolerate a high mud flow through their feeding structures. This may account for the fact that in the Palaeozoic, when brachiopods were a major part of these epifaunal communities, there appears to have been a much lower correlation between an epifaunal community and the sediment on which it rested than there is at present time (Ziegler, 1965).

Rocky bottom communities consist mainly of suspension feeders and grazers; they are a mixture of epifauna and of infaunal boring animals. In the geological record they are restricted to beds immediately above depositional breaks and to hardgrounds. As far as we understand them at present, the animals inhabiting rocky bottoms have always formed quite distinct communities from those on areas of unconsolidated sediments.

Carnivores and scavengers feed on other animals. They can thus never be as abundant as suspension feeders and deposit feeders. Today, certain worms, echinoderms, gastropods and fish are the most important carnivores on the sea floor; similar forms are also present as far back as the Mesozoic. The direct evidence for early carnivores is very meagre; before fish became common in the sea and before gastropods developed radulae, soft-bodied worms, arthropods, medusoids and sea-anemones appear to have been the most probable carnivores and scavengers. If a carnivore species becomes specialized in its choice of food, it will naturally become a member of the animal community in the habitat where its food is situated. Unspecialized carnivores may, on the other hand, feed on members of several communities and thus extend over broader areas than any one species of the suspension feeders or deposit feeders on which they prey.

It can be concluded from the above discussion that Mesozoic, Tertiary and modern bottom-dwelling communities are more intimately linked with sediment type than those in the Palaeozoic (except for some early Palaeozoic communities which were dominated by the deposit-feeding trilobites). All communities appear to be influenced to some extent by depth of water, but this influence has nothing to do with hydrostatic pressure. There is no question of depth control, but depth is correlated with greater food supply in shallow environments, and with uniform conditions on deeper sea floors. It will be observed that in this book the community names reflect this change with time. Those from the Ordovician to the Carboniferous are often named after a characteristic genus or species, while those in the Cambrian and from the Jurassic onwards are normally named after the habitat in which they occur.

Plankton and nekton may occur in all depths of water. The chief controls which govern their distribution are temperature, salinity and water currents. Their abundance can vary greatly with the distribution of dissolved nutrients in the water. Many migrate long distances during their life span, and seasonal changes in distribution are common in many groups. From the palaeontological point of view, plankton and nekton are not always very reliable indicators of the environment in which they are found fossil; many may drift for long distances after death. For example, *Nautilus* shells have been recovered from Madagascar and Japan, though living forms are only known between Australia, the Philippines and the Fiji Islands.

Salinity is largely independent of water depth or sediment. In most open seas the salinity is nearly constant at about 35 parts of dissolved salts to a thousand parts of water. But in isolated or semi-isolated seas and lagoons this may change a great deal: much of the Baltic Sea, for example, has a salinity of less than 10 parts per thousand, while in areas subject to high evaporation rates the salinity may increase to over 40 parts per thousand or much more.

When there is a change in salinity (either upwards or downwards) only the more tolerant (euryhaline) species survive. Most marine species are not very tolerant of changes in salinity (they are stenohaline), so that the effect of any marked change is reflected by a reduction in the number of species present. But this does not necessarily mean that there are fewer animals; many brackish water regions contain an abundance of individuals of a very few species. It is often possible to determine salinity tolerance in extinct species. If a sedimentary formation changes from normal marine deposits into brackish water deposits, a series of fossil collections may then show a progressive reduction in the number of species present. In the Middle Jurassic of England, for instance, it is possible to deduce that ammonites are among the most stenohaline animals, and that certain oysters are among the most euryhaline.

In addition to the marine faunas, there are some aquatic organisms which have become adapted exclusively to fresh water. These are usually stenohaline, that is, they are restricted to fresh water; but some species are euryhaline and can extend their range into brackish environments, where they may occur with euryhaline marine species.

STRATIGRAPHIC NOMENCLATURE

With the development of stratigraphy, it has been found useful to develop a standard method of nomenclature. The Cambrian **System** consists of rocks deposited during the Cambrian **Period**. The system consists of rocks; while the period is the time during which these rocks were formed. Both terms have an age connotation.

Systems are recognized and defined by the fossils they contain. If the fossils within a system can be shown to change regularly with time, systems can be subdivided into series, stages and zones, each of which represent the rocks laid down during shorter intervals of time respectively (as recognized by the fossils).

The systems are grouped together into eras. After the Precambrian, there are three eras: Palaeozoic, Mesozoic and Cenozoic (Table 1). The Cenozoic Era can be divided into the Tertiary (Palaeocene to Pliocene) and the Quaternary.

17

Introduction

Table I

Age (in millions of years) to base of:

CENOZOIC	Quaternary	2
	Pliocene	5
	Miocene	26
	Oligocene	38
	Eocene	55
	Palaeocene	67
MESOZOIC	Cretaceous	138
	Jurassic	200
	Triassic	245
PALAEOZOIC	Permian	285
	Carboniferous	360
	Devonian	411
	Silurian	438
	Ordovician	519
	Cambrian	570

PRECAMBRIAN

In addition to the above terms, which all have a time connotation, geologists have developed a hierarchy of rock-types (lithostratigraphical) terms for bodies of rock which have distinctive characters, but which may not necessarily all be of the same age everywhere they occur. Groups, formations and members are the terms used. For example, the Swerford Member is a sand unit in the Chipping Norton Formation of Oxfordshire; the Chipping Norton Formation is a unit of sands and limestones which occurs at the base of the Great Oolite Group of central England; and the Great Oolite Group can be recognized throughout Britain and northern France as a group of limestones, clays and sands occurring below the Oxford Clay Group.

Fossil sequences do not always reflect time. If, in a particular region, the sea was getting progressively deeper, a succession of fossil communities will be present which reflect progressively deeper environments. And if deep water conditions occur in some parts of the region earlier than in other parts, the deep water fauna

18

will appear earlier in these areas. Before the development of palaeo-ecology in recent years, this progression of communities was not often recognized, and many zonal schemes based on benthic fossils have had to be revised. On the other hand, zones based on pelagic animals (graptolites, ammonites and globigerinid foraminifera) have stood the test of time very well; slowly migrating communities do not pose the same problems when zones are based on these free-swimming and drifting animals. The revision of zonal schemes based on benthic animals involves a search for changes that are independent of changing environments. These can be found by examining the evolution of particular genera. Many (possibly most) benthic lineages appear to change very little once they have been established. But evolution with time can be demonstrated in some benthic lineages (for instance, the progressively weaker ribs in the Silurian brachiopod *Eocoelia* discussed in the Silurian chapter) and in several pelagic groups (like the Mesozoic ammonites).

AN OUTLINE OF THE FOSSIL RECORD

Radioactive isotopes of rubidium and uranium decay slowly through time, and measurements of very small quantities of these isotopes show that some igneous rocks in western Greenland are about 3,800 million years old (Moorbath, 1975). These are the oldest dated rocks on the Earth's surface. Some inclusions in these very old rocks appear to have been derived from sedimentary rocks, suggesting that water has existed on the Earth's surface for more than 3,800 million years. Other isotopic work suggests that the Earth's crust is about 4,600 million years old. If this estimate is correct, the first 800 million years of the Earth's history have not yet been deduced from the rocks on its surface.

The oldest fossils known occur in rocks of about 3,400 million years old; they appear to be microscopic prokaryotes. Later in these Precambrian times (at about 2,700 million years) layered larger, dome-shaped structures (stromatolites) appear, which were formed by blue-green algae. Apart from a few burrows, possibly made by worms, these are the only fossil groups known prior to about 700 million years before the present. Thus, during the greater part of these Precambrian times, it is stromatolites which dominate the fossil record.

During the later part of the Precambrian (about 700 to 570 million years) the impressions of some soft-bodied coelenterates and annelid worms occur (the chief characteristics of the common animals and plant fossils are set out in the Classification of Organisms on page 30). Thus animals existed for some considerable time before they developed hard parts at the base of the Cambrian.

Introduction

The Cambrian system was first named by Adam Sedgwick after the rocks of Wales, which he mapped and described during the nineteenth century. Like all systems, Cambrian beds can be recognized by the fossils which they contain. The base of the Cambrian is usually defined by the first appearance of trilobites, hyolithids or archaeocyathids.

The Ordovician system was first defined by Charles Lapworth (1879), who had to clear up a controversy between Sedgwick and Sir Roderick Murchison. The original Cambrian system of Sedgwick contained many beds which were included by Murchison in his Silurian system. To resolve the overlap, Lapworth introduced the Ordovician (like the Silurian, it was named after an old British tribe). He defined it as the beds lying between the base of the Arenig Series and the base of the Llandovery Series; and he thus restricted the Silurian to the younger beds of Murchison's original system, and the Cambrian to the older beds of Sedgwick's system.

To the geologist working in Wales, Lapworth's new Ordovician system was easy to recognize, as unconformities exist above and below it. These mark local episodes of uplift and folding, but, fortunately, there were also faunal changes which can be detected at many sites around the world. Not only did several new trilobite families appear at the base of the Ordovician, but also the first families of bryozoans, the first strophomenide brachiopods and some other new brachiopod families, the first crinoids, the first graptoloids (excluding the dendroids), and the first agnathan fish.

During the Ordovician the brachiopods became steadily more abundant relative to the trilobites, and, except in some carbonate environments (where colonial corals were making their first appearance or where algae were common), the brachiopods dominated most Silurian shelf communities.

Three main types of biological change occur near the base of the Silurian:

1. Many Ordovician shelf animals became extinct; some of this may be attributable directly to the lowering of sea level because of the formation of ice caps.

2. Most Silurian faunas were cosmopolitan. The provincialism seen in the Ordovician disappeared except for a polar fauna in parts in Gondwanaland. When the faunas mixed, competition increased, so that some well adapted animals increased their geographical range, while others died out.

3. In the Lower Silurian, when the shelf seas spread over the continental margins (and in some cases over whole continents), many habitats became available for recolonization. The shallow communities at this time showed many large changes in composition; some habitats were occupied by several different genera in succession.

In eastern Wales and western England (where Murchison did his most important work), the base of the Silurian is marked by an unconformity. It now appears that perhaps all shallow water sequences have a break in sedimentation at this time. Certainly in the better studied sections of Europe (Ziegler, Rickards and McKerrow, 1974) and North America (Berry and Boucot, 1970) there appears to be no continuous sedimentation from the Ordovician into the Silurian within a shallow shelf sequence. But these breaks do not mean that there was a widespread folding at the end of the Ordovician, for most deep water sequences show a continuous succession of beds. A likely explanation for these observations is to be found in Africa and South America, where terrestrial glacial deposits are present over a wide area (Sheehan, 1973). A major Ashgill ice age could account for a temporary lowering of sea level sufficient to produce the break seen in the basal Silurian shallow water areas. The lowering of sea level was, in places, accompanied by the erosion of a well marked shelf. Certainly in the British Lower Palaeozoic it is only in the Lower Silurian that a sharp shelf margin can be detected (Hancock et al. 1974).

In England, the marine Silurian beds are followed by the river and brackish water deposits of the Old Red Sandstone. Murchison had doubts as to where to draw the upper boundary of the Silurian, and there has been debate about the matter ever since. It would now appear that non-marine environments developed in several parts of Britain before the end of the Ludlow Series, and the transition into these can certainly not be used to mark a synchronous event. Moreover, work in Poland and Czechoslovakia has recently shown that there is a succession of beds in many areas which are younger than the Ludlow but older than the basal Devonian of the Ardennes (currently the area where the base is defined). We follow Berry and Boucot (1970) in recognizing the Pridoli as the top series in the Silurian system. The Downtonian beds, which contain freshwater fish and have often been mapped as part of the Old Red Sandstone, are of Pridoli age and are thus included in the Silurian.

It was during the latest Silurian that the first land plants (psilophytes) developed, followed almost at once by myriapods and insects. By the Upper Devonian the amphibia had also colonized low-lying terrestrial semi-aquatic environments.

Devonian benthic marine animals were not very different from those in the Silurian, but among the brachiopods spiriferides became more common, and productoids and terebratulides appeared for the first time. A more significant change is seen in the pelagic faunas: *Monograptus* became extinct just before the end of the Lower Devonian. This extinction was nearly coincident with the rise of the ammonoids. The earliest true ammonoid families developed in the Lower and Middle Devonian; from then until the

Late Carboniferous these ammonoids serve as the best zonal indicators in stratigraphy.

The base of the Carboniferous system, although named on account of its coals, was originally defined by the return of the sea over much of the Old Red Sandstone continent of north-west Europe. Like most of such large marine transgressions, the sea did not spread everywhere at once. The base of the system is now defined by goniatitic ammonoids in marine facies, though in many freshwater successions it is still not possible to place the boundary precisely. In many marine environments, the most common fossils were brachiopods (especially productoids and spiriferides), crinoids and rugose corals. Among the echinoderms, not only did the crinoids flourish, but the blastoids too reached their acme. Marine bivalves became more common.

The Carboniferous saw the spread of deltaic conditions over a large part of northern Europe and North America, which resulted at times in widespread brackish water and freshwater conditions. The non-marine faunas included bivalves (myalinids and anthracosiids), branchiopods and fish; and the marginal marine faunas included lingulids, arenaceous foraminifera and worms. Coal-producing forests occurred throughout Carboniferous times, but they were more common in the later parts of the period; the commonest plants were lycopods and horsetails, but some gymnosperms were also present. In the Late Carboniferous and Permian, fusulinid foraminifera became important in marine environments; they were often a major component of the fauna and they provide good zonal indices for this part of the stratigraphic column.

The Late Carboniferous mountain-building episode (the Hercynian orogeny) has been interpreted as the result of southern Europe colliding with northern Europe (McKerrow and Ziegler, 1972). The westward continuation of this fold belt lies in the southern Appalachians which is considered to be the result of a collision between Africa and North America. The Uralian orogeny also took place in the Permian, uniting Siberia with the Russian platform. The consequence was that, by the end of the Permian, the largest continents of the world were fused to form Pangea, a single supercontinent (Briden et al. 1974). This large continent was only sporadically covered by shelf seas. Those seas near the margins of Pangea had normal marine faunas, for example those covering the Perm (later called Molotov) region north-east of Moscow, parts of southern Europe, Mexico and western America, Indonesia and the Himalayas, while inland seas, like the Zechstein Sea of northern Europe, only had euryhaline species which could tolerate the increase in salinity. Bryozoans, bivalves and productoids were among the commonest of these animals that could tolerate life in the Zechstein Sea. The normal marine forms were mostly descendants of the same groups that were present in the

Carboniferous.

The Late Permian or Early Triassic saw the extinction of the following marine animals:

1. Many families of foraminifera, including the fusulinids.

2. The tabulate and rugose corals.

3. The cystoporate, trepostomate and most of the cryptostomate bryozoans.

4. Orthide brachiopods, and many strophomenides and spiriferides.

5. Many ammonoid families.

6. All the trilobites, and many ostracode families.

7. The blastoids and many crinoid families.

8. Many fish families, including acanthodians and primitive crossopterygians.

Most of these groups gradually decreased in numbers and diversity during the Permian, but some appear to have died out suddenly at the very end of the period (Harland et al. 1967; Kummel and Teichert, 1970). Some groups became very restricted geographically before they became extinct.

The reasons for this massive extinction of faunas was perhaps due to increase in competition. By the end of the Permian, shelf seas only extended over about fifteen per cent of the area covered in the Early Permian. The geographic extent of shallow marine benthic species therefore would have become reduced, and the probability of their extinction increased (Schopf, 1974).

On land much, but not all, of the present northern hemisphere was desert, while, by contrast, India and parts of the present day southern hemisphere (Gondwanaland) were, at least at times, covered by ice. Glaciation started in the Early Carboniferous of South America and Africa, but did not reach India or Australia until the late Carboniferous; and glacial tillites continued through much of the Permian of Australia (King, 1958). Over this large area the glaciation was followed by the *Glossopteris* flora (Chaloner and Lacy, 1973) again earlier in South America and Africa, and later in India and Australia. The *Glossopteris* distributions were one of the first pieces of evidence to suggest Gondwanaland existed as a single continent. At the end of the Permian, apparently independently of climate (as it had become warm in South America early in the Permian while remaining cold in Australia), many Pteridospermopsida (the class containing *Glossopteris*) became extinct. Also on the land, the amphibians decreased in diversity throughout the Permian, while the synapsid reptiles showed extinction of many families at the very end of this period (Harland et al. 1967). The pattern of extinctions on land is thus as complex as that in marine life. The causes of extinction of the land flora and faunas may also be due to competition in the extremes of climate we know existed. But not all life was going through hard times in the Permian; many new insect suborders appeared throughout the period.

The Triassic system was named after the threefold division present in Germany. During this period, the sea spread over the margins of Pangea, and especially along the borders of the Tethys Ocean which extended from Indonesia to Spain between Asia and Gondwanaland. This increase in area allowed some marine animals to diversify, but they did so slowly. The appearance of the new groups was spread out through Triassic time. They included: scleractinian corals, several new ammonoid families, actinopterygian fish. Although few new orders appeared, bivalves became more common. On land, the Triassic saw some new amphibians and many varied reptiles. The first mammals appeared in the Late Triassic.

The Jurassic system was named after the Jura Mountains of eastern France. In much of north-west Europe, there was a major transgression of the sea around the Triassic-Jurassic boundary; though, like many such transgressions, it occurred over a moderate length of time. Ammonites dominated the open seas of the Jurassic; while the benthos was characterized by an abundance of bivalves (especially oysters) and burrowing worms and crustacea (represented more by their burrows than by their skeletons). Echinoids diversified greatly, with the appearance of many new families.

The Cretaceous system is named after the widespread development of chalk deposits seen in much of Europe north of the Alps. This is composed entirely of marine organisms of which the protistid coccoliths make up a large part. In general, the higher Cretaceous taxa were very similar to those present in the Jurassic.

The close of the Cretaceous saw the extinction of many reptiles: some families became reduced gradually during the Upper Cretaceous, but others disappeared suddenly at the very end of the period (Romer, 1966). Most of the reptiles, (like the majority of dinosaurs and the pterosaurs) were land animals, but some (like the ichthyosaurs and the sauropterygians) were marine. The Mesozoic mammals diversified gradually during the Late Cretaceous, but this development was small compared with the great radiation seen in the early Tertiary; both may be linked with the contemporaneous development of angiosperms.

The Late Cretaceous saw the extinction of marine animals, including many foraminifera families, both pelagic (*Globotruncata*) and benthic. Some bivalves (like the rudists) died out at the end of the Maastrichtian (the top Cretaceous stage), but others did not become extinct until the Eocene. The belemnites showed a similar pattern, with some groups continuing until the Late Eocene. Ammonoids decreased from twenty-two families at the base of the Upper Cretaceous to eleven in the Maastrichtian; all families died out before the Tertiary (Hancock, 1967). Many echinoid families died out, and some groups of bony fish became extinct. As in the Permian, many Upper Cretaceous animals became restricted in their geographical distribution, so that they have different upper limits in different areas.

After these extinctions (or in some cases while the Cretaceous groups were decreasing in numbers and diversity), many other animal groups showed evolutionary radiation. These included the pelagic foraminifera (*Globorotalia* and its descendants), some benthic foraminifera, gastropods, echinoids and bony fish (teleosts) and, most markedly, the mammals.

Many explanations have been put forward for these changes, and no doubt the critical factors are complex and may vary from one group to another. For the marine animals, the key factor was possibly a reduction in the area of shelf sea after the widespread Cretaceous transgressions; the sea had probably covered more of the continent in the Upper Cretaceous than at any other time in the Earth's history. This reduction in living space would (as is postulated for the Permian) cause increase in competition, and hence extinction of the less successful groups. On land, there are several possible (and not necessarily exclusive) reasons for the reptilian extinctions; the three most popular are:

1 The rise of the angiosperms, which could have necessitated changes in feeding habits for many large herbivorous dinosaurs. If the herbivores could not obtain enough to eat (some of them must have had to consume many hundreds of pounds of foliage each day), then carnivores which preyed on them would also suffer.

2 Uplift of many land areas, reducing the amount of swamp-covered areas where the lush vegetation was situated (Romer, 1966).

3 A change in climate, produced by the same uplift, possibly coupled with the northward movements of all the northern hemisphere continents (Smith and Briden, 1977); this would have resulted in cooler conditions over many land areas, to which the cold-blooded reptiles would be unable to adapt.

During the Cenozoic, most of the major changes in faunas and floras are a reflection of climate. The climatic changes that we see in the fossil record of Europe and North America took place in two ways: there was a slow northward drift of these two continents throughout the Tertiary, which resulted in progressively cooler climates; and, especially in the Pleistocene, there were more rapid world-wide fluctuations in temperature. It is still not yet clear why these Pleistocene fluctuations took place (and they are probably still in action), but they provide a very convenient method for dating sediments during the past few million years. Each cooler period is marked by a cooler fauna and flora, and although the composition of these fossil assemblages depends on the latitude of the collecting site the fluctuations in temperature can be determined, so that the hot and cold peaks can be correlated over large areas.

PRESERVATION OF FOSSILS

Fossils are the remains of plants and animals found buried in sedimentary rocks. To be preserved, an organism must leave some permanent record in the rock such as a shell, a burrow or some other impression. But it is also necessary for the sediment in which fossils occur to be preserved; this can only happen by subsidence of the area where the sediment is deposited.

The process of subsidence does not affect all habitats equally. In upland areas the remains of the inhabitants tend to have been eroded or to have disintegrated before burial. Most of the rocks in which fossils are found were originally laid down in seas or in low-lying land areas. There is thus a strong bias in the geological record against the preservation of terrestrial organisms, especially those that lived in upland areas.

Examination of a physical map of Britain today shows that south-east England between the Wash and the Thames estuary is low-lying; here fluviatile sediments (interrupted by the occasional marine incursion like the 1951 surge) are accumulating in areas of subsidence like the Fens. Further east, sediments have been accumulating in the North Sea since the Carboniferous. Though neither the Fens nor the North Sea have had a continuous record of subsidence, both areas contrast with the greater part of the British Isles which have been generally characterized by uplift, especially in Wales, western Ireland and western Scotland, during most of the Tertiary. These areas are being eroded by rain and rivers. Any small areas of sedimentation are likely to be very temporary; all the sediments resulting from the erosion of upland Britain are carried out into the surrounding marine or fenland areas. The past sedimentary record is very similar. The majority of sediments are marine or deposited in deltas, estuaries or the lower reaches of rivers.

Most fossils are the hard parts of animals. Many soft-bodied animals are only known from rare impressions (often preserved in unusual conditions) or from burrows or other traces, which can seldom be assigned with confidence to any particular species. So a second bias in the fossil record is that towards animals with shells or bones. The fossils illustrated in this book are those that are most easily found; both types of bias apply to the examples given.

Some readers may think that there are far too many fossil names presented, but this has been done so that individual fossils can be looked up in other reference books if the reader wishes to find out more about them. Names should be used as reference indices, and not as long lists to learn by heart. Although many names are given, it must be stressed that the fossils in this book are only a very small sample of the more common genera. Rare fossils, many microfossils, and those that occur in unusual environments with a restricted geographic range are all omitted.

The major divisions of the animal kingdom are called phyla (singular: phylum). Most phyla have some members which secreted calcareous skeletons (Table II), and all phyla, except some groups of worm-like organisms, have some members preserved as fossils.

Calcium carbonate comes in two distinct crystalline forms: calcite and aragonite. Both crystal types occur in fossils, and in addition organically formed calcite may have a low or a high magnesium content. Aragonite and high-magnesium calcite are both more unstable than low-magnesium calcite if exposed to air or to fresh water (as will happen when rocks are uplifted to form land). This means that many benthic foraminifera, scleractinian corals, and various molluscs are often only preserved as moulds, or, if the mould is filled in, replaced by another mineral. Some sand formations may even have all their calcareous fossils removed by solution.

The hard parts are primarily for support in the Protista, the Porifera, the Coelenterata, the Hemichordata and Chordata: they also serve for protection in the Bryozoa, the Brachiopoda, the Mollusca, the Arthropoda and in some Echinodermata; but in many animals they perform several additional functions, notably in brachiopods, molluscs, arthropods and chordates where they often serve as muscle attachments. In these latter groups a careful study of the hard parts can give the palaeontologist a lot of useful information about soft parts of the animal that are not preserved fossil and on how extinct animals lived (functional morphology).

Burrowing animals frequently have thinner shells than have epifauna living exposed on the sea floor. Burrowers are, however, less likely to be destroyed by currents or free-swimming predators and scavengers. Mechanical abrasion can seriously affect the preservation of some epifauna. Experiments have been conducted by tumbling shells in a barrel with pebbles and noting how soon they were ground down; bryozoans and some calcareous algae did not last long compared with gastropods, bivalves or corals (Chave,

Phylum	Calcareous skeleton	Other materials forming hard parts
Protista	most Foraminiferida	protein and agglutinated Foraminiferida; siliceous and strontium sulphate Radiolaria
Porifera	calcareous sponges	siliceous sponges
Archaeocyatha	all	—
Coelenterata	corals and pennatulaceans	chitinophosphatic conulariids
Annelida	serpulid polychaetes	protein jaws
Bryozoa	ectoprocta	—
Brachiopoda	all Articulata and some Inarticulata	chitinophosphatic Inarticulata
Mollusca	most molluscs	the outer shell layer is commonly protein
Arthropoda	calcareous reinforcement of protein in some groups	all have a hard protein covering
Echinodermata	all	—
Conodontophora	—	calcium phosphate
Hemichordata	—	protein pterobranchs and graptolites
Chordata	—	cartilage and bone (impregnated with calcium salts); scales and teeth of dentine

Table II

Composition of the hard parts of animals.

1964). Fossils preserved in coarse sediment may have been treated in a similar manner, with selective mechanical destruction of the less robust material.

TRANSPORTATION

In certain conglomerates and in all turbidites it is often obvious that shells have been transported, but in many shelly beds the degree of transportation may not be obvious until some careful statistical work has been carried out on the fossil populations.

Schafer (1972) has described five types of situations in which marine fossils may be preserved. At one extreme, there is a shallow water reef assemblage, where all the animal skeletons are preserved in their life positions, and there are no clear bedding planes. In moderately turbulent shallow water environments, some burrowing organisms are preserved in place, but the majority of the epifauna is transported. With high turbulence, nothing is preserved in position of growth. In quieter environments, there may be some areas where a large proportion of the benthic fauna is in place, but the nekton and plankton may make up a large proportion of the fossils. Finally, in very tranquil areas with no bottom currents, the sea floor may become stagnant and the only fossils are from the nekton and plankton.

If a collection of a living species is measured (say the length or some other character which changes with growth), most characters show variations in growth rates with ontogeny. For example, most mammals stop growing after maturity is reached; there will thus be a skewed distribution in the measurements because there will be more adult-sized individuals than juveniles in a population. In fact, different animals will show widely different size distributions according to the relationships between recruitment, growth and mortality (Craig and Oertel, 1966). By contrast, the size distributions of a transported population will be normal (provided it has not been carried wholesale in a density current); the shell size distribution will depend on the speed of the depositing current for a particular shape of shell (Boucot, 1953; Craig and Hallam, 1963; McKerrow et al. 1969).

Although this type of statistical work has not been done on many fossil communities, there is one other supporting fact to suggest that the majority of those described in this book, do, in fact, represent animals which lived in the same habitat: the same communities occur repeatedly. Although the different contributors have often approached the study of fossils in different ways, the communities described have at least this characteristic in common: they have been observed occurring many times by the author concerned, usually over a wide geographical area.

Classification of organisms

Taxa rare as fossils are not always included in the following classification.

Until recently, it has been the custom to divide organisms into two kingdoms: the animal kingdom and the plant kingdom; the animals generally feeding on other organisms, and the plants generally producing their energy by photosynthesis. This division has become increasingly difficult to apply to many simple organisms, and some biologists (e.g. Whittaker, 1969) think that the primary division should be defined on the nucleus:

Eukaryotic organisms, with definite nuclei in their protoplasm, like most large plants and animals.

Prokaryotic organisms, with no compact nucleus, but nuclear material dispersed in the protoplasm.

PROKARYOTA (Earlier Precambrian to Present)

The prokaryotes include some of the **blue-green algae** and the **bacteria.** The blue-green algae have chlorophyll and are capable of photosynthesis. The bacteria, in the main, lack chlorophyll and must obtain their energy from sources other than light; they have a large range of chemical powers. Both groups have been found in early Precambrian rocks which are much older than the rocks containing the first definite eukaryotes. The blue-green algae are considered to have been the world's first source of free oxygen produced by photosynthesis; they were thus necessary precursors to the development of the first animals.

Most **stromatolites** were (and are) formed by blue-green algae but some other groups can also form these mounds (Walter, 1972). Crowded, erect filaments on their upper surface trap mud particles, which become consolidated to form domes with a layered internal structure. Similar algae can also form **oncoliths**, spherical layered structures like large ooliths.

EUKARYOTA (Precambrian to Present)

Eukaryotes can be divided roughly according to their methods of obtaining energy:

plant-like: photosynthesis using chlorophyll, light and carbon dioxide;

animal-like: devouring plants or other animals (dead or alive) and oxidizing their substance;

fungus-like: absorbing and degrading organic substances (rare as fossils).

These three divisions become blurred in many single-celled organisms, most of which are microscopic; these have been artificially grouped in the Protista Kingdom. Some of the varied groups or protists are members of the Animal and Plant Kingdoms.

THE PROTISTA KINGDOM

A Protists with purely plant-like nutrition (photosynthesis):

DIATOMACEA (Cretaceous to Present)

Diatoms have brown protoplasm enclosed in a rigid silica skeleton. They occur as plankton and benthos in marine and fresh water. They are important primary producers of the marine food chain.

B Protists with characters intermediate between animals and plants; some are capable of swallowing solids, but most depend on photosynthesis; capable of locomotion by means of flagella:

COCCOLITHOPHORIDA (Jurassic to Present)

Organisms with the envelope of the cell covered by calcareous discs (2 to 30 microns in diameter). Coccoliths are pelagic; marine or fresh water.

SILICOFLAGELLATA (Cretaceous to Present)

Organisms with a skeleton composed of hollow bars of silica.

DINOFLAGELLATA (Silurian to Present)

A varied group of organisms, many of which have cellulose plates (6 to 100 microns); some may be the origin of **acritarchs** (which are known from the late Precambrian and later systems).

The **Zooxanthella** may be related; they are round yellow-green cells which live associated with hermatypic corals and some other marine invertebrates.

31

C Protists with purely animal-like nutrition (including the **Proto-zoa**); mostly feeding on organic matter; no photosynthesis. Most soft-bodied, but one phylum includes animals with hard parts.

PHYLUM SARCODINA (Cambrian to Present)

class ACTINOPODA (Cambrian to Present)

subclass RADIOLARIA (Cambrian to Present)
Marine pelagic protozoans with hard parts of silica or strontium sulphate. As silica does not dissolve as readily as calcite in very deep water, radiolarian deposits are commoner in very deep water sediments.

class RHIZOPODEA (Cambrian to Present)

order Foraminiferida (Cambrian to Present)
Protoplasm of body supported by a skeleton usually calcareous. Mostly in marine or brackish water. Mostly benthic, but some pelagic; pelagic foraminifera (e.g. *Globigerina*) are important in Cretaceous and Tertiary stratigraphy.

THE PLANT KINGDOM

ALGAE (other than prokaryotes and protists)
Nearly all aquatic. Most are soft and not easily preserved as fossils; but a few shallow marine forms (which are often abundant in shelf environments) have calcium carbonate in or on their cell walls, and the freshwater **Characeae** (Silurian to Present) may have the whole plant, or just the outer layer of the egg cell, calcified.

BRYOPHYTA (Devonian to Present)
Mosses and **liverworts.**

VASCULAR PLANTS
Land plants with woody tissue.

PTERIDOPHYTA (Silurian to Present)

Plants that produce spores asexually in alternate generations; the spores then grow into very different small plants, which reproduce sexually to form a new generation of larger plants.

PSILOPHYTES (Upper Silurian to Upper Devonian)
The earliest known land plants, with a simple organization.

LYCOPODS (Lower Devonian to Present)
Plants with simple leaves borne singly. Include varied and large forms in the Carboniferous (e.g. *Lepidodendron*). At present fewer and smaller (the **club mosses**).

CALAMITES (Devonian to Present)
Plants with simple leaves in circlets. Reached climax in the Carboniferous (e.g. *Calamites*); now represented only by the **horsetails**.

FERNS (Devonian to Present)
Plants with leaves elaborately branched. Still important today.

GYMNOSPERMAE (Devonian to Present)

Plants producing seeds fertilized by pollen carried by wind or insects.

CONIFERS (Carboniferous to Present)
Plants with simple needle leaves. Abundant in the Mesozoic and still important.

CYCADS (Permian to Present)
Plants with leaves like feather-palms. Abundant in the Mesozoic, less common now.
The Gymnosperms also include the **Pteridosperms** (e.g. *Glossopteris*) the **Cordaites** and the **Ginkgos**.

ANGIOSPERMAE (Cretaceous to Present)

Flowering plants, with the young seed enclosed in an ovary, and the pollen received on a special organ, the stigma. May be preserved as wood, leaves, seeds, fruit and pollen.

THE ANIMAL KINGDOM

PHYLUM PORIFERA

Sponges consist of cells which are only poorly organized into tissues. Water is drawn through many small inlets on the surface of the sponge and ejected through fewer exits. Benthic, mostly in shallow marine environments, but some in deep water and a few in fresh water. The marine **stromatoporoids** (Cambrian to Cretaceous), commonly thought to be coelenterates, might be included in the Porifera (Hartman and Goreau, 1970). Many so-called tabulate corals (e.g. *Chaetetes*) may also be sponges.

class DEMOSPONGEA (Cambrian to Present)

Sponges with complex canal systems. Skeleton of spongin and/or siliceous spicules; some groups, including the **lithistids**, have the spicules fused to form a rigid frame. Shapes very variable. Mostly epifaunal, but some are borers, e.g. *Cliona.*

class HYALOSPONGEA (Cambrian to Present)

Sponges with a siliceous skeleton; usually with a large exhalent cavity. Formerly called Hexactinellida, but this class does still include the family **Hexactinellidae.**

class CALCISPONGEA (Cambrian to Present)

Sponges with a skeleton of small calcite spicules. Degree of fusion of the spicules variable. Shape of individuals and complexity of canal systems very variable.

PHYLUM ARCHAEOCYATHA (Cambrian)

Confined to the Lower Cambrian and early Middle Cambrian. Skeleton in the form of a calcareous cup, usually solitary. The walls and some internal plates are porous, like sponges, but other non-porous structures may be present (like corals). Shallow marine, benthic.

PHYLUM COELENTERATA

Animals with a nervous system; often with stinging cells. The body has a single cavity with only one opening (no separate anus). Free-swimming coelenterates (**medusoids**) have marginal tentacles; benthic forms have tentacles round mouth on the upper surface, and usually have radial symmetry. Includes many soft-bodied forms (e.g. jellyfish), but also the corals, which are important as fossils.

class PROTOMEDUSAE (Late Precambrian to Ordovician)

Jellyfish-shaped animals with radial lobes. Probably free-swimming, marine.

class HYDROZOA (Late Precambrian to Present)

A varied group of soft- and hard-bodied marine and freshwater coelenterates. It includes *Ediacaria*, a medusoid from the Late Precambrian. Many are colonial, including the coral-like order **Milleporina** (Cretaceous to Present) which occur on coral reefs and, unusually for hydrozoans, have a calcareous skeleton.

class SCYPHOZOA (Cambrian to Present)

Mainly soft-bodied, marine **jellyfish**, which are seldom preserved as fossils. This class also includes the order **Conulariida** (Cambrian to Triassic) which have the body protected by a cone or a four-sided pyramid of chitinophosphatic material; like the jellyfish, conulariids have a 4-rayed symmetry, and are thought to have been entirely marine and free-swimming or attached.

class ANTHOZOA (Late Precambrian to Present)

Benthic coelenterates, with the central cavity radially partitioned by fleshy mesenteries. The mouth is surrounded by retractable tentacles. Entirely marine.

subclass OCTOCORALLIA (Late Precambrian to Present)

Colonial anthozoans with eight tentacles and eight mesenteries in each individual polyp. The skeleton is usually of calcareous spicules, sometimes it is a rigid calcareous framework; these latter forms common as fossils from the Cretaceous to the Present. Includes the order **Pennatulacea** (the **sea pens**), which have unbranched individuals anchored by a calcareous stalk, or possibly (in the case of the Late Precambrian *Charnia* and *Rangea*) by a circular calcareous disc. Most octocorallia are benthic shallow marine animals, but some have been recovered from the sea floor at very great depths.

subclass ZOANTHARIA (Cambrian to Present)

Includes **sea anemones** and **corals.** Only the orders containing corals are listed here.

order Tabulata (Cambrian to Permian)

Colonial corals in which the individual polyps had their tubes partitioned by horizontal plates (tabulae), while the vertical radiating plates (septa), seen in most other coral groups, were rudimentary or absent. Some of this extinct group may, in fact, not be coelenterates but sponges.

order Rugosa (Ordovician to Permian)

Solitary or colonial corals, usually with alternating longer and shorter septa, showing bilateral or radial symmetry. Tabulae and other internal plates were usually present. Colonial corals were restricted to shallow water, but the solitary corals occurred over a wider depth range. The name is derived from rugae (broad ribs) which occur on the external walls of many of these corals.

order Heterocorallia (Carboniferous)
A small group of very elongate solitary corals.

order Scleractinia (Triassic to Present)
Solitary or colonial corals with radial septa in successive cycles (starting with 6 in the first cycle). Abundant small plates or rods between the septa. Scleractinias are the most abundant corals alive today. Some are dependent for their existence on the presence of large numbers of single-celled algae (dinoflagellates or zooxanthellae) in their tissues. These (hermatypic) corals have a maximum depth range of about 90m, and include most of the modern reef builders. Those corals not dependent on the light-requiring algae (ahermatypic corals) can live in all depths of water, but most prefer depths of less than about 500m in areas with slow sedimentation rates.

PHYLUM ANNELIDA and other phyla of worm-like animals

Occasional impressions of soft-bodied segmented annelids occur from the Late Precambrian (Ediacara fauna) to the present day. More frequently some hard protein jaws of certain annelids (mostly **polychaetes**) may be preserved fossil; these are known as **scolecodonts** and range from the Cambrian. Most scolecodonts are marine, like the majority of present day polychaetes. Worm burrows, trails and the calcareous tubes are more abundant; many of these cannot be assigned to a phylum, but some traces are characteristic of particular groups (like the lining of **terebellid tubes** by cemented shell fragments). The calcareous tubes of **serpulids** can also be attributed to a particular group of polychaete worms. Some polychaetes which crawl on the sea floor leave characteristic trails; others, like the **sabellids**, form mucus-lined burrows.

The paucity of fossils worms contrasts with their abundance in modern marine environments; they include free-swimming, crawling, sessile and burrowing animals. They also have a variety of feeding habits; in addition to the filter-feeding majority, they include carnivores and scavengers.

The **sipunculid worms** (Cambrian to Present) are assigned to a separate phylum from the annelids. Their bodies are not segmented, and most of them narrow anteriorly; the anterior extension is retractable and bears tentacles. Sipunculids are mostly infaunal deposit feeders.

PHYLUM BRYOZOA

Small colonial animals with a curved ridge bearing tentacles (the lophophore) which set up currents for feeding and respiration. The vast majority of the phylum (the Ectoprocta) have an external calcareous skeleton.

subphylum ECTOPROCTA (Ordovician to Present)

The skeleton may be encrusting (for example on a molluscan or brachiopod shell), branching (like a small coral), fan-shaped or nodular. Some ectoprocts contribute to reef formation. They are mostly sessile, usually on the sea floor, but some can be attached to floating seaweeds, and a few are borers.

PHYLUM BRACHIOPODA

Marine animals with a lophophore (similar to the Bryozoa) protected by a bivalved shell. They are normally attached to the sea floor by a pedicle which emerges from the larger valve, but some brachiopods lose the pedicle and rest (on one valve) directly on the sea floor. Brachiopods use their lophophore to circulate water in the cavity between the valves, where they can extract organic matter in suspension or in solution. They have a free-swimming larval stage (of from one to twenty days) which allows them to disperse. Most brachiopods are confined to shelf seas, but a few live in very deep water.

class INARTICULATA (Cambrian to Present)

Brachiopods with no hinge teeth to hold the valves in place. The shell often chitinophosphatic, but can be calcareous. Most live on the sea floor attached by a pedicle, but some (e.g. *Lingula*) burrow, and others lose the pedicle and become cemented (e.g. *Crania*).

class ARTICULATA (Cambrian to Present)

The majority of brachiopods are in this class. They all have a calcite shell. Most articulates have two teeth in the pedicle valve which fit into sockets on the smaller (brachial) valve and hold valves in place; a few groups have lost their teeth.

order Orthida (Cambrian to Present)

The earliest articulate brachiopods; probably ancestral to the rest of the class. The calcite shells generally have radial ribs or striations and a broad hinge line. Most orthides have a large triangular pedicle opening, situated in a distinct interarea.

order Strophomenida (Ordovician to Jurassic)

Most strophomenides have no functional pedicle in the adult, most forms have large valves, one of which is convex; an adaptation for resting on the sea floor. Other adaptations for loss of pedicle are seen in cemented forms and in those (like *Chonetes* and *Productus*) which develop spines. This order is the largest in the Brachiopoda, with nearly 400 genera.

order Pentamerida (Cambrian to Devonian)

Biconvex shells, usually smooth, with a spondylium. The pentamerides possibly had functional pedicles, but the thickened shell near the umbones could have allowed some to be stable on the sea floor without the need for a pedicle.

order Rhynchonellida (Ordovician to Present)

Biconvex shells with a pointed umbo. Pedicle usually functional. Shell normally with strong radial ribs, and with a marked fold anteriorly.

order Spiriferida (Ordovician to Present)

Spiriferides usually with a functional pedicle and the lophophore supported by a spiral brachial skeleton.

order Terebratulida (Devonian to Present)

The lophophore is supported by a calcareous loop, and those with a long loop usually have dental plates. The umbo is truncated by a large pedicle opening. This is the most abundant brachiopod order still alive today. The **Thecideidina** (Triassic to Present) are a group of small, cemented brachiopods which may be related to the Terebratulida.

PHYLUM MOLLUSCA

Animals with a body composed of a head and/or a foot, a visceral mass containing the internal organs, and a sheet of tissue (the mantle) which can secrete a calcareous shell. A space between the mantle and the visceral mass, known as the mantle cavity, contains gills and acts as a respiratory chamber. Most molluscs have a free-swimming larval stage.

class MONOPLACOPHORA (Cambrian to Present)

Mollusca with a single valve containing several paired muscle scars. Shallow marine benthos in the Palaeozoic; living at abyssal depths today.

class POLYPLACOPHORA (Cambrian to Present)
Marine mollusca, normally with a row of calcareous plates; commonly called the **chitons.**

class APLACOPHORA (Present)
Worm-like animals without shelly plates.

class SCAPHOPODA (Devonian to Present)
Benthic, marine mollusca with an external tubular tapering curved shell, open at both ends. They live partly embedded in the sediment, with the narrow end of the shell extending up into the water. Scaphopods feed on micro-organisms and organic debris.

class GASTROPODA (Cambrian to Present)
Mollusca usually with a single valve, and no internal septa. Most gastropods crawl on their foot, but in some the foot is adapted for swimming. Gastropods may be marine, terrestrial, or freshwater.

subclass PROSOBRANCHIA (Cambrian to Present)
Gastropods with mantle cavity opening to front. Shell rarely absent.

order Archaeogastropoda (Cambrian to Present)
Mostly with paired gills, each gill having two rows of filaments. Mainly algal feeders.

order Mesogastropoda (Ordovician to Present)
Usually with a single gill having one row of filaments. Often with a siphon. Include benthic grazers, suspension feeders and carnivores; some are pelagic.

order Neogastropoda (Cretaceous to Present)
Gastropods with an inhalent siphon emerging through a siphonal canal on the aperture margin. Mainly carnivorous; radula with few teeth (usually not more than three) in a row.

subclass OPISTHOBRANCHIA (Carboniferous to Present)
Mantle cavity opens at side or rear of body. Shell often reduced or absent. Includes benthic forms and the pelagic **Pteropods.**

subclass PULMONATA (Jurassic to Present)
No normal gills; the mantle cavity acts as a lung. Mostly terrestrial; a secondary gill occurs in some freshwater forms.

class CEPHALOPODA (Cambrian to Present)
Marine mollusca with a single-valved shell partitioned by septa, which are traversed by a fleshy tube, the siphuncle. All known modern cephalopods are carnivorous.

subclass NAUTILOIDEA (Cambrian to Present)
Siphuncle variable in position, but not inflated between septa. No closely packed calcareous cones within siphuncle. Curvature of shell variable. Septal sutures with smooth curves.

subclass ENDOCERATOIDEA (Ordovician and ? Silurian)
Posterior part of siphuncle filled by close packed plates. Straight or slightly curved shell. Septal sutures straight or slightly flexured.

subclass ACTINOCERATOIDEA (Ordovician to Carboniferous)
Straight shells, with siphuncle inflated between septa. Other internal structures may also be present. Septal sutures straight or slightly flexured.

subclass BACTRITOIDEA (Ordovician to Permian)
Straight or gently curved shells, with siphuncle in contact with ventral wall. Septal suture with a strong V-shaped lobe.

subclass AMMONOIDEA (Devonian to Cretaceous)
Cephalopods with angular or more complex septal sutures. Normally with ventral siphuncle and planospiral shell. Includes the **goniatites** (Palaeozoic), **ceratites** (Triassic) and **ammonites** (Jurassic and cretaceous).

subclass COLEOIDEA (Devonian to Present)
Most fossil coleoids are **belemnites** with a cigar-shaped calcite guard, which contains a conical phragmacone (with septa and ventral siphuncle) at one end. The belemnites became extinct in the early Tertiary (Hancock, 1967). The modern coleoids (**cuttlefish, squids** and **octopus**) have no guard.

class BIVALVIA (Cambrian to Present)
Marine or freshwater molluscs with two calcareous valves joined at the hinge with a flexible ligament. Bivalves have very varied modes of life; they include fixed forms (by byssus or cementation) and mobile forms (free-living, sessile, burrowing and boring); they include filter feeders and detritus feeders.

subclass PALAEOTAXODONTA (Ordovician to Present)
The **nuculoids,** with simple gills and many small teeth (taxodont) are mostly shallow burrowers.

subclass CRYPTODONTA (Cambrian to Ordovician to Present)
  Toothless burrowers with simple gills.

subclass PTERIOMORPHIA (Ordovician to Present)
  This large group includes bivalves with more complex gills. They are varied, but the fossil record suggests that they may have a common origin.

  order Arcoida (Ordovician to Present)
    Forms with both adductor muscles equal (isomyarian), and with a characteristic area between the hinge line and the umbo. Usually with many small teeth.

  order Mytiloida (Devonian to Present)
    The **mussels** are usually byssally attached and have a reduced anterior adductor muscle (heteromyarian).

  order Pterioida (Ordovician to Present)
    Although descended from byssally fixed forms (and thus heteromyarian or monomyarian), many pterioids have developed other modes of life including the free-living **pectinids** and the cemented **oysters.**

subclass PALAEOHETERODONTA (Cambrian to Present)
  This group includes the earliest known bivalves (Middle Cambrian). Most forms have the adductor muscles about the same size (isomyarian), and are equivalved.

  order Modiomorphoida (Cambrian to Present)
    Includes the **actinodonts,** which have a few elongate teeth radiating from the umbo; they may be ancestral to most other bivalve groups.

  order Unionoida (Devonian to Present)
    Freshwater bivalves, smooth except for growth lines. Characterized by a thick outer coating of horny material (periostracum) covering the calcareous part of the shell.

  order Trigonoida (Devonian to Present)
    Marine bivalves with an angular posterior margin, and a few strong teeth near the umbo.

subclass HETERODONTA (Ordovician to Present)
  The name refers to the fact that the teeth may include a group near the umbo (cardinal teeth) as well as lateral teeth. The adductor muscles are usually equal in size.

41

order Veneroida (Ordovician to Present)
Active, nestling or burrowing heterodonts. Teeth well developed.

order Myoida (Carboniferous to Present)
Burrowing heterodonts with siphons, weak teeth and thin shell.

order Hippuritoida (Silurian to Cretaceous)
Thick-shelled bivalves, mainly attached by one valve. This group includes equivalved early forms, but the later members include the **rudists**, which have one large fixed conical valve and a small free valve on top. The rudists formed extensive reefs in low latitudes during the Cretaceous.

subclass ANOMALODESMATA (Ordovician to Present)
Burrowing, byssate or cemented bivalves with weak teeth (and never any lateral teeth).

class ROSTROCONCHIA (Ordovician to Permian)

order Conocardiacea (Ordovician to Permian)
A peculiar group of mollusca in which the two valves have become fused.

phylum uncertain

class CALYPTOPTOMATIDIDA (Cambrian to Permian)
The **hyolithids**; simple conical or pyramidal shells, with the aperture usually closed by an operculum. Near the base of some Lower Cambrian transgressive sequences, hyolithids occur in shallow marine beds under those containing the first local occurrence of trilobites; and for this reason (only) some authors consider that the hyolithids may extend into the Precambrian.

class CRICOCONARIDA (Ordovician to Carboniferous)
The **tentaculitids** and **cornulitids**; conical shells with exterior ornament of transverse rings.

PHYLUM ARTHROPODA
Aquatic, terrestrial and aerial invertebrates, with a segmented body and jointed legs (hence the name) covered by chitin. Growth takes place through moults, so the chitin skeleton shows no growth lines.

subphylum TRILOBITOMORPHA (Cambrian to Permian)

In addition to the trilobites, this subphylum includes the class **TRILOBITOIDEA**, which are largely represented by fossils from the Middle Cambrian Burgess Shale of British Columbia. They appear to be a varied group, but to have similar appendages to trilobites.

class TRILOBITA (Cambrian to Permian)

Marine arthropods with a dorsal skeleton divided longitudinally into three lobes. The skeleton is also divided laterally into three parts: the head shield (cephalon), the thorax (which shows distinct segments) and the tail (pygidium). Each segment carried a pair of ventral appendages for locomotion and respiration. The appendages are not often preserved fossil, as their covering is not reinforced by calcite like the dorsal skeleton. Many trilobites were deposit feeders, living on the sea floor. Common trilobite traces include *Cruziana* (a linear trail with oblique scratch marks) and *Rusophycus* (an excavation of a resting place).

subphylum CHELICERATA (Cambrian to Present)

Terrestrial and aquatic arthropods, with the front pair of appendages developed as pincers. The body is divided into a cephalothorax (with six segments) and an abdomen (with up to twelve segments).

class MEROSTOMATA (Cambrian to Present)

Aquatic arthropods which include the **king-crab** (*Limulus*) and its relatives, and also an extinct group, the **eurypterids,** which include a species of *Pterygotus* which is over 1.8m long and is the largest known arthropod.

class **ARACHNIDA** (Silurian to Present)

Air-breathing **chelicerates,** which include the **scorpions** and the **spiders.**

subphylum CRUSTACEA (Cambrian to Present)

Arthropods with two pairs of antennae in front of the mouth. Body generally covered with a hard carapace. Respiration by gills. Some classes are rare or **absent** as fossils, and are omitted here. The **cycloidea** (Carboniferous to Triassic) have limpet-like shells and are perhaps related to the **Crustacea.**

class BRANCHIOPODA (Devonian to Present)

Crustacea in which the carapace may form a single dorsal plate or a bivalved shell usually about 5mm long. The branchiopods (literally gill-feet) use their limbs for breathing, and also for swimming and feeding; most swim or crawl; some burrow in mud. Many prefer fresh or brackish water to normal marine environments.

class OSTRACODA (Cambrian to Present)

Small, bivalved crustacea with a calcified carapace hinged along dorsal margin, covering a body with few segments. Freshwater or marine; mostly benthic. Benthic ostracodes do not have a pelagic larval stage, and thus many species are unable to cross deep water.

class COPEPODA (Miocene to Present)

Elongate segmented carapace with prominent central articulation. Copepods are the most abundant of modern marine animals, but known early forms were all lacustrine.

class CIRRIPEDIA (Silurian to Present)

Adults permanently fixed, often with calcareous plates (the **barnacles**).

class MALACOSTRACA (Cambrian to Present)

A varied group of highly developed crustacea; usually with a thorax of eight segments and an abdomen of six or seven segments, most of which bear appendages. Includes the **decapods** (**lobsters**, **crabs** and **shrimps**), the isopods and the amphipods. Mostly shallow marine, but some freshwater or terrestrial.

subphylum MYRIAPODA (Silurian to Present)

The **centipedes** and **millipedes**; terrestrial arthropods. The head has a single pair of antennae.

subphylum HEXAPODA (Devonian to Present)

The **insects** develop wings and have six walking legs. They are all air breathing.

PHYLUM ECHINODERMATA

Marine animals usually with a pentamerous symmetry, though some early forms have no radial symmetry, and some later echinoderms develop a superimposed bilateral symmetry. Skeleton composed of calcite plates.

subphylum HOMALOZOA (Cambrian to Devonian)

The **carpoids**; echinoderms without radial symmetry.

subphylum CRINOZOA (Cambrian to Present)
Benthic echinoderms, with a clear pentamerous symmetry. This group includes the **crinoids** (which are usually attached by a stalk to the sea floor). Other crinozoans (e.g. some **cystoids** and **blastoids**) have no stalks. Arms usually provide food to the mouth.

subphylum ASTEROZOA (Ordovician to Present)
The **starfish**, typically with five arms. Some primitive forms are suspension feeders, but most are carnivorous. **Asteroids** have arms which merge towards the centre, while **ophiuroids** have long arms which are distinct extensions from a central disc.

subphylum ECHINOZOA (Cambrian to Present)
Free-living echinoderms, which include the Lower Cambrian **Helicoplacoidea** (which have only one ambulacrum), the **Edrioasteroidea**, the leathery **Holothuroidea**, and the **sea urchins** (**Echinoidea**). Many advanced echinoids develop a bilateral symmetry, which is especially associated with the burrowing deposit-feeding heart urchins.

PHYLUM CONODONTOPHORA (Cambrian to Triassic)
Small tooth-like and plate-like structures of calcium phosphate (**conodonts**) belonging to an unknown group of extinct marine animals.

PHYLUM HEMICHORDATA
A varied but small group, probably related to the Chordata, but without a notochord.

class ENTEROPNEUSTA (Present)
Worm-like animals with no external skeleton.

class PTEROBRANCHIA (Ordovician to Present)
Fixed colonial organisms. Skeleton of protein. Lophophore attached by a stalk to an internal protein tube (the stolon).

class GRAPTOLITHINA (Cambrian to Carboniferous)
Colonial marine animals with a skeleton of protein.

order Dendroidea (Cambrian to Carboniferous)
Graptolites with a protein stolon, include benthic and pelagic forms.

order Graptolidea (Ordovician to Lower Devonian)
Graptolites with no preserved stolon; probably all pelagic.

45

PHYLUM CHORDATA
Animals with a notochord (a cylindrical sheath forming a flexible support for the back) or a backbone.

subphylum CEPHALOCHORDATA (Present)
Fish-like animals without bones or fins, but with a notochord (e.g. *Amphioxus*).

subphylum UROCHORDATA (Permian to Present)
The **sea squirts** or **tunicates.** Notochord only present in the larvae, which are active swimmers. Adults marine; benthic or pelagic; some colonial.

subphylum VERTEBRATA (Ordovician to Present)
Chordates with a skeleton of cartilage or bone.

class AGNATHA (Ordovician to Permian; also Present)
Vertebrates without jaws; they include Palaeozoic **ostracoderms**, with external plates, and the modern soft-bodied **cyclostomes** (**hagfishes** and **lampreys**). Most ostracoderms appear to have lived in rivers and freshwater lakes, but some were marine.

class PLACODERMI (Devonian and Carboniferous)
Jawed fishes with heavy external plates. Marine and fresh water.

class CHONDRICHTHYES (Devonian to Present)
Sharks with a cartilaginous skeleton. Mostly marine.

class OSTEICHTHYES (Silurian to Present)
This large class is split into several classes by some authors. It includes the Palaeozoic **spiny sharks** (Acanthodii), the **lungfish** and the **Crossopterygians** (some of which were ancestral to the first amphibians) and all the other bony fishes. The **actinopterygian** bony fishes include the Chondrostei, the Holostei and the Teleostei. The **chondrosteans** were common in the Upper Palaeozoic and Triassic. The **holosteans** were the dominant bony fishes in the Mesozoic. During the Cretaceous the **teleosts** became more common, and were the dominant fish of the Cenozoic. The ear-bones of bony fishes are known as **otoliths.** Mostly marine, but many freshwater.

class AMPHIBIA (Devonian to Present)
The adults are land animals, but the eggs are typically laid in water and the early life stages are usually aquatic.

class REPTILIA (Carboniferous to Present)

> Mainly land animals, which develop from an egg that can be laid on land, though some are marine. They reached their maximum development in the Mesozoic. Reptilian teeth are usually simple cones.

subclass ANAPSIDA (Carboniferous to Present)

> Includes the earliest reptiles and also the turtles and the mesosaurs. The **mesosaurs** were stream-lined swimmers, in contrast to the **turtles**, which nevertheless were (and are) also mobile aquatic animals.

subclass LEPIDOSAURIA (Permian to Present)

> Includes the **lizards,** the **snakes** and the lizard-like **rhynchosaurs.**

subclass ARCHOSAURIA (Permian to Present)

> The ancestral **thecodonts** gave rise to the **dinosaurs,** which included many varied types which were dominant on the land during the Mesozoic. This subclass also includes the **crocodiles** and the winged **pterosaurs.**

subclass EURYAPSIDA (Permian to Cretaceous)

> Mostly aquatic or amphibious, this subclass includes the early and relatively unspecialized **nothosaurs,** the long-necked **plesiosaurs** with paddle-like legs, the short-necked **pliosaurs,** and the fish-shaped **ichthyosaurs.**

subclass SYNAPSIDA (Permian to Jurassic)

> These terrestrial **mammal-like reptiles (paramammals)** include the **pelycosaurs,** which had large dorsal spines.

class AVES (Jurassic to Present)

> **Birds** typically have feathers, and show several skeletal differences from the reptiles (e.g. hollow bones). Except for some early forms, the birds have no teeth. They do not have a very complete fossil record.

class MAMMALIA (Triassic to Present)

> Fossil mammals can be distinguished from the reptiles by many details of their skeleton, including the teeth which are among the more durable parts preserved fossil. Mammalian cheek teeth typically have several cusps. Fossil mammals are rare until the Palaeocene, that is, until after the extinction of the dinosaurs at the end of the Cretaceous.

THE COMMUNITY DIAGRAMS

The convention used in the construction of the diagrams has been devised to show the relations between the rock containing fossils as observed and the mode of life of the organisms illustrated (interpretation).

Most diagrams have three sections: the sea, a block of soft sediment, and a block (protruding at the base) of sedimentary rock. The latter block represents the sediment as it now occurs in the field, and it is capped by a bedding plane which shows the fossils as they may be seen today.

Above this projecting fossil bed, the sediment is shown as it may have existed shortly after deposition, and the infauna in this soft sediment is shown, with soft parts, as it may have been when the animals were alive. The upper part of each diagram represents the sea, and the organisms in the sea (nekton and plankton) and on the sea floor (epifauna) are also reconstructed to show how they may have lived. Some empty shells may also be shown if they form a significant part of the habitat.

It is important to stress that the reconstructions of the soft parts and of the mode of life of extinct forms are interpretations, which are based mainly on comparisons with the animals' modern relatives. For some extinct groups, however, the mode of life has to be deduced from a study of the function of those hard parts preserved, a study of the traces left on the sea floor (e.g. trilobites), or a study of quite unrelated animals which appear to resemble the fossils.

The key to understanding these diagrams is the large time gap present above the projecting fossiliferous bed: the fossil bed and the sediments below it are as they appear today, while the sediments and the organisms above this bed are interpretations showing how they must have appeared at the time the bed was laid down and the animals were alive.

Precambrian

The most obvious fossils in Precambrian sediments are structures, called stromatolites, built up in layers by blue-green algae, but many microfossils have also been reported; some of these latter are rods and spheres which were perhaps formed by algae or bacteria. Traces of animals are absent until relatively late Precambrian times.

Stromatolites appear very early in the geological record and continue to the present day. They are laminated calcareous structures built up from fine sedimentary material (usually calcium carbonate) accumulating on blue-green algae (see Fig. 39). Modern filamentous algae trap fine-grained sediment particles with sticky mucosal coverings coating minute algal strands. Sediment which has been trapped in this way is then bound by further growth of the algae. The oldest stromatolites may be around 3,350 million years old (Muir and Grant, 1976); from this time onwards they are the dominant fossils in the Precambrian and they have been used as a basis for correlation (Walter, 1972; Hoffman, 1974). The oldest known water-laid sediments are about 3,800 million years in age, so some primitive forms of life appear to have existed not very long (geologically speaking) after the earliest direct record of water on the earth's surface.

The evidence from a sedimentary basin about 1,800 million years old, in Northern Canada, suggests that stromatolites had a greater depth range in the Precambrian than in later times; they appear to have lived in deep water as well as in shallow water (Hoffman, 1974). Some individual stromatolite colonies are very large: over 10m high. Today, stromatolites are restricted to very shallow water, and to some hypersaline (very salty) lagoons. This restriction is due to the action of grazing animals, like some gastropods, which can destroy the laminations faster than the blue-green algae can form them, and to destruction by burrowers such as bivalves and worms. The algae can therefore only flourish today where these destructive animals are absent (Garrett, 1970); thus at present they are found mainly in hypersaline lagoons — they have a greater tolerance of salinity than most animals — and on tidal flats exposed for long periods, where grazing animals are absent and burrowers are few. These observations help to explain why stromatolites were more abundant in the Precambrian (before the grazers and burrowers evolved) than at later times.

Plants do not necessarily need free oxygen, but animals do: the two universal necessities for animal life are water and oxygen. Early evidence of water is seen in the oldest sediments (3,800 million years old) which were deposited by water currents, but it is probable that free oxygen was not produced until much later in the Earth's history. After the time when plants first developed (about 3,400 million years ago), there was probably a very long period before the amount of free oxygen in the atmosphere and dissolved in seawater reached a level which could support the first primitive animals (Cloud, 1976). The oldest red (oxidized) sediments are about 1,800 million years old, and occur in many parts of the world.

Apart from a few possible burrows, perhaps made by annelid worms, the first reliable records of fossil animals occur in late Precambrian rocks, younger than about 700 million years. The Pound Quartzite of Ediacara in South Australia has yielded soft-bodied coelenterates, annelid worms and some other animals of uncertain affinities (Glaessner and Wade, 1966); many of them were possibly free-floating (Cloud, 1968). Elements of the Ediacara fauna have been recorded in south-west Africa (in beds considered to be of Lower Cambrian age), in England (in Charnian beds of about 680 million years), in Siberia (670 million years) and in south-east Newfoundland. The fauna thus appears to span a long period of time (680 to 570 million years), but there are some doubts about the exact age of many of the beds in which it occurs (Glaessner, 1971).

It is not known when animals first developed. They might have had a long history prior to the development of the Ediacara fauna, or they might have developed rapidly shortly before this fauna appeared. There is also uncertainty about the exact relations between the Ediacara fauna and those animals present in the Early Cambrian. Were the varied Lower Cambrian faunas all descended from the Ediacara coelenterates and annelids? Or were they descended from soft-bodied ancestors which developed slowly over a long period of time before the Ediacara animals developed? In the absence of any direct evidence, we can only speculate. Cloud (1968, 1976) summarizes most of the relevant facts and concludes that some rapid late Precambrian adaptive radiation in animals is the best guess in the light of our present knowledge.

The Appearance of Animals with Hard Parts

Many geologists have postulated that some critical geochemical change or some physical event on the Earth may have been responsible for the appearance of animals with hard parts at the base of the Cambrian. However, there is no evidence of any significant differences between most Cambrian sediments and those that were laid down at the end of the Precambrian. Nor can the late Precambrian glaciations be held directly responsible, for it now appears that these ice ages occurred at slightly different times on different continents, so that there may not ever have been a single major glacial event.

A biological change seems more probable. If we study the trilobites, archaeocyathids, brachiopods, sponges and molluscs which appear at or near the base of the Lower Cambrian, we can observe that they all have one feature in common: the hard parts are external. Though the skeletons clearly support the animal and act as attachments for muscles, the most important feature which they all possess in common is that they protect the animals. Very few Early Cambrian shelled animals were burrowers or active swimmers. The variety of forms present in the Early Cambrian suggests that the immediate soft-bodied ancestors must have been quite diverse. Though there may have been a late Precambrian evolutionary radiation, this variety was greatly enhanced when animals with hard parts appeared. The hard part could perhaps have been developed in response to the evolution of the first carnivores (Hutchinson, 1961), the only change required at the start of the Cambrian being for some scavenging worm to start eating animals that were still alive instead of those that were already dead. All the other developments could have stemmed from this change. In this connection, it is relevant to note that animal groups (like the graptolites, fish and corals) where the skeleton is not for external protection, do not appear until after the Lower Cambrian.

Stanley (1976) considered that the known Cambrian faunas could all be descended from the animals represented in the Ediacara fauna. He pointed out that, during the first part of Early Cambrian time (the Tommotian), the only groups additional to those in the Ediacara fauna were: hexactinellid sponges, archaeocyathids, inarticulate brachiopods, gastropods, hyolithids, and a few problematical groups. The remaining Early Cambrian taxa (including the trilobites and the articulate brachiopods) appeared in the later parts of the Early Cambrian.

Stanley concluded that two basic factors could have been responsible for the widespread development of hard parts in these different groups: first, protection from predators; and second, the increase in efficiency and performance brought about by the development of hard parts, which allowed many taxa to undergo adaptive radiation and relatively rapid diversification. The hard parts act as body supports and as muscle attachments, and it is difficult to conceive of a trilobite or a brachiopod functioning without their hard parts. Their soft-bodied ancestors must have been very different types of animals.

Cambrian

Cambrian time represents perhaps 80 million years, and is important as providing the first rocks containing abundant fossils of animals with hard parts (see Precambrian chapter). However, few Cambrian fossil assemblages have been described from Britain and north-west Europe; they are thus represented in this book by only four sea floor reconstructions designed to give a few glimpses of Cambrian life rather than to be a comprehensive sample.

The Cambrian system takes its name from the Cambrian mountains of Wales, where it was first recognized by Adam Sedgwick. The Harlech Dome in North Wales consists of a thick sequence of largely deep-water sediments in which fossils are hard to find, and similar rocks are also found in the tectonically deformed Caernarvonshire Slate Belt. However, sediments deposited under relatively shallow water, in which fossils are more common, occur as small outcrops in south Wales, particularly near St David's, in the Welsh Borderland and in the English Midlands, particularly near Nuneaton (Rushton, 1974). Our four reconstructions are typical of these shallower water facies.

Cambrian fossils are rare in Scotland, but molluscs and trilobites from near Durness, Sutherland, also occur in rocks of the same age in Spitzbergen, Greenland, eastern Canada and Arctic Canada, indicating, as in the Ordovician, a northern continent separated by the Iapetus Ocean from England, Wales and Scandinavia (Fig. b).

1 Cruziana-Lingulella Community

In many localities throughout the world, Cambrian sediments show cross-bedding, ripples and other signs of deposition under relatively shallow water. In many places the only fossils are tracks and trails, which were partly due to the work of sabellid worms, and partly to the various activities of trilobites (Crimes, 1970; Seilacker, 1967). These sediments seldom contain the remains of the trilobites themselves, which suggests that these arthropods, like some today, spent only part of their time in shallow-water areas, and retreated,

52

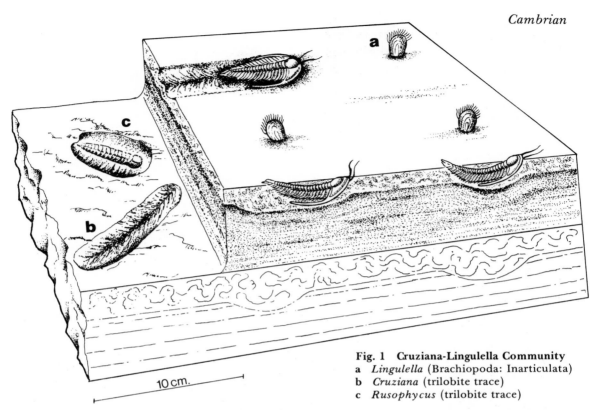

Fig. 1 Cruziana-Lingulella Community
a *Lingulella* (Brachiopoda: Inarticulata)
b *Cruziana* (trilobite trace)
c *Rusophycus* (trilobite trace)

10 cm.

perhaps with the tides, to deeper waters at other times. Trilobites have an exoskeleton which is commonly preserved as a fossil and consists of a thick dorsal body cover of calcite set in a framework of protein, and a very thin ventral skin of scleroprotein (which also covers the limbs), which is only preserved under exceptional conditions. The dorsal cover includes segments which are best developed in the middle part of the trilobite; in some trilobites the relative movement of these segments was slight, but in others there was considerable flexibility between adjacent segments so that the animal could roll up (like a woodlouse) for protection. On the ventral side, where protection was less important, the cover was very thin, but the appendages had sufficient strength for burrowing, swimming and feeding. Trilobites were mostly deposit feeders, but a few were probably carnivores (Bergström, 1973).

In the *Cruziana-Lingulella* Community the trilobite burrows take various forms. Some are elongate traces probably made while the animal was ploughing through the sea floor, and these forms are known as *Cruziana*. (In palaeontology it is more convenient to name animal traces by a different set of generic and specific names from those for the animals themselves; this allows a stable nomenclature to develop, even though it may be quite uncertain which trace belongs to which trilobite species). *Cruziana* is characterized

53

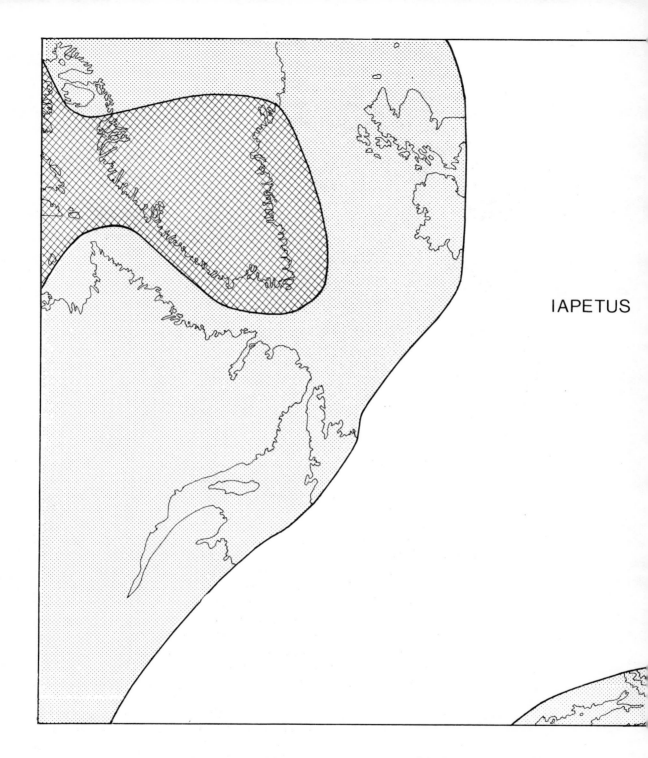

IAPETUS

Fig. b. *Geography of the North Atlantic region in Cambrian times (oceans after McKerrow and Ziegler, 1972; land after Cowie, 1974, p. 150). The trilobite faunas of North America are distinct from those of northern Europe;*

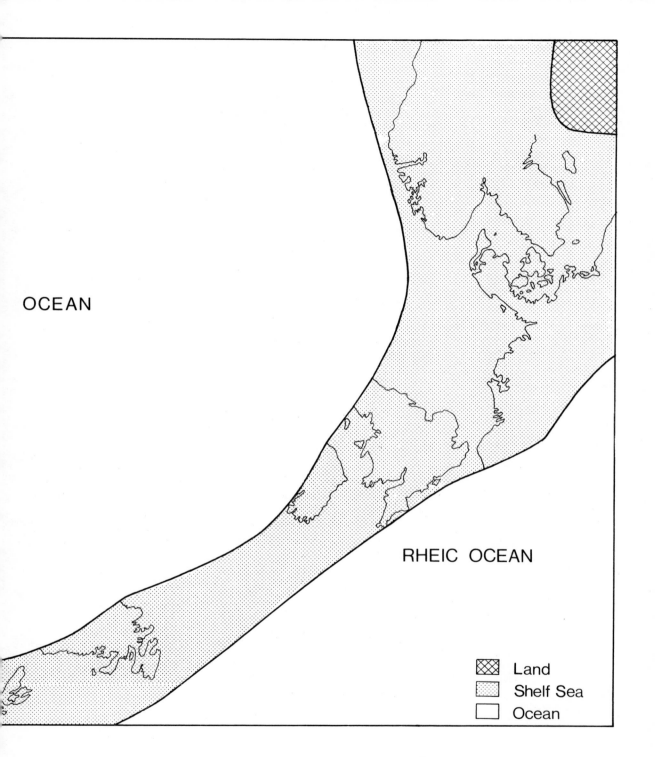

OCEAN

RHEIC OCEAN

Land
Shelf Sea
Ocean

Community 8 is North American, but the remainder of the Cambrian and Ordovician communities illustrated are from northern Europe. The details of regions away from outcrops of Cambrian rocks are speculative.

by a double set of oblique grooves which were made by a set of appendages on each side of the trilobite. Extra oblique grooves, presumably made by further appendages, can also be present. Sometimes longitudinal grooves can also be seen, which are thought to have been produced by rigid spines dragging over the sea floor. *Cruziana* is the result of a trilobite in motion through the sediment, but sometimes trilobites dug resting places in the sediment; they then produced an oval trace, known as *Rusophycus*. Leg impressions are also known where trilobites appear to have disturbed the sediment surface only gently.

Fossils other than trilobite traces are rare in this community, apart from the inarticulate brachiopod *Lingulella*, which can be very abundant in some areas, often crowded along bedding planes. *Lingulella* occurs thus in the Upper Cambrian rocks of Wales, where shallow marine environments became widespread during the deposition of the Merioneth Series. *Lingulella* is generally not so elongate as its later relative *Lingula* (which survives with little change in its hard parts from the Ordovician to the present day). *Lingula* has a burrowing mode of life unique among brachiopods. Early Cambrian *Lingulella* is not known to occur in burrows, but may nevertheless have lived partially buried in the sediment.

2 Shallow Water Shelly Community

In slightly more open sea environments, Cambrian bottom-dwelling animals become more diverse than in the *Cruziana-Lingulella* Community. Trilobites are the dominant group of fossils. *Paradoxides* was a large trilobite (up to 40cm long), which lived in a range of marine environments; *Solenopleura* was smaller and had fewer segments than *Paradoxides* and it also had a larger tail; *Bailiella* was very similar to *Solenopleura*, but had no eyes; nor did *Agnostus*. Many blind trilobites appear to have had long and successful evolutionary histories; so eyes were clearly not essential to all these animals. Trilobite eyes often covered a wide field of vision (Clarkson, 1966, 1966a). They were probably primarily useful in the detection of predators, but we have little idea what the predators were. However, in general, the majority of the blind trilobites seem to have been forms living in deeper water, where the amount of light available is small; shallower water blind forms like *Bailiella* must have relied on other senses for protection.

In addition to the trilobites, hyolithids and brachiopods are usually present in this community. Hyolithids are conical calcareous shells, sometimes with a covering plate called an operculum, which were probably inhabited by some epifaunal mollusc. They do not appear to have been closely related to any living molluscan class,

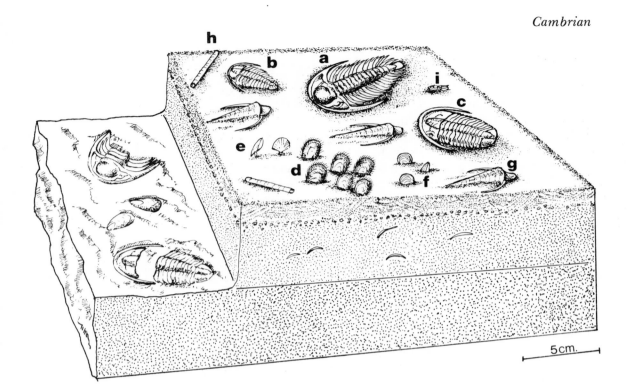

Fig. 2 Shallow Water Shelly Community
a *Paradoxides* (Arthropoda: Trilobita)
b *Bailiella* (Arthropoda: Trilobita)
c *Solenopleura* (Arthropoda: Trilobita)
d *Lingulella* (Brachiopoda: Inarticulata)
e *Billingsella* (Brachiopoda: Articulata: Orthida)
f *Micromitra* (Brachiopoda: Inarticulata)
g *Hyolithes* (Caliptoptomatida)
h *Hyolithellus* (uncertain affinities)
i agnostid (Arthropoda: Trilobita)

and it is not certain how the animals functioned. The Cambrian brachiopods are mainly inarticulate forms (with no hinge teeth) and have a dark phosphatic shell. Most inarticulate brachiopods are epifaunal, being attached to the sea floor by their pedicle, but a few, like *Micromitra*, had no pedicle and probably rested on the sea floor. Some articulate brachiopods also occur, differing from inarticulates in having a tooth and socket system which keeps the two valves in place. Articulate brachiopods all have a shell made of calcite. Many later forms developed a variety of internal plates and other calcite structures, but most Cambrian articulates, like *Billingsella*, have no internal hard parts other than teeth and sockets, and *Billingsella* is probably ancestral to the great majority of later articulates. All brachiopods, both inarticulate and articulate, are filter feeders. The trilobites, hyolithids and brachiopods can all occur in sands, shales or limestones; some species were tolerant of varied shallow marine environments, whilst other species are known only from one particular kind of rock.

Calcareous algae, which can be of any size from microscopic to a meter or so in diameter, are generally restricted to clearwater areas. *Girvanella* is a small intertwined tube-like form, which is locally common in Britain.

3 Deeper Shelf Community

In those deeper parts of the shelf sea floor which were still well oxygenated, the Cambrian bottom-dwelling faunas show an increase in diversity. Apart from the especially high diversities associated with shallow coral reefs, this increase in number of species with increase in depth is generally characteristic of benthic communities from the Cambrian to the present day. In the deeper parts of the shelf fifteen or more genera of trilobites may occur at a single locality. Since food is scarcer at great depth, a greater variability and size range of animals is developed to exploit to the full the different types of possible feeding strategy available.

Trilobites are the most abundant group of animals in the deep shelf areas of the Cambrian. The four genera illustrated all occur in the Tremadoc, a time division which is placed at the end of the Cambrian in Britain, but at the start of the Ordovician in some parts of the world. *Euloma* is similar to its older relative *Solenopleura*. *Agnostus* is a small trilobite (usually less than 1cm long) which is often found with its tail folded up under its head, though it was probably a deposit feeder like many other trilobites and may only have curled up when threatened. Although nothing is known of the limbs of *Agnostus*, there is good reason to believe that *Agnostus* swam near the sea floor, in the same way as many arthropods today. *Asaphellus*, like *Agnostus*, had a head and tail equal in size, but it is a larger trilobite than *Agnostus*, with a flattened rim to its carapace. It probably lay flat on the muddy sea floor, and the rim could have helped prevent it from sinking into the mud. *Platypeltoides* is similar to *Asaphellus*, but it has no flattened rim and is smoother. This smooth outline may have assisted the animal to burrow through the sediment.

Other bottom-dwelling organisms present in this well-oxygenated deeper shelf include brachiopods and molluscs. Articulate brachiopods are nearly all restricted to shallower areas during the Cambrian, but inarticulate forms, not very different from those found in shallow water, occur sporadically. Two groups of univalved molluscs can be present in small numbers: the hyolithids, such as *Hyolithes*, whose straight conical shells are generally more abundant in shallow-water sediments, and the gastropods, which have coiled shells. Plano-spiral bellerophontids are common in the Cambrian and Ordovician; those with thick shells probably crawled on their foot along the sea floor, just like many modern marine snails.

The earliest graptolites, which were bottom-encrusting forms, occur near the top of the Middle Cambrian. However graptolites were not widespread until planktonic forms evolved. One grapto-

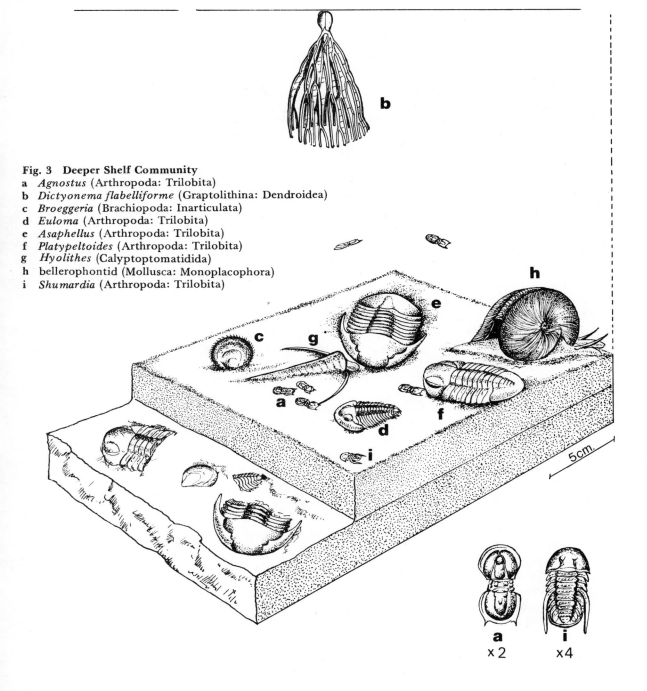

Fig. 3 Deeper Shelf Community

a *Agnostus* (Arthropoda: Trilobita)
b *Dictyonema flabelliforme* (Graptolithina: Dendroidea)
c *Broeggeria* (Brachiopoda: Inarticulata)
d *Euloma* (Arthropoda: Trilobita)
e *Asaphellus* (Arthropoda: Trilobita)
f *Platypeltoides* (Arthropoda: Trilobita)
g *Hyolithes* (Calyptoptomatidida)
h bellerophontid (Mollusca: Monoplacophora)
i *Shumardia* (Arthropoda: Trilobita)

lite with a world-wide distribution in Tremadoc time was *Dictyonema flabelliforme*, which either floated by itself or may have been attached to some other floating object, such as seaweed, by a long thread.

4 Olenid Community

The Olenid Community is characteristic of relatively stagnant bottom conditions under a variety of water depths. When bottom currents are absent, excess plant material accumulates on the sea floor, and the action of bacteria leads to oxygen deficiency and a high sulphide content in the bottom sediments. This environment occurs particularly in seas which were wholly or partially isolated from the main oceans, like the Black Sea today, and is represented by rocks in places throughout the stratigraphical column. The sediments are usually bituminous shales; currents are seldom strong enough to transport any coarser material. In these adverse conditions the diversity of bottom dwellers is never high, regardless of water depth. In the Cambrian the fauna of stagnant environments usually consisted of just a few trilobites; only olenids and some agnostids appear to have adapted enough to be able to tolerate these conditions, such as the small agnostid *Lotagnostus*, and the larger *Olenus, Peltura* and *Ctenopyge*. Characteristically only one or two of the forms are common at any one locality.

Agnostids can occur in various environments, but the other trilobites are only common when bottom circulation is reduced. The olenid family is characterized by having a broad head and a small tail, and in most olenids, such as *Peltura* and *Ctenopyge*, the eyes are situated in the normal place, one each side of the glabella, on slightly raised areas on the upper surface. However, in some other olenids the eyes are spherical and lie far back, or down on the sides of the head, much increasing the all-round field of vision. *Ctenopyge* has many prominent spines, and probably swam just above the sediment surface. None of these trilobites burrowed; there was probably an abundance of organic food on the sea floor. Their problem would not have been a shortage of food, but a shortage of oxygen for respiration. Compared with other trilobite groups, olenids have a relatively large number of segments and, since each segment would have had its own gill branch, this would no doubt have assisted them to obtain enough oxygen.

Few other fossils occur in this environment during the Cambrian, but some flattened masses with lath-shaped components can be found; these may be algal filaments (Taylor and Rushton, 1971). Olenid communities have also been described from the Ordovician (Fortey, 1975).

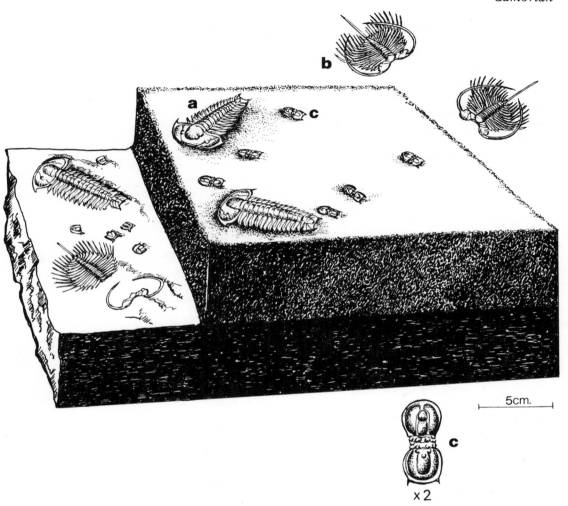

Fig. 4 Olenid Community
a *Peltura* (Arthropoda: Trilobita)
b *Ctenopyge* (Arthropoda: Trilobita)
c *Lotagnostus* (Arthropoda: Trilobita)

Ordovician

The Ordovician period probably saw a greater expansion within the animal kingdom than any other similar time interval. The trilobites, which were the dominant marine animal group of the Cambrian, continued to diversify during the Ordovician but, in proportion to the other animals present, the trilobites no longer had their former dominance. However, new developments, like large raised eyes in phacopids and numerous long spines in odontopleurids, suggest that some trilobites spread out into new ecological niches. The same is true for the inarticulate brachiopods, which continued in about the same relative abundance as in the Cambrian.

By contrast, the articulate brachiopods increased greatly in abundance, in numbers of genera, and in the colonization of the sea floor in areas of deeper water. The brachiopods were, and are, all suspension feeders, circulating water containing organic matter through their mantle cavities in order to extract their food. This food mainly originated from marine phytoplankton, which was most abundant in shallow waters. There is little doubt that in the Lower Palaeozoic (as at the present day) there was a gradual decrease in the quantity of food for suspension-feeding benthos with increase in the distance from shore. Perhaps the gradual colonization of the deeper environments by the brachiopods during the Ordovician may be related to some increase in efficiency of the brachiopod feeding mechanisms. Although the most primitive articulates (the orthides) probably had a fairly simple lophophore (the orthide lophophore is not preserved fossil) they diversified to a great extent. Other articulate groups to develop from the orthides in the Ordovician were the rhynchonellides, the spiriferides and the pentamerides; probably all of these had more advanced lophophores than the orthides. The strophomenides developed in a different way: many of these lost the use of their pedicles in the adult and developed curved shell shapes which allowed them to rest on soft substrates without sinking.

In the Cambrian, most of the graptolites belong to the order Dendroidea. The first dendroids were sessile, but in the Late Cambrian (Tremadoc) *Dictyonema flabelliforme* and other pelagic forms appeared, enabling fast world-wide dispersal, useful for zonal dating. The Early Ordovician saw a great expansion of pelagic graptolites with the rise to dominance of the order Graptoloidea.

The molluscs are more varied in their modes of life than most other phyla. The Cambrian bivalves were mostly deposit feeders, but in the Lower Ordovician considerable diversification took place and the teeth (which are of value in classification) show that most modern bivalve orders appeared at this time. Some of these new forms developed their gills so that they could feed on material suspended in the seawater. This adaptation is frequently accompanied by a byssal mode of attachment. Deposit-feeding forms never had a byssus and usually nestled in the sand (they could not burrow to any great depth, as bivalves did not develop siphons until the Upper Palaeozoic).

The gastropods were probably mainly bottom-dwelling scavengers; they only diversified to a limited extent during the Ordovician. The third major group of molluscs are the cephalopods; during the Lower Palaeozoic they are mainly represented by the nautiloids. Most Palaeozoic nautiloids had straight shells (orthocones), but in the Ordovician curved forms appeared and also forms with other shapes. The nautiloids also showed diversification in the shape of the aperture and in many internal features. Most of this nautiloid radiation occurred quite quickly in the Lower Ordovician. Nautiloids (like *Nautilus* today) all have shells divided by calcareous plates into gas chambers; modern cephalopods are known to be able partially to fill these chambers with liquid in order to adjust their buoyancy. It is probable that most nautiloids swam near the sea floor scavenging organic material. Like most modern scavengers, the nautiloids were seldom as abundant as the suspension feeders (brachiopods and bivalves), but in some Ordovician carbonate environments (like parts of Scandinavia and North America), where perhaps algae were present in quantity, nautiloids were the most abundant animals with shells.

Many bryozoan orders appeared for the first time at the base of the Ordovician. Bryozoans are similar to the lophophore-bearing brachiopods in many ways, except that they are colonial; like the brachiopods, bryozoans are often more abundant in shallower waters where perhaps the food supply for these suspension feeders was greatest. In the Palaeozoic, bryozoans were sufficiently abundant to form the major proportion of certain reefs.

There are two large groups of Palaeozoic corals: the Tabulata and the Rugosa. The tabulates appeared in the Late Cambrian (Tremadoc); they were all colonial corals with very simple internal structures, but with a wide variety of colony shapes. Solitary rugose corals appeared for the first time in the Arenig in shallow carbonate environments. Rare compound rugose corals are first known in the Caradoc, but they do not become common until the Silurian.

The echinoderms are mainly represented by Crinozoa in the Ordovician. Of these, the most widespread group is the Crinoidea, which first appeared in the Cambrian, but became abundant in the Ordovician. Crinoids were either cemented to a hard substrate or anchored in sediment by a stem made up of numerous cylindrical plates. The cup at the head of the stem contained the mouth, digestive tract and reproductive organs; it was supplied with food by currents set up along the arms (usually five, ten or twenty in number). Crinoids, like the other Ordovician suspension feeders, were usually more abundant in shallow water environments. Other groups of Crinozoa can also be locally very abundant; for example, the cystoids are a major component of some Middle and Upper Ordovician limestones, particularly in North America and Scandinavia. Echinoids are very rare indeed during the Ordovician.

The Agnatha are a primitive group of fish-like animals: some fragments which occur in the Arenig may belong to this order, but more complete fossils assigned to this group did not appear until the Caradoc. True fish (Pisces of Tarlo, 1967) did not develop until the Silurian.

The Ordovician can be divided into five series:

Ashgill	(after a stream in the English Lake District)	Upper
Caradoc	(after a hill in Shropshire)	Middle
Llandeilo	(after a town in Dyfed)	
Llanvirn	(after a farm in Dyfed)	Lower
Arenig	(after a mountain in Gwynedd)	

The Ordovician System was first defined by Lapworth (1879) to resolve the differences between Sedgwick (who wished to include it in his Cambrian System) and Murchison (who claimed it for his Silurian System). Lapworth defined the base of the Ordovician as the base of the Arenig Series in north Wales; this is still the normal practice in Britain (and the one followed in this book), though many geologists in Europe and North America include the underlying Tremadoc in the Ordovician.

Many British geologists regard the Caradoc Series as part of the Upper Ordovician, but for the purposes of this book we group the series as above. During the Arenig and Llanvirn, brachiopods, bryozoans and molluscs were largely confined to the shallower parts of the shelf. In the Llandeilo and Caradoc these three groups expanded in numbers and in genera and the crinoids became important; there was also an ecological expansion of these bottom dwellers into the deeper parts of the shelf, a region hitherto mainly inhabited by trilobites. In the Late Ordovician, this process continued more slowly, with slightly deeper areas being colonized by bottom dwellers. By the end of the Ashgill, evolution of the common brachiopod and trilobite genera, sometimes accompanied by a change of ecological niche, had produced a new set of animal communities.

The Ordovician period lasted from about 519 to 438 million years ago: this interval represents over half of the total time of the Lower Palaeozoic (which lasted from about 570 to 411 million years ago).

The faunal provinces seen in the British Cambrian rocks continue into the Ordovician (Fig. c), with Scottish and most American bottom-dwelling faunas remaining very different from those in the equivalent communities in England and Wales. This difference is due to the great width of the Iapetus Ocean which extended from west of Norway, along the border between Scotland and England, through the centre of Ireland and of Newfoundland, and down the northern Appalachians to Connecticut. In the Arenig, the ocean was wide enough to prevent most animals from crossing. The only groups which were widespread on both sides are *Dictyonema* (which had crossed in the Tremadoc), some related pelagic dendroid graptolites and some conodont-bearing animals (conodonts are microscopic structures probably representing parts of the feeding apparatus of some unknown extinct animal). Many conodontophorids are geographically restricted and were probably nekton or benthos (Barnes and Fahraeus, 1975), but the widespread distribution of some forms suggests that they belong to pelagic animals; they were certainly among the first to be able to cross freely the gradually closing Iapetus Ocean. Correlation by means of the less restricted conodonts (Bergström and Cooper, 1973) shows that few graptolites crossed this ocean until the appearance in Wales of *Didymograptus bifidus* (whose ancestors are known only in North America) at the start of the Llanvirn. Though some graptolites could cross freely after the Llanvirn, many others remained provincial until later in the Ordovician.

Brachiopods and trilobites were both bottom-dwelling groups, but with free-swimming larval stages. There are occasional records of certain genera crossing the Iapetus Ocean prior to the Caradoc, but the vast majority of forms were different on both sides of the ocean. The agnostid trilobites, for example, seem to appear more often on both sides of the ocean than other groups: it may be that they were pelagic, or had longer-lived larval stages than most other trilobites. In the Caradoc, many brachiopod genera are common to both sides of the ocean, although the species are in most cases different. This suggests that while there was some connection between the species, the ocean still formed a barrier to free migration, and that after the original crossing further mixing was prohibited.

It has been suggested that oceanic islands may have played an

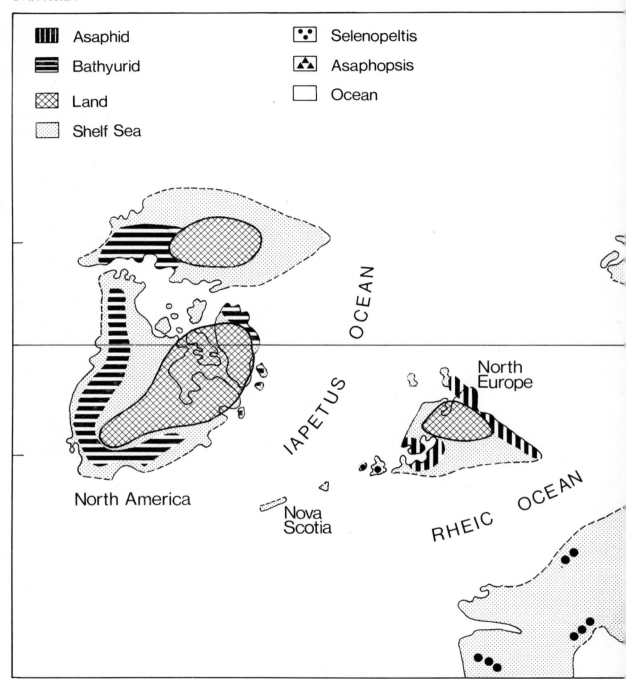

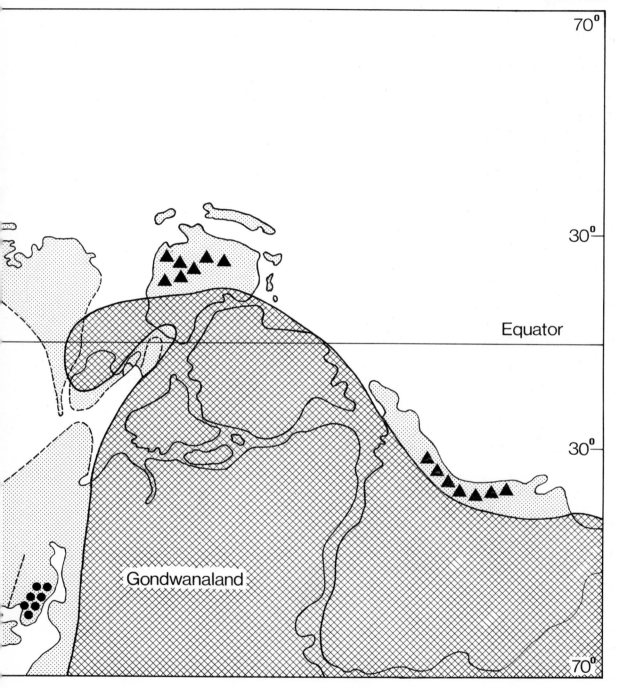

Fig. c. *The world during the Ordovician (land areas and positions of continents after Barnes and Fahraeus, 1975, and McKerrow and Ziegler, 1972; Lower Ordovician trilobite provinces after Whittington and Hughes, 1973).*

67

important role at this time, both as links between the continents where the pelagic larvae could settle and as evolutionary centres where the ecological pressures would be less than on the continents. Some unusual Ordovician brachiopod associations have been found in volcanic areas in the northern Appalachians (Neuman, 1972) where some forms occur earlier than on the adjacent continental areas.

During the Llandeilo, the volcanic rocks in the Lake District may have been part of an island arc off the continental areas of England; these volcanic rocks are followed by Caradoc sediments which contain brachiopod and trilobite faunas similar to those of the Welsh Borderland. Later in the Caradoc, and in the Ashgill, genera appear which previously were present at Girvan (in southern Scotland) on the other side of the Iapetus Ocean from the Lake District. By the end of the Ashgill nearly all brachiopod and trilobite species are common to both sides, showing that the ocean was now narrow enough for the pelagic spat to cross freely. This distance was probably around 2,000 to 3,000km (McKerrow and Cocks, 1976).

Many small bivalved ostracodes are different from most bottom-dwelling marine crustaceans in that they do not have a pelagic larval stage. Ostracodes are still alive today and it is known that they can lay their eggs on the sea floor from which their young hatch out directly. They have thus no means of crossing very deep water, no matter how narrow this water may be. The Iapetus Ocean did not close completely in Britain nor the northern Appalachians until the Devonian, but it closed slightly earlier to the west of Norway. It is not until late Silurian times that the ostracode faunas are similar on both sides; this is thus probably the time when the Iapetus Ocean first closed between Norway and Greenland. When the Iapetus Ocean finally closed altogether in the Lower Devonian, the freshwater fish faunas became the same on both sides. This record of progressive mixing of different animal groups thus provides the best documentation of the closing Iapetus Ocean. North America and the land around the Baltic Shield (of which England was a part) were only two of the four or five large continents in the world during the Ordovician (Hughes, 1973); other oceans opened and closed at different times, but so far the palaeontological record of these has not been fully studied.

The examples that follow are only a small selection of Ordovician communities. Apart from the Lower Ordovician Carbonate Community (8), they are drawn from Wales and the Welsh Borderland. Similar communities occur to the east of the Appalachians, but the majority of North America is in a different faunal province. Bretsky (1970) and Walker (1972) have described some Ordovician brachiopod communities from the North American faunal province east of the Appalachians.

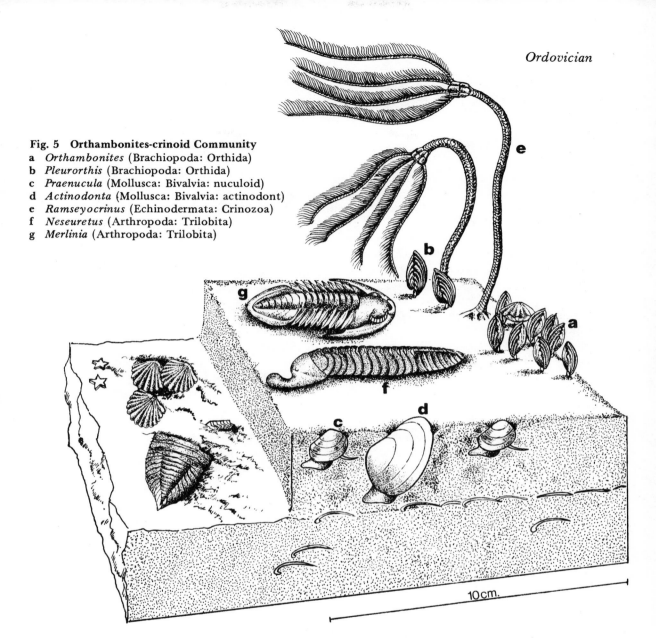

Fig. 5 Orthambonites-crinoid Community
a *Orthambonites* (Brachiopoda: Orthida)
b *Pleurorthis* (Brachiopoda: Orthida)
c *Praenucula* (Mollusca: Bivalvia: nuculoid)
d *Actinodonta* (Mollusca: Bivalvia: actinodont)
e *Ramseyocrinus* (Echinodermata: Crinozoa)
f *Neseuretus* (Arthropoda: Trilobita)
g *Merlinia* (Arthropoda: Trilobita)

10cm.

Lower Ordovician   5   Orthambonites-crinoid Community

In the Lower Ordovician most bottom-dwelling invertebrates, with
the exception of the trilobites, lived in shallow water. Since these
were mostly suspension feeders, like brachiopods, molluscs and
echinoderms, they preferred the turbulent environments where the
food supply was most abundant. The trilobites, being chiefly
deposit feeders, were not so restricted. Unlike later benthic communities, Ordovician communities were thus more diverse in shallow water than in the deeper parts of the sea. They also differed

69

from those of the Cambrian, where suspension feeders made up a smaller proportion of the total fauna. But trilobite diversities were different: in clastic sediments the diversity of trilobites generally increased with depth during the Lower Palaeozoic, but the most diverse of all communities were in certain shallow water carbonate environments.

A typical example of a rich Lower Ordovician (Arenig) shallow water fauna occurs in south-west Wales (Bates, 1969). At this time, the brachiopods were dominated by impunctate orthids like *Orth-ambonites* and *Pleurorthis,* but before the end of the Lower Ordovician many other groups of articulate brachiopods made their first appearance. Both these genera have strong radial ribs which may have served to strengthen the shell, but Rudwick (1964, 1970) has shown that strong ribs also help to keep sand grains out of the brachiopod interior while admitting a sufficient flow of water for feeding and respiration. Ribbed brachiopods are thus especially common in turbulent sandy environments. These genera also have large pedicle openings and were probably anchored by strong pedicles to shell fragments on the sea floor.

The bivalves include both suspension feeders and deposit feeders; unlike most modern bivalves in this environment they had not developed ribs. *Praenucula* (a deposit feeder) and *Actinodonta* (probably a filter feeder) are the commonest genera of the five which have been recorded in this community; they have moderately strong teeth which serve to keep the two valves interlocked.

In addition to the brachiopods and bivalves, the echinoderms, gastropods and bryozoans are also present in the epifauna. The echinoderms include crinoids like *Ramseyocrinus* which can occur in large numbers, cystoids, and asterozoans.

The trilobites were only represented by a few genera in this shallow environment where sands with little organic matter were the dominant sediments; most trilobites, being deposit feeders, were associated with shales which contain a higher proportion of organic matter. *Neseuretus* and *Merlinia* are two genera which were apparently able to feed in sand and silt: they are the only trilobites present at many localities.

6 Asaphid Community

In deeper water, provided it was not stagnant, the trilobites were more abundant and diverse while the suspension feeders decreased greatly in importance. The trilobites included such varied forms as the large *Asaphus, Ogygiocaris* and nileids and the small blind tri-nucleids. Many asaphaceans have flattened margins to their shell which probably assisted them on soft muddy substrates. *Asaphus* is distinguished from *Ogygiocaris* by having a wider axial region.

Fig. 6 Asaphid Community
a *Asaphus* (Arthropoda: Trilobita)
b *Ogygiocaris* (Arthropoda: Trilobita)
c *Homalopteon* (Arthropoda: Trilobita)
d trinucleid (Arthropoda: Trilobita)

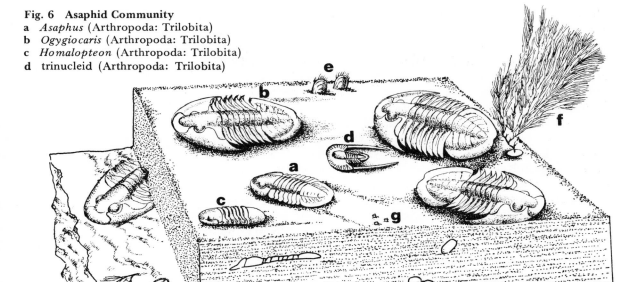

e *Lingulella* (Brachiopoda: Inarticulata)
f *Callograptus* (Hemichordata: Dendroidea)
g *Conotreta* (Brachiopoda: Inarticulata)

Homalopteon, a nileid, is smooth and has much larger eyes than *Asaphus* or *Ogygiocaris*; the smooth outline to the carapace may have been an adaptation for easier movement through the mud on the sea floor. The trinucleids have long genal spines extending back from the sides of their heads; when the small posterior part of this animal rolled up, these spines still projected backwards and would have assisted the animal to remain stable even in soft mud. Different species of trinucleids can be distinguished by variations in the ornament round the front of the head.

A few articulate brachiopods occurred in deeper water environments during the Lower Ordovician, but the deeper brachiopods are mostly inarticulates like *Lingulella* and *Conotreta*, which are both small compared with the trilobites. Their sparseness, as compared with the Cambrian *Lingulella* Community (1), and their size may reflect the small amount of food available in suspension near

71

the sea floor in this deeper environment. Both genera had a functional pedicle, but *Conotreta* probably lived with the flat part of
the larger pedicle valve resting on the sea floor.

Most Ordovician graptolites were pelagic but some, like *Callograptus,* were attached to the sea floor. *Callograptus* resembles its
relative *Dictyonema flabelliforme* (Community 3) in many details,
including the large number of branches; its attached mode of life
was similar to that of the earlier Middle Cambrian graptolites from
which *Dictyonema flabelliforme* and the later pelagic graptolites
were probably derived. The animals which inhabited the numerous
small cups on this colony were probably suspension feeders.

It is unlikely that there were many active carnivores in the
Ordovician, but scavenging worms, cephalopods and other animals
were perhaps present, though only the cephalopods are preserved
as fossils. Scavengers, like predators, are normally considerably less
abundant than the animals on which they feed. The commonest
fossil scavengers in the Lower Ordovician were nautiloids: some of
these were straight orthocones but others had a gently curved shell.
All the nautiloids have gas chambers which may have been partly
full of liquid so that, even when they had a thick heavy shell, the
overall density of the animal was probably close to that of seawater.

Fig. 7 Triarthrus Community
a *Triarthrus* (Arthropoda: Trilobita)
b *Hypermecaspis* (Arthropoda: Trilobita)

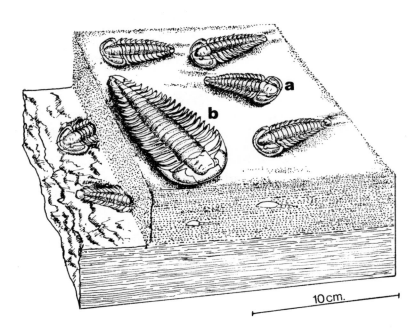

10 cm.

7   Triarthrus Community

In muddy environments, where the water near the sea floor was stagnant and poorly oxygenated, the fauna was sometimes restricted to just two trilobite genera, *Triarthrus* and *Hypermecaspis*. Both are related to the Cambrian *Olenus* (Community 4) and have numerous segments. Since each segment carried a gill branch (for breathing) an increase in gill branches may have increased the oxygen intake.

In Wales and other areas with Ordovician marine deposits isolated from the main oceans, stagnant water seems to have been locally quite common. There are many extensive sedimentary formations in which only *Triarthrus* and *Hypermecaspis* occur as fossils.

8   Carbonate Community

In the Lower Ordovician, carbonates are much more abundant in North America than in England and Wales; they are present in Nevada, New York, western Newfoundland, Greenland and Spitzbergen. In Britain, carbonates similar to those of North America are represented by the Durness Group of north-west Scotland which was probably part of the North American continent at this time, as its faunas are similar. It is probable that most of North America lay in low latitudes during the Early Ordovician. Some carbonates are present around the Baltic Shield; these may also have been near the equator during this time, but their faunas are more similar to the rest of northern Europe. The differences were thus due to isolation, and not to climate. The Baltic area, like England and Wales, appears to have been separated from North America by a wide Iapetus Ocean.

Similar genera of stromatolites, gastropods, trilobites, brachiopods and nautiloids, occur in many Lower Ordovician limestones of Scotland and North America. The stromatolites are the commonest fossils in many of the Durness Group carbonates; they are layered dome-shaped structures built up by miscroscopic algae and bacteria. Today they are restricted to very shallow water and it is probable that after the appearance of grazing invertebrates the maximum depth for fossil stromatolites was about 45m (Walter, 1972). Often associated with the stromatolites are large genera of gastropods like *Maclurites* and *Ceratopoea* which probably grazed on them.

The sediment surrounding the stromatolites may consist of algal debris and pebbles of calcareous mud which contain fragmented trilobites and brachiopods. The most common trilobites are *Hystricurus* and *Bathyurellus*. The low profile of both genera

Fig. 8 Carbonate Community
a *Apheorthis* (Brachiopoda: Orthida)
b *Finckelnburgia* (Brachiopoda: Orthida)
c *Maclurites* (Mollusca: Archaeogastropoda)
d operculum of *Ceratopoea* (Mollusca: Archaeogastropoda)
e *Piloceras* (Mollusca: Endoceratoidea)
f *Tritarthiceras* (Mollusca: Nautiloidea)
g *Hystricurus* (Arthropoda: Trilobita)
h *Bathyurellus* (Arthropoda: Trilobita)

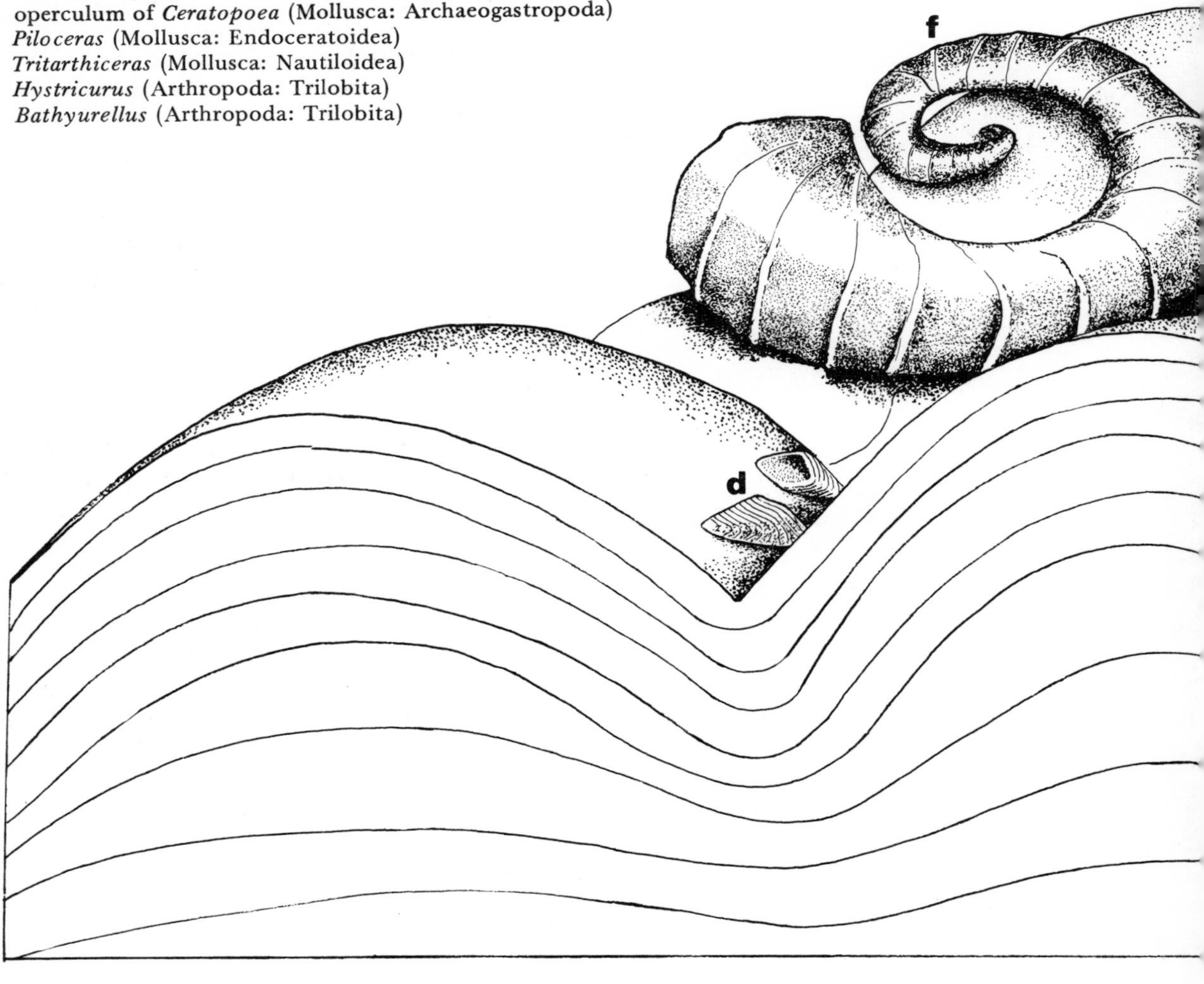

suggests that they were bottom dwellers; they could have moved over soft sediment supported on their long genal spines. Like most trilobites they were probably deposit feeders. Brachiopods are less common than the trilobites. *Apheorthis* and *Finckelnburgia* are two orthid brachiopods which occur as single species clumps between the stromatolites.

The nautiloids and other primitive cephalopods reached their maximum development in the Early Ordovician. Some of the gently curved forms, like *Piloceras*, and some more markedly

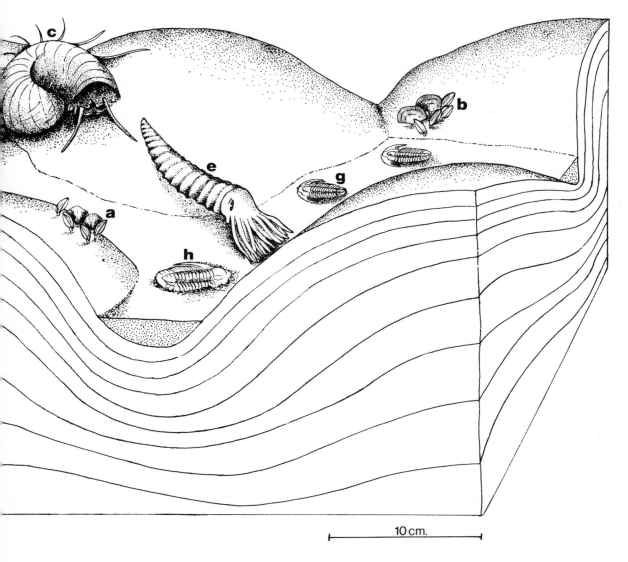

|⊢————— 10 cm. —————⊣|

coiled forms, like *Tritarthiceras,* may have grazed on the sea floor
or may possibly have been buoyant enough to swim.

The Durness Group seldom has more than six or seven different species at any one locality. This is a much lower diversity than
equivalent carbonates in Nevada, where sponges and other animal
groups occur. But even in Nevada, the Early Ordovician carbonates
have a lower diversity than most later carbonate environments.
Corals and bryozoans, which became abundant in many later carbonates, were very rare indeed during the Early Ordovician.

9   Graptolite Assemblage

Pelagic graptolites were abundant throughout the Ordovician and we show as an example a Lower Ordovician reconstruction. Graptolites were colonial hemichordates which formed a protein cup around each individual. These cups are normally arranged linearly along branches. Each colony is attached to a float, either directly or by a thread (the nema).

There are two main types of graptolite: the Dendroidea (e.g. *Dictyonema*) which usually have two types of cup and always have an inner cylindrical canal system (the stolon), and the Graptoloidea which have only one type of cup and never have a stolon preserved fossil. It is the stolon that is the main indication of the graptolites' zoological affinities; a similar structure is present in certain other hemichordates, which are primitive relatives of the vertebrates. The hard parts of a graptolite have been built up in two layers; the outer layer appears to have formed subsequently to the inner layer, suggesting that it was deposited by some outer covering of soft tissue and that the skeleton was thus internal. An internal skeleton is another indication of affinities between hemichordates and graptolites.

Our reconstruction shows two graptolite genera, *Goniograptus* and *Cryptograptus*, covered by a thin film of soft tissue. Both genera belong to the Graptoloidea, the order in which all the cups are the same size and in which no inner stolon is seen.

Like many early Ordovician genera, *Goniograptus* had numerous branches, while *Cryptograptus* had only two. A thin protein thread extends upwards from between the branches of *Cryptograptus*, which, in our reconstruction, is shown at the centre of a spiral ribbon of soft tissue which acted as a float.

Diverse graptolites are abundant in all marine Ordovician deposits where the rate of sedimentation was low. Most species only persisted for a relatively short time, though there were some exceptions. Since most graptolites were free-floating they have a wide distribution. Graptolites are thus the most useful Ordovician fossils for stratigraphical correlation.

10   Dinorthis flabellulum Community   Middle Ordovician

Many genera in shallow marine environments (equivalent to those of the *Orthambonites*-crinoid Community (5) of the Lower Ordovician) became more restricted in their environmental distributions during the later Ordovician. It thus becomes possible to distinguish two shallow water benthic communities: one which was probably restricted to coastal areas not far below the low tide level, and the

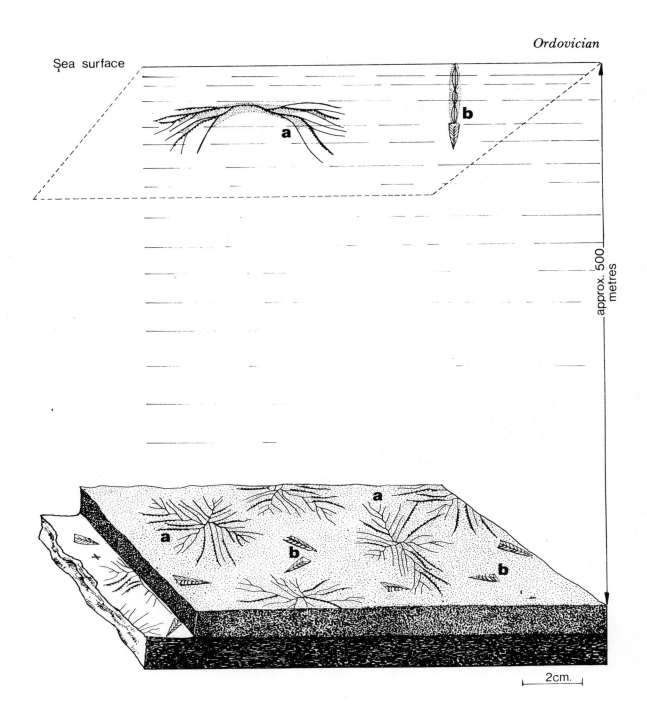

Fig. 9   Graptolite Assemblage
  a   *Goniograptus* (Hemichordata: Graptoloidea)
  b   *Cryptograptus* (Hemichordata: Graptoloidea)

other which occurred in slightly deeper conditions (perhaps in up to 15m of water); in addition to depth, the type of substrate could also be important.

The *Dinorthis flabellulum* Community is one of these near-shore communities. It is common in coarse sandy sediments which display a wide variety of sedimentary structures indicative of very shallow water; in particular it can be found close to the base of transgressive sedimentary sequences (for example the base of the Caradoc of Shropshire).

Only two articulate brachiopods (*Dinorthis* and *Harknessella*) were at all common in shallow water high energy marine environments. They were both orthides occupying similar niches, although they seldom occur together in quantity. Some orthides are punctate, like *Harknessella*, with numerous very small pores extending through the shell at right angles to the surface; others, like *Dinorthis,* are impunctate (with no pores). In living punctate brachiopods (like *Terebratula*) there is no obvious function for these punctations; nevertheless it is clear that their development was significant in many brachiopod groups and they are of importance in the recognition and classification of many brachiopod orders and families. *Dinorthis* and *Harknessella* are distinguished by various internal features and also by the stronger ribs in *Dinorthis*. They both had large pedicle openings and were probably anchored vertically on the sea floor.

The inarticulate brachiopods were represented by the lingulid *Palaeoglossa.* The lingulid family developed the ability to burrow during the Ordovician and can sometimes be found in its vertical life position. By contracting its pedicle a lingulid could obtain the protection of its burrow, and by extension of the pedicle the shell margin could be raised above the sea floor for feeding. This family is unique among brachiopods in having a burrow; this could be an adaptation for environments with periodic sedimentation. *Lingula* itself is the only modern brachiopod capable of surviving temporary emersion and fluctuations in salinity.

Bryozoans are colonial organisms which often secrete a calcareous skeleton for support and protection. This is made up of numerous small cups, each containing an individual; it may take the form of an encrustation (often on a brachiopod shell) or the skeleton may rise above the surface (rather like some small coral) in a dome or with cylindrical branches. The individuals are suspension feeders; they have a lophophore rather like brachiopods. Bryozoans first appear as fossils in the Lower Ordovician and by the Caradoc they were locally abundant in shelf environments. Some, such as *Hallopora*, were sometimes the dominant constituents of the *Dinorthis* Community.

By the Middle Ordovician, early representatives of most of the major modern bivalve groups had developed. These included the

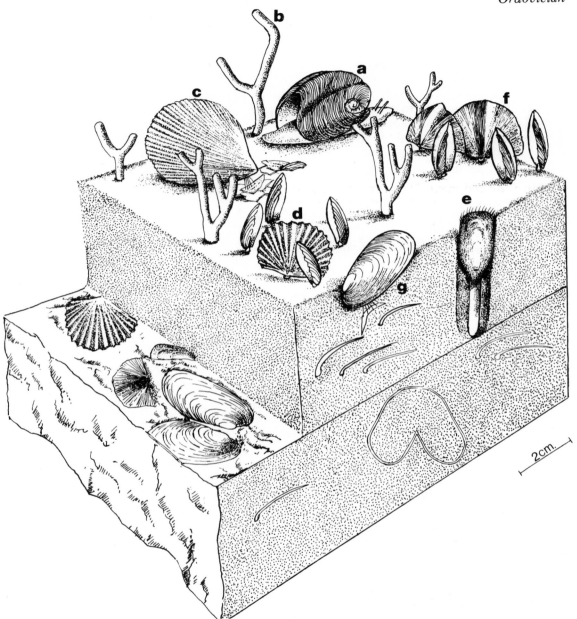

Fig. 10 Dinorthis flabellulum Community
a *Plectonotus* (Mollusca: Monoplacophora)
b *Hallopora* (Bryozoa: Ectoprocta)
c *Ambonychia* (Mollusca: Bivalvia: Pterioida)
d *Dinorthis flabellulum* (Brachiopoda: Orthida)
e *Palaeoglossa* (Brachiopoda: Inarticulata)
f *Harknessella* (Brachiopoda: Orthida)
g *Modiolopsis* (Mollusca: Bivalvia: Modiomorphoida)

79

Ordovician

shallow infaunal deposit-feeding nuculoids which were more common in fine-grained clastic sediments. Byssate bivalves (e.g. *Modiolopsis, Lyrodesma* and *Ambonychia*) appeared in the Middle Ordovician: as they were immobile they must have been filter feeders. *Ambonychia* has fine ribs and *Modiolopsis* is smooth. Other groups present include the shallow, burrowing actinodonts and the deeper-burrowing early anomalodesmatids: these were probably also filter feeders. None of them had well developed siphons, so they either lived on the surface of the sediment or could only burrow to a small extent, since, having no spihon, they would need to have the posterior part of their shells extending above the sediment. Detritus-feeding monoplacophorans are sometimes also present (such as *Plectonotus*).

11 Sowerbyella-Dalmanella Community

In Middle Ordovician shallow water environments with finer clastics *Dinorthis flabellulum* became rarer and several other articulate brachiopods occur (including other species of *Dinorthis*). These were dominated numerically by two genera: *Sowerbyella* and *Dalmanella*. *Dalmanella* is a punctate orthide like *Harknessella*. *Sowerbyella* belongs to a very different group of brachiopods (the strophomenides) which have curved valves with a relatively small space between them for the soft parts of the animal and the mantle cavity. The inner layers of the strophomenide shell contain calcareous spicules (known as pseudopunctations) which are best seen on weathered specimens. *Sowerbyella* had no functional pedicle and lived resting on its convex valve. This position was unstable in fast currents; most fossil *Sowerbyella* are found convex side upwards, having been turned over after death. If they were flipped over while alive it is likely that quick muscle contraction could have righted the shell (Cocks, 1970).

Other, but much less abundant, brachiopods include the strongly ribbed biconvex *Oxoplecia* and the large strophomenide *Kjaerina*. Crinoids such as *Balacrinus* were very abundant at many localities. Both the crinoid stems and the large strophomenide may have provided attachment areas for the pedicles of *Dalmanella*.

Bryozoa in this community are less robust than those in the *Dinorthis flabellulum* Community: they are abundant in places and can include six or seven different species.

Several different genera of scavenging gastropods are present. The commonest Archaeogastropod family is the Pleurotomariidae, which includes *Lophospira*. This family all have a notch on the outer part of their apertures which allowed a symmetrical current to flow over their two gills (many later gastropods have no sinus and only one gill). From the Upper Palaeozoic onwards many

Fig. 11 Sowerbyella-Dalmanella Community
a *Sowerbyella* (Brachiopoda: Strophomenida)
b *Dalmanella* (Brachiopoda: Orthida)
c *Kjaerina* (Brachiopoda: Strophomenida)
d *Oxoplecia* (Brachiopoda: Orthida)
e *Balacrinus* (Echinodermata: Crinozoa)
f *Lophospira* (Mollusca: Archaeogastropoda)

10 cm.

gastropods were predators, drilling through the shells of their prey by means of their radulae. No Lower Palaeozoic shells are known which have been drilled by gastropod radulae and it is probable that most of these Ordovician genera were scavengers.

Trilobites were never very common in Middle Ordovician shallow water environments but a few forms are occasionally present.

12   Diverse brachiopod Community

In the middle to deep shelf areas there was a large increase in the diversity of the bottom-dwelling faunas, especially among brachiopods, during the Middle Ordovician. The articulate orthides and strophomenides are the two dominant orders. The large strophomenides are represented by five or six genera; most of them lay free on the sea floor on the gently curved convex valve (in some genera this was the pedicle valve, in others the brachial valve). One member of this group was *Leptaena*, in which the part of the shell nearest the umbo has strong concentric ribs which helped to keep it stable; the distal parts of the shell were bent sharply upwards to allow the water intake to be clear of the sea floor.

The orthides include both punctate forms (like *Dalmanella* and *Onniella*) and impunctate forms (like *Platystrophia*). At most localities several genera of each type are usually present.

The inarticulate brachiopods are less common than the articulate forms but they are sometimes quite diverse; some collections may contain as many as fifteen articulate and five inarticulate genera.

Although the trilobites continued to diversify until near the end of the Ordovician, this expansion in genera is not reflected by much increase in the number of individuals. Except in the deepest communities, trilobites are always outnumbered by the brachiopods. The trilobites show a variety of different forms. They include the large smooth *Brongniartella*, *Calyptaulax* which has large eyes, and *Platylichas* which has a crenulated head and tail.

The scavengers include bellerophontid monoplacophorans and orthocones.

In the Middle Ordovician, ostracodes became abundant for the first time. They are small crustaceans (often only one or two mm long) with a bivalve shell, which swim near the sea floor. They occur in most marine sediments but are more noticeable in clays and other fine grained sediments. Crinoids, bivalves and bryozoans are usually less common in the deeper parts of the shelf, in contrast with their abundance in the shallow water environments.

The straight calcareous cones of *Tentaculites* were common in this environment. Their zoological affinities and their mode of life are both uncertain, but it is likely that *Tentaculites* is related to

Fig. 12 Diverse Brachiopod Community
a strophomenid (Brachiopoda: Strophomenida)
b *Leptaena* (Brachiopoda: Strophomenida)
c *Onniella* (Brachiopoda: Orthida)
d *Platystrophia* (Brachiopoda: Orthida)
e *Brongniartella* (Arthropoda: Trilobita)
f *Calyptaulax* (Arthropoda: Trilobita)
g *Playtlichas* (Arthropoda: Trilobita)
h bellerophontid (Mollusca: (Monoplacophora)
i orthoceratid (Mollusca: Nautiloidea)
j *Tentaculites* (Cricoconarida)

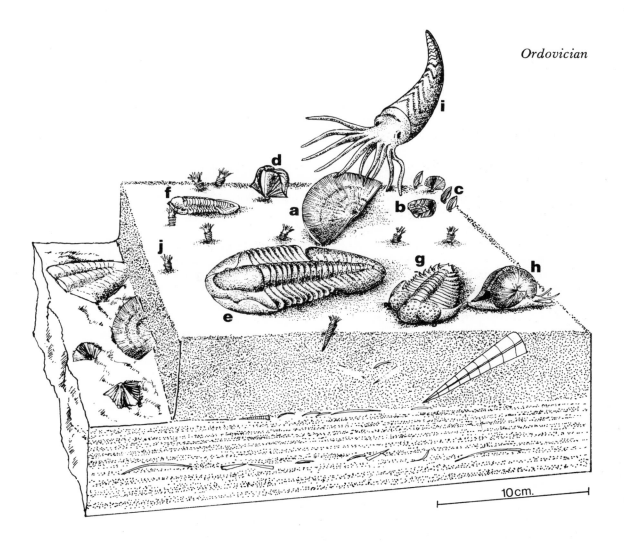

the molluscs and, as it is most common in shallow water deposits, it is likely to have been epifaunal rather than free-swimming or floating. It occurs mostly in siltstones and sandstones; it is thus unlikely to have been a burrowing deposit feeder. Our reconstruction, showing it resting partially buried, takes account of the possible function of the strong circular ribs which may have helped to stabilize it in the sediment.

This community occurs in Wales and England, but in Scotland (notably the Girvan district) different genera are present in similar environments. These faunal differences occur on either side of the Iapetus Ocean (see Fig. d). This old ocean was wide enough during most of the Ordovician to prevent the pelagic spat of trilobites and brachiopods from migrating across, so although some species got across in the Late Ordovician, for most of this period the bottom-dwelling faunas were quite distinct on either side.

13 Trilobite-Onniella Community

Some quiet water or deeper-shelf environments of the Middle Ordovician include many diverse types of trilobites and a few species of brachiopods. The reconstruction shown is based on the (type) Caradoc sections of the Onny River in Shropshire. The two most common trilobites are the trinucleid *Onnia* and the calymenid *Onnicalymene*. The trinucleids had a large convex head with long genal spines which supported the whole animal. The thorax and tail were small in comparison with the head and appear to have been suspended above the surface of the sediment. Appendages below the thorax probably assisted in food collection. *Onnicaly-mene* is typical of the calymenid family in having strongly grooved thoracic segments which enabled it to roll up into a ball (in the same way as a modern woodlouse); it is sometimes found fossil in this attitude. The calymenids are one of the Ordovician trilobite groups which survived into the Silurian.

Although most trilobite communities have a high diversity of genera, only three other forms are illustrated here. *Chasmops* has large eyes which are compound, like all arthropod eyes, but the individual cells are unusually large so that they can be seen easily by means of a hand lens on well preserved specimens. These eyes are raised in semicircular arcs which gave *Chasmops* excellent all-round vision. *Remopleurides* was a swimming trilobite with long narrow eyes; this suggests that though it had good vision laterally it could not see forwards or backwards like *Chasmops*. *Lonchodo-mas* had extremely long genal spines and also a long rostral spine in front of the head. These spines served to support the trilobites above the fine grained sediment; many other trilobites with genal spines are found in this community, which normally occurs in mudstones and siltstones.

The most characteristic brachiopod in this community is the orthide *Onniella* (see Fig. 12). Another brachiopod, which is found locally in large numbers (but not shown here) is the small (4mm) strophomenide *Chonetoidea;* it has a functional pedicle and may either have been attached to algae on the sea floor or have been epiplanktonic, attached to floating material.

Bivalves and gastropods, usually with thin shells, occur sometimes in the trilobite-*Onniella* Community.

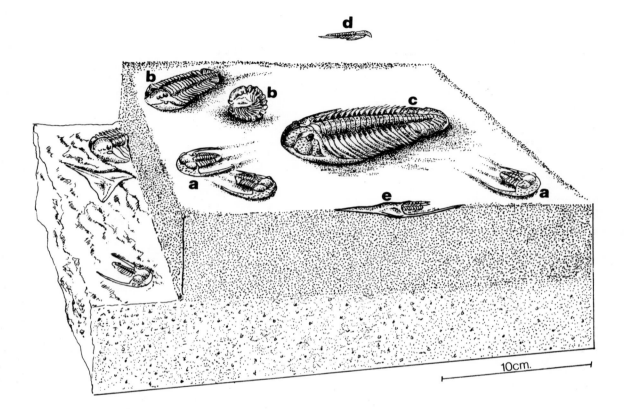

Fig. 13 Trilobite-Onniella Community
a   *Onnia* (Arthropoda: Trilobita)
b   *Onnicalymene* (Arthropoda: Trilobita)
c   *Chasmops* (Arthropoda: Trilobita)
d   *Remopleurides* (Arthropoda: Trilobita)
e   *Lonchodomas* (Arthropoda: Trilobita)

Upper Ordovician   **14   Restricted Hirnantia Community**

During the late Ashgill many assemblages have a very restricted fauna which often consist of only two dominant genera, with perhaps one or two other less common brachiopods together with a few crinoid and bryozoan remains.

85

Fig. 14 Restricted Hirnantia Community
a *Hirnantia* (Brachiopoda: Orthida)
b *Eostropheodonta* (Brachiopoda: Strophomenida)
c bryozoan (Bryozoa: Ectoprocta)

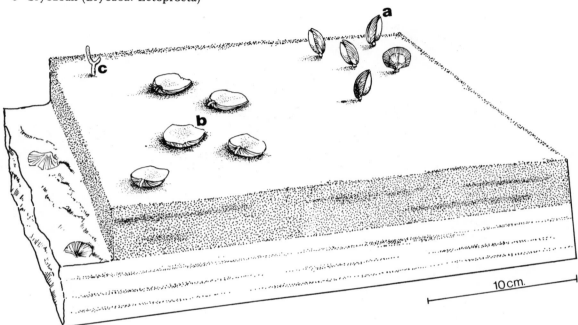

Hirnantia is a large biconvex dalmanellid with weak ribs and distinctive muscle scars. The other brachiopod often occurring in large numbers in this community is *Eostropheodonta*, a large flat strophomenide. *Eostropheodonta* is an early member of a family which becomes common in the Silurian. In addition to the two teeth in the pedicle valve, this family, the stropheodontids, developed denticulations along their straight hinges: these assisted in holding the two valves together when the shell was open for feeding.

The crinoids are usually fragmented and may constitute a fairly large proportion of the sediment. Many areas of the sea floor which were receiving little clastic sediment were probably covered by large numbers of these attached animals. This restricted community also occurs in algal limestones; the algae grow in concentric layers around a nucleus and form subspherical objects 2 or 3mm in diameter (much like large ooliths) and denote growth under shallow clear water; however, not all Palaeozoic crinoidal and algal limestones have a restricted fauna. Trilobites are very rare or absent from this community; this may be due to the substrates having been unsuitable for these sediment feeders.

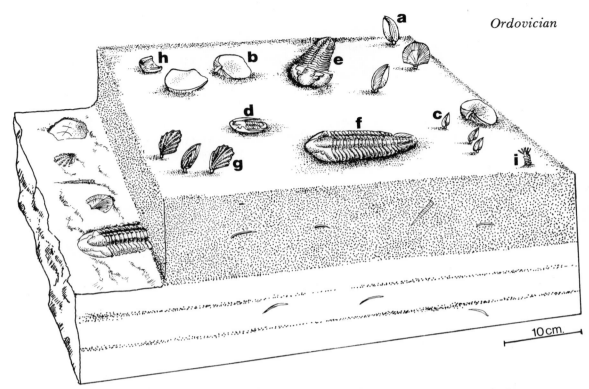

Fig. 15 Diverse Hirnantia Community
a *Hirnantia* (Brachiopoda: Orthida)
b *Eostropheodonta* (Brachiopoda: Strophomenida)
c *Dalmanella* (Brachiopoda: Orthida)
d *Phillipsinella* (Arthropoda: Trilobita)
e *Dalmanitina* (Arthropoda: Trilobita)
f *Brongniartella* (Arthropoda: Trilobita)
g *Plectothyrella* (Brachiopoda: Rhynchonellida)
h *Leptaena* (Brachiopoda: Strophomenida)
i *Tentaculites* (Cricoconarida)

15 Diverse Hirnantia Community

Hirnantia and *Eostropheodonta* also occur in beds with a large number of different brachiopods and other benthic animals. Amongst the other brachiopods are *Plectothyrella* (a ribbed rhynchonellide), *Leptaena*, *Dalmanella* and its relative *Kinnella*. As many as eighteen different brachiopod genera are present in some beds. Trilobites may also be important; as many as six or seven different forms can occur, including *Phillipsinella*, *Dalmanitina* and *Brongniartella*, although the trilobites are heavily outnumbered by the brachiopods. There are also a few bivalves, gastropods, ostracodes and tentaculitids.

This community is very widespread around the world, particularly at the very end of the Ordovician Period. It has been recorded in Wales, Scotland, the Lake District and Ireland, the northern Appalachians, north Africa, Scandinavia, central Europe and Burma. It would appear that few Ashgill oceans were wide enough to prevent the migrations of the faunas. Wales and Scotland, for example, were separated by the Iapetus Ocean at this time (see Fig. d) and Bohemia and north Africa were separated from Wales by the Rheic Ocean. All modern brachiopods and modern crustacea have pelagic larval stages when the young can drift with ocean currents for periods ranging from three to fifty days. It is probable that the young of these Ordovician brachiopods and trilobites went through the same pelagic larval stages, so that only wide oceans with few oceanic islands could act as barriers to migration.

In parts of north Africa, muddy siltstones containing examples of the diverse *Hirnantia* Community are found interbedded with glacial deposits. At this time there was an ice sheet covering much of Africa and South America. Thus the community could tolerate cold conditions in high latitudes, but its wide distribution suggests that it was not confined to polar regions.

16 Christiania-Sampo Community

The Ashgill benthic communities, like those of all ages in open marine flat-bottom conditions, show an increase in diversity with depth. The *Christiania-Sampo* Community characteristically contains fifteen to thirty different brachiopod genera (including *Dinorthis* and *Rafinesquina*, and ten to twenty other forms of bottom-dwelling organisms, chiefly trilobites, bivalves, gastropods, crinoids and bryozoans as well as rarer tentaculitids, sponges and other groups. Because of the great diversity in this community it is not easy to designate characteristic genera but the strophomenides *Christiania* and *Sampo* occur (often in quantities of five per cent or less) in most collections.

Christiania is a small (1cm) smooth brachiopod which probably rested loose on its convex pedicle valve in a similar manner to the Jurassic bivalve *Gryphaea.* In the Caradoc, *Christiania* is absent in England and Wales, but it spread across the Iapetus Ocean from North America and Scotland during the early Ashgill and is found consistently in the deeper water Ashgill shelf environments of Wales.

The trilobites are very varied (only *Tretaspis* and *Brongniartella* are shown) and can sometimes make up twenty per cent of the total fauna. The trilobites, being deposit feeders, were much more dependent on the type of substrate than were the suspension-feeding brachiopods and thus the trilobite genera present in this community vary much from one locality to another, depending on the local substrate.

Fig. 16 Christiania-Sampo Community
a   *Sampo* (Brachiopoda: Strophomenida)
b   *Christiania* (Brachiopoda: Strophomenida)
c   *Dinorthis* (Brachiopoda: Orthida)
d   *Tretaspis* (Arthropoda: Trilobita)
e   *Rafinesquina* (Brachiopoda: Strophomenida)
f   bellerephontid (Mollusca: Monoplacophora)
g   *Brongniartella* (Arthropoda: Trilobita)
h   crinoid (Echinodermata: Crinozoa)

10 cm.

89

17   Tretaspis Community

This community is dominated by trilobites. Up to twenty genera may be present at any one locality, but they are often distributed sporadically throughout the rock and large numbers are not easy to collect. The commonest genera are *Tretaspis, Lonchodomas, Cybeloides* and *Remopleurides*.

Tretaspis is a trinucleid with small eyes (many of its close relatives appear to have been blind) and long genal spines. *Lonchodomas* has a rostral spine projecting forwards from its head. These spines may have been either for protection or to prevent the animal sinking into the muddy sediment; since there were few large predators in the Ashgill the latter explanation seems more likely. In his reconstruction of this community, Bergström (1973) shows *Cybeloides* completely covered by sediment apart from its eyes. These were elevated above its dorsal surface, perhaps to project through the sediment as Bergström thought, or perhaps just to allow the animal good all-round vision. *Remopleurides* had large semicircular eyes; it clearly had all-round vision, as the many cells (which make up all arthropod eyes) face in all directions.

In addition to the trilobites, there were some other groups occurring less commonly in the *Tretaspis* Community; these include two ostracode genera, an asterozoan, and tubes which may have been made by burrowing worms. Some brachiopods are present but they are not abundant; they include the orthacean *Skenidioides,* three different plectambonitaceans, and a small subspherical spire-bearer. All of these brachiopods are closely related to genera which occur in the Silurian *Clorinda* Community; they represent the earliest of the deeper water brachiopod faunas. Prior to the Caradoc, brachiopods are largely confined to shallower water environments.

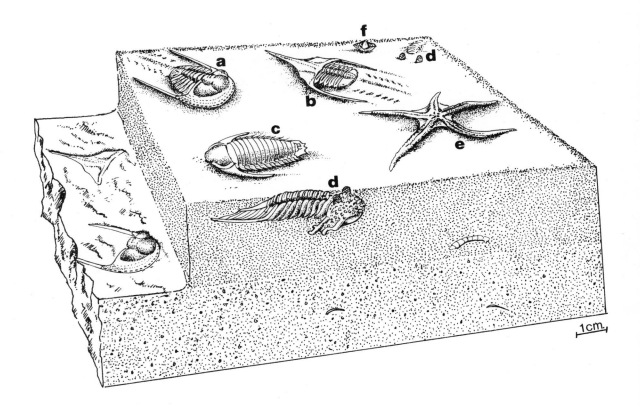

Fig. 17 Tretaspis Community
a *Tretaspis* (Arthropoda: Trilobita)
b *Lonchodomas* (Arthropoda: Trilobita)
c *Remopleurides* (Arthropoda: Trilobita)
d *Cybeloides* (Arthropoda: Trilobita)
e asterozoan (Echinodermata: Asterozoa)
f *Skenidioides* (Brachiopoda: Orthida)

Fig. d. *The distribution of the late Llandovery marine benthic communities of Wales and the Welsh Borderland (after Ziegler, 1965; Ziegler et al. 1968a). The fact that these communities were parallel to the shore (on the east and south) and to the shelf margin (on the north-west) suggests that they are related to the depth of water. Dotted lines indicate the boundaries between communities.*

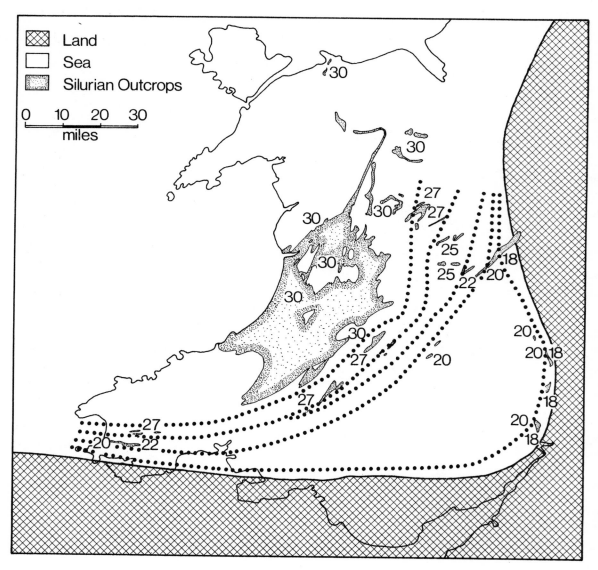

Silurian

The marine faunas in the Silurian were dominated by benthic brachiopods and pelagic graptolites. Although trilobites occur in many environments, they were far less important than in the Ordovician. Molluscs, bryozoans, ostracodes and crinoids are also present, but only common in some habitats. Towards the end of the Silurian, the land was just beginning to be colonized by plants, but animals with hard parts were absent from the land, although some arthropods and worms may have become terrestrial before the end of the period.

The Iapetus Ocean (Fig. e), which in the Ordovician separated the Canadian Shield (including Scotland) from the Baltic Shield (to which England was attached), became progressively narrower during the Silurian. Therefore, in their drifting (pelagic) larval stages, the brachiopods and trilobites were able to cross the ocean freely, and the faunas became the same on both sides of the ocean. In the Lower Silurian these brachiopods and trilobites were world-wide in their distribution; no ocean was wide enough to act as a permanent barrier to their pelagic larvae. Most ostracodes, however, do not have any pelagic larval stage — their eggs hatch out on the sea floor — and, as a result, even a narrow ocean, like the Iapetus in the Silurian, was a barrier to their migration.

The Lower Silurian brachiopods of Britain were one of the first groups of fossils from which communities have been described. Ziegler et al.(1968, 1968a) have concluded that these communities are more closely related to varying depths of water than to varying types of sediment. The evidence for this is, first, a consistent community sequence from the shore to deep water as shown by maps of community distributions at any one time (Fig. d), second, the fact that in the transgressive Lower Silurian of the Welsh Borderland different local successions show the same vertical sequence of communities (from shallow to deep), and, third, that like modern bottom marine communities, there is often a general progressive increase in the diversity of animal species with depth.

Because many communities occur in a variety of clastic sediments it appears that other factors also affected the distribution of brachiopods in the Silurian. Brachiopods are suspension feeders,

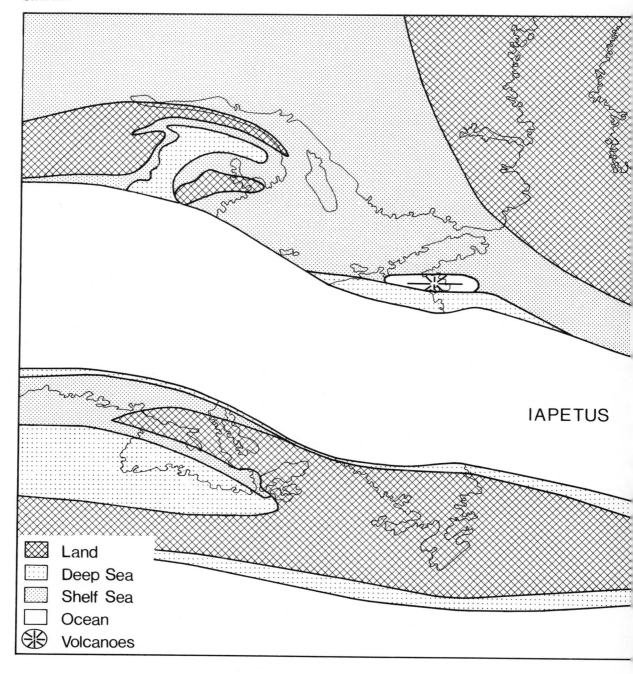

Fig. e. *Geography of the North Atlantic region in Silurian times (after McKerrow and Ziegler, 1972; Ziegler, 1970). The shorelines and volcanic areas are shown for late Llandovery or early Wenlock time. The Welsh Basin (Fig. d)*

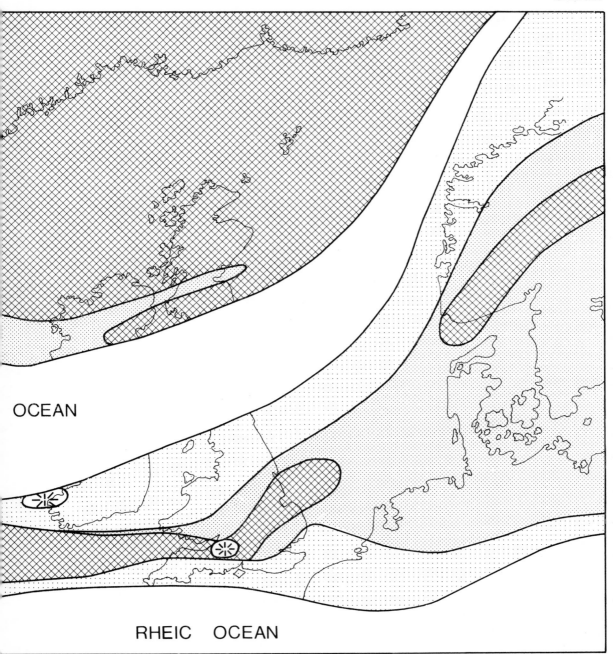

OCEAN

RHEIC OCEAN

lies to the north-west of the Mendip volcanics shown in southern England. The Iapetus Ocean was narrow enough to permit free migration of brachio-pods and other benthos with pelagic larvae.

attached to objects on the sea floor by pedicles, or lying loose on one of their valves. Their food supply originates from primitive microscopic plants, which may have been more abundant in shallow water. Many brachiopod species decrease in size away from the shore, suggesting that a corresponding decrease in food supply could have controlled their size. The food supply may also have been a significant factor in the control of the distribution of the communities, with only the more efficient feeders thriving in deep water.

Food supply alone cannot explain the diversification of Silurian brachiopods into communities. There must have been some advantages for those brachiopods which lived in the deeper water communities, and stability of environment and in particular constancy of sedimentation rates and of temperature would be among such benefits.

Brachiopods make up as much as 80 per cent of the preserved Silurian benthos in clastic marine sediments. Apart from competition for food, there was probably little interaction between one brachiopod and its neighbours on the sea floor. Each brachiopod community therefore did not form a discrete unit; there is little correlation between the ecological distribution of any two species, and in areas with an even sea floor there is a continuous gradation of changes in the brachiopod distribution from the shore to deep water, with few natural breaks.

The Silurian is divided into four Series, named Llandovery, Wenlock, Ludlow and Pridoli. The lower three were originally defined by Murchison, and correspond approximately to major lithological groups in the Welsh Borderland. The Llandovery takes its name from the town in Dyfed, where the Lower Silurian is rich in brachiopods; the Wenlock and Ludlow are named from Much Wenlock and Ludlow, which are towns in Shropshire. The name Pridoli comes from an area in Czechoslovakia, where marine conditions were continuous from the Silurian into the Devonian. The Pridoli Series corresponds to the Downtonian freshwater sediments of Wales and the Welsh Borderland. Recent estimates of dates from radioactive isotopes suggest that the Silurian lasted for about 27 million years, from 438 to 411 million years ago. The period is sub-divided into 34 graptolite zones, which are on average each a little less than a million years long. Since the classic work of Lapworth (1878) the graptolites have been the chief basis of Silurian time correlations, especially on a world-wide scale. Other animal groups which have been used to correlate the Silurian include conodonts, ostracodes, acritarchs and brachiopods. These have been especially useful in shallow water deposits where graptolites are uncommon.

In time correlation, brachiopods are most reliable when the evolution of individual species has been studied. Assemblages of

Fig. f. *The evolution of the brachiopod* Eocoelia *during the late Llandovery (after Ziegler, 1966). The diagrams on the right show moulds of the interior of the pedicle valve. The early species had strong ribs and no grooves on its teeth; the later species show gradual changes towards smooth forms with strong grooves on their teeth. Most Silurian benthic genera do not show such well marked evolutionary changes, over a period of time lasting several million years, but the few examples that do (like Eocoelia) are of great significance in dating the beds in which they occur. Eocoelia extends from the western United States to the U.S.S.R. and the changes are synchronous over this large area.*

brachiopods reflect environment rather than time. Very similar assemblages can occur in different places at different times in a similar environment, and are thus virtually useless for correlation. But the evolution of a genus reflects genetic changes through time within a similar habitat. Most Silurian brachiopod lineages show no detectable change with time, but those lineages which do show changes are very useful in correlation. The most useful are *Eocoelia* (Fig. f (after Ziegler 1966)), which lived in shallow

Lower Wenlock		*Eocoelia angelini*		
Upper Llandovery	C6	*Eocoelia sulcata*		
	C4–5	*Eocoelia curtisi*		
	C3	*Eocoelia intermedia*		
	C1–2	*Eocoelia hemisphae- rica*		

water (Upper Llandovery — Lower Wenlock), *Stricklandia,* which was in deeper water (Lower Llandovery — Lower Wenlock) and *Resserella sabrinae* which was also in deeper water (Wenlock and Lower Ludlow). The age relationships of these lineages and the graptolite zones are known by means of the occasional graptolites which have been recorded amongst shelly faunas. Fig. f shows the changes in *Eocoelia* from strongly ribbed forms at the base of the Upper Llandovery to virtually smooth forms by the Lower Wenlock, and through the same time interval there were also changes in the dentition between the two valves. Most of the localities in Fig. d have been dated by *Eocoelia* and *Stricklandia*, but those in the deeper water areas have been correlated by graptolites.

In the Wenlock and Ludlow Series, the successors of the Llandovery Communities form similar gradational sequences from the shore to deep water. These Upper Silurian Communities include many genera which survived from the Llandovery in the same general habitat, but there are also many differences. For example, the *Eocoelia* and *Stricklandia* lineages became first rare and then died out during early Wenlock time, and genera which are less abundant in the lower Silurian occupied the vacant habitats and became much more common.

In the case of the *Clorinda* Community, the two genera *Clorinda* and *Dicoelosia* occur together through much of the Silurian, but *Dicoelosia* is more prevalent during the Wenlock and Lud-

Time ↓ Ecogroup →	*Lingula*	*Eocoelia-Salopina*	*Pentamerus-Sphaerirhynchia*	*Stricklandia-Isorthis*	*Clorinda-Dicoelosia*	*Visbyella*	Graptolite
Pridoli and Ludlow	*Lingula*	*Salopina*	*Sphaerirhynchia*	*Isorthis*	*Dicoelosia*	*Visbyella*	Graptolite
Wenlock	*Lingula*	*Salopina* / *Eocoelia*	*Sphaerirhynchia* (sometimes *Homoeospira* in clastics)	*Isorthis*	*Dicoelosia*	*Visbyella*	Graptolite
Late Llandovery	*Lingula*	*Eocoelia*	*Pentameroides* / *Pentamerus*	*Costistricklandia* / *Stricklandia*	*Clorinda*	Marginal *Clorinda*	Graptolite
Early Llandovery	*Lingula*	*Cryptothyrella* / *Pentamerus*		*Stricklandia*	*Clorinda*		Graptolite

Table III *The Silurian ecogroups and chief animal communities and their evolution with time*

low. A few species change from one habitat to another with time, but the great majority of Silurian brachiopod lineages remain in the same habitat throughout the period.

The succession of communities in one habitat (with component genera changing through time) must be distinguished from a single community (where the component genera are always similar). We use the term COMMUNITY for an association of genera which live in a single habitat and occur in roughly similar proportions, and we use the word ECOGROUP for successive communities in a similar habitat. Within a single community there are often local variations in the presence and proportion of genera. For example the *Eocoelia* Community may sometimes include an ASSEMB-LAGE (a group of organisms) rich in rhynchonellids, and at other times rhynchonellids may be less common. Whenever a fossil collection is made from one site it is an assemblage that is collected. Similar assemblages are grouped into communities, and communities occupying similar habitats at different times can be classified together into ecogroups (Table III).

The Silurian Beds of Britain vary greatly from place to place (Table IV) and a large number of formations (each with a local name) are present. For details of these and their stratigraphical thickness in each area see Cocks et al. (1971) and Ziegler et al. (1974), which are papers dealing with the correlation of the whole of the British Silurian and some other countries.

Time Division	Wales	Welsh Borderland	Scotland and Lake District
Pridoli and Ludlow	Turbidites in Lower Ludlow, then no record in North Wales. Regression in SE Wales from *Dicoelosia* to *Lingula* Communities, then freshwater (Old Red Sandstone).	Progressive shallowing from *Dicoelosia* to *Lingula* Communities, then freshwater (Old Red Sandstone).	Graptolite shales in Lake District becoming shallow at top. No record in Southern Uplands. Freshwater or very shallow marine (Stonehaven) in Midland Valley.
Wenlock	Turbidites and graptolitic shales in north, irregularly shallower to south.	Deeper water at first, becoming shallower with Wenlock Limestone at top.	Mostly turbidites in Southern Uplands and Lake District. Non-marine in Midland Valley.
Upper Llandovery	Marine basin continues, with thick turbidities at Aberystwyth. Shallower to south and SE with *Pentamerus* to *Clorinda* Communities.	Marine transgression southeastwards, with progressive deepening from *Lingula* to graptolitic Communities.	Mostly thick turbidites in Southern Uplands, and graptolitic shales in Lake District. Regression from *Clorinda* to *Lingula* Communities in east Midland valley.
Lower and Middle Llandovery	Marine basin with deep water in North Wales, shallower in Pembrokeshire.	Land	Deepwater deposits across Southern Uplands and Lake District. Marine transgression in Girvan, *Lingula* through deeper brachiopod communities to graptolites.

Table IV *The history of the chief Silurian areas of Britain.*

THE LINGULA-BIVALVE ECOGROUP (18)

In the marine areas closest to the coast, and also in lagoons and estuaries where the salinity was slightly reduced, the commonest fossils are bivalves, gastropods and *Lingula*. This association is seen throughout the Silurian. Although there were few major changes with time, the composition of individual assemblages varies greatly; in particular gastropods and ostracodes are sometimes very common, and sometimes absent. The ecogroup is stable through time, and only includes one community, the *Lingula* Community.

It is important to realize that *Lingula* can occur at any depth. It is not found only in this ecogroup. Individuals which were present in deeper environments can often be distinguished by their smaller size, and they are seldom very common.

18 Lingula Community

Where the salinity was normal, *Lingula* occurred with more dominant articulate brachiopods (rhynchonellids and strophomenids), many of which are also present in the *Eocoelia-Salopina* ecogroup. In other assemblages *Lingula* is accompanied by burrowing and epifaunal bivalves. If the water was slightly less saline than the open sea, then the number of species became reduced, the fauna became very restricted and in extreme cases may consist of *Lingula* alone, as in the Skomer Volcanic Group of Pembrokeshire, or of a single species of bivalve.

Even in normal salinity, the *Lingula* Community is not diverse, but its constituents represent a wide range of adaptive types. As in the other Silurian marine communities suspension feeders are dominant, but here they include as many bivalves as brachiopods. The suspension feeders are dominated by bivalves (including the byssally attached *Pteronitella* and *Modiolopsis*) and brachiopods; the latter include the burrowing *Lingula*, and *Stegerhynchus* which was attached to other shells or to seaweed by its pedicle. Other suspension feeders include stick bryozoans and cornulitids. *Stegerhynchus*, like some other rhynchonellids throughout the Upper Palaeozoic and Mesozoic, was euryhaline, probably more so than any other articulate brachiopod.

Deposit feeders include burrowing bivalves like *Palaeoneilo* and *Lyrodesma*. Some muddier beds containing this community lack fine bedding because of disturbance by burrowers (bioturbation); it is not known which animals made these burrows, but arthropods or some group of worms seem the most likely. The gastropods present were probably either grazers or scavengers, since Silurian shells seldom show signs of attack by gastropods

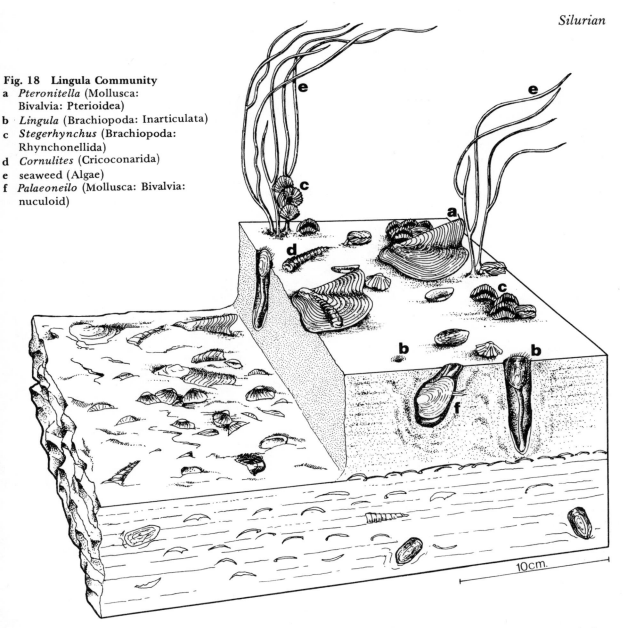

Fig. 18   Lingula Community
a   *Pteronitella* (Mollusca:
     Bivalvia: Pterioidea)
b   *Lingula* (Brachiopoda: Inarticulata)
c   *Stegerhynchus* (Brachiopoda:
     Rhynchonellida)
d   *Cornulites* (Cricoconarida)
e   seaweed (Algae)
f   *Palaeoneilo* (Mollusca: Bivalvia:
     nuculoid)

with rasping radulae and the only known predators were cephalopods and starfish. Abundant grazing gastropods in some localities suggest the presence of enough seaweed to support the population. Seaweed may also be linked with the local abundance of *Stegerhynchus* in some areas. This brachiopod has a large pedicle opening, suggesting that it was attached throughout its life, and the seaweeds probably provided the best anchorage.

The *Lingula* Community has been recorded from Llandovery and from Downtonian beds in the Welsh Borderland.

101

THE EOCOELIA-SALOPINA ECOGROUP (19–21)

In the Upper Llandovery the *Eocoelia* Community occurs to the seaward side of the *Lingula* Community, and many brachiopods, especially rhynchonellids and *Eostropheodonta*, are common to both communities. Similarly on the seaward edge of the *Eocoelia* Community, the *Eocoelia* and *Pentamerus* Community constituents may also be found together. In the Lower and Middle Llandovery the *Eocoelia-Salopina* ecogroup is represented by the *Cryptothyrella* Community, and during the late Silurian by the *Salopina* Community.

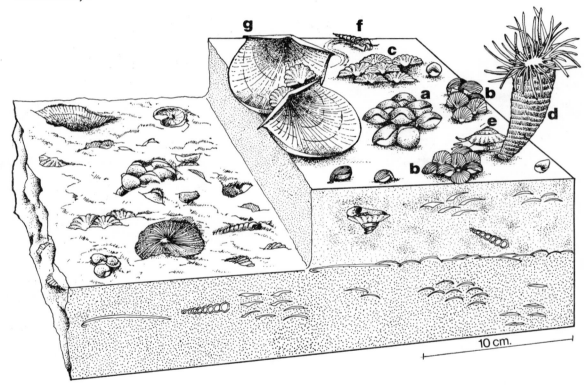

Fig. 19 Cryptothyrella Community
a *Cryptothyrella* (Brachiopoda: Spiriferida)
b dalmanellid (Brachiopoda: Orthida)
c *Zygospiraella* (Brachiopoda: Spiriferida)
d streptelasmatid coral (Coelenterata: Rugosa)
e *Liospira* (Mollusca: Archaeogastropoda)
f *Loxonema* (Mollusca: Mesogastropoda)
g *Eostropheodonta* (Brachiopoda: Strophomenida)

19 Cryptothyrella Community

The Ordovician-Silurian boundary was marked by a glacial period at least as severe as that of the Pleistocene, which, together with the resultant lowering of sea level, appears to have caused the extinction of many of the Ordovician marine benthos. At the very base of the Silurian, marine faunas are both scarcer and less diverse than later; and the Lower and Middle Llandovery saw a steady increase in the number of different animal types, as well as a gradual evolution of their ecological distribution into the markedly depth related communities of the Upper Llandovery and later. The depth range occupied by the *Eocoelia* Community, and perhaps also the shallower part of the *Pentamerus* Community in the Upper Llandovery is taken up by the *Cryptothyrella* Community in the Lower and Middle Llandovery. Four brachiopod genera are dominant, *Cryptothyrella* (a smooth atrypoid), *Zygospiraella* (a ribbed atrypoid externally similar to *Eocoelia*) and two different dalmanellids. Also present are the strophomenid *Eostropheodonta, Tentaculites* and stick bryozoans. Most of the habitats appear to be directly comparable to those of the *Eocoelia* Community, and many of the brachiopods in the *Cryptothyrella* Community are closely similar in external shape to other brachiopods present in the *Eocoelia* Community. The *Cryptothyrella* Community occurs in Lower Llandovery beds of the Girvan area, Ayrshire.

20 Eocoelia Community

This community is dominated by *Eocoelia* (whose range is late Middle Llandovery to Lower Wenlock), with the rhynchonellid *Ferganella* and dalmanellids the two commonest associated forms. *Eostropheodonta* or *Leptostrophia*, bivalves, gastropods, and *Tentaculites* are common in places. The rhynchonellids and the bivalves are especially common in the shallower part of the community depth range. Only one or two species of trilobites are common in the *Eocoelia* Community. In contrast to the suspension-feeding brachiopods, the trilobites and many of the molluscs are related to particular substrates. Those molluscs which are characteristic of particular sediments may also have been deposit feeders like the trilobites.

Although the *Eocoelia* Community can be found in any clastic sediment, even fine muds, it is most often found in sands. It is of interest that the two most abundant forms in this community, *Eocoelia* and *Stegerhynchus*, are both strongly ribbed. Most brachiopods only need to open their valves by 1 or 2mm to feed; the

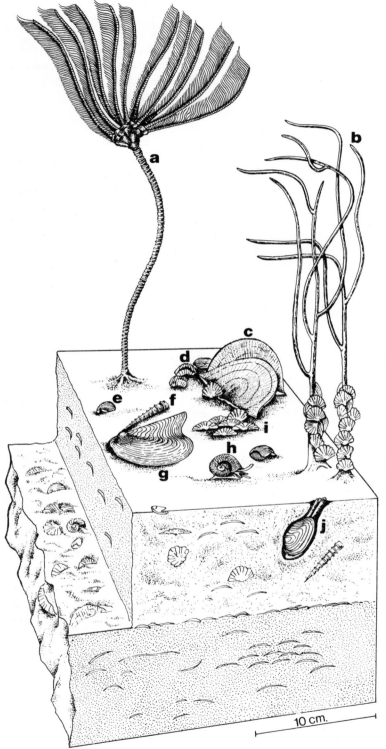

Fig. 20 Eocoelia Community
a   crinoid (Echinodermata: Crinozoa)
b   seaweed (Algae)
c   *Eostropheodonta* (Brachiopoda: Strophomenida)
d   *Stegerhynchus* (Brachiopoda: Rhynchonellida)
e   *Salopina* (Brachiopoda: Orthida)
f   *Tentaculites* (Cricoconarida)
g   *Pteronitella* (Mollusca: Bivalvia: Pterioidea)
h   *Poleumita* (Mollusca: Archaeogastropoda)
i   *Eocoelia* (Brachiopoda: Rhynchonellida)
j   *Palaeoneilo* (Mollusca: Bivalvia: nuculoid)

10 cm.

irregular opening provided by ribs acted as a sieve which excluded sand grains from the interior when the two valves were open (Rudwick, 1964).

The majority of collections from this community contain only four or five common species (more than three per cent), while some assemblages are almost completely dominated (over 90 per cent) by *Eocoelia* itself. Both today and in the Silurian an increase in species diversity is often related to an increase in water depth. We do not know of any *Eocoelia* Community in which salinity changes can be proved to be the cause of low diversity.

The *Eocoelia* Community occurs in many shallow water sands and shales in the Welsh Borderland and in the Girvan area.

21   Salopina Community

The Upper Silurian *Salopina* Community occupies the same habitat as the Upper Llandovery *Eocoelia* Community. The small finely ribbed dalmanellid *Salopina* is rare in the Llandovery and becomes more abundant as *Eocoelia* disappears during the Wenlock. Rhynchonellids and bivalves are still important constituents of the community. A species which occurs locally in huge numbers is *Microsphaeridiorhynchus nucula*; this can be distinguished from most other Silurian rhynchonellids by its small number of ribs. In the Ludlow the brachiopod *Protochonetes* also becomes an important element in the *Salopina* Community. *Protochonetes* has small spines sticking outwards from its hinge line and these may have served as an anchor and counterweight, so that the curved feeding edge away from the spines would be held stable above the substrate. The practical advantage of this arrangement would have been great, since brachiopods can be overwhelmed and killed by quantities of sediment. *Protochonetes* occurs with *Microsphaeridiorhynchus nucula* and little other fauna in some beds just below the transition from marine (Ludlow) to non-marine (Downtonian) deposits in the

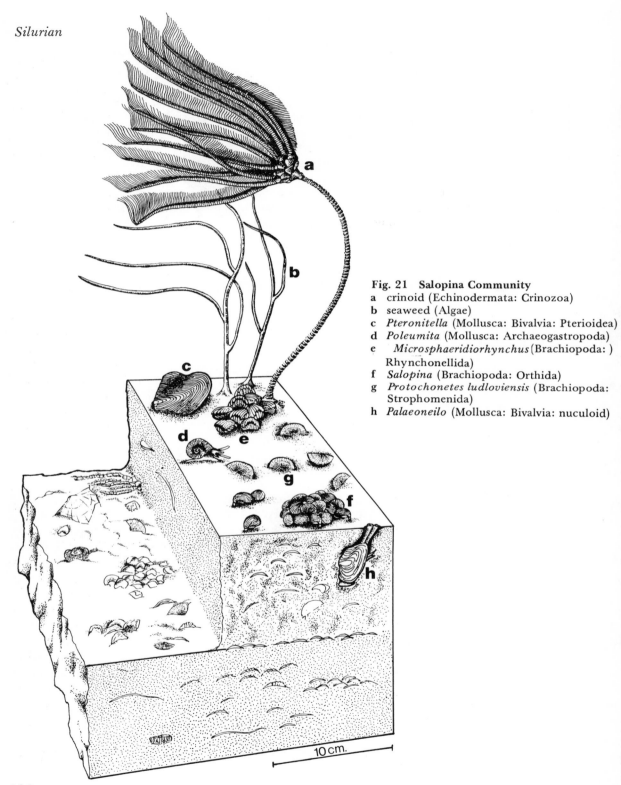

Fig. 21   Salopina Community
a   crinoid (Echinodermata: Crinozoa)
b   seaweed (Algae)
c   *Pteronitella* (Mollusca: Bivalvia: Pterioidea)
d   *Poleumita* (Mollusca: Archaeogastropoda)
e   *Microsphaeridiorhynchus* (Brachiopoda:
     Rhynchonellida)
f   *Salopina* (Brachiopoda: Orthida)
g   *Protochonetes ludloviensis* (Brachiopoda:
     Strophomenida)
h   *Palaeoneilo* (Mollusca: Bivalvia: nuculoid)

10 cm.

Welsh Borderland. This distribution suggests that *Protochonetes* may have resembled the rhynchonellids in being euryhaline.

The *Salopina* Community is normal in sandstones or siltstones, but it can occur in clays. Though most of the genera present are suspension-feeding brachiopods and bivalves, bioturbation (from burrowing bivalves, worms or trilobites) is present in places. The *Salopina* Community occurs in the late Ludlow beds of the Welsh Borderland.

THE PENTAMERUS-SPHAERIRHYNCHIA ECOGROUP (22—24)

Pentamerus is unknown from the Lower Llandovery and no community representing the *Pentamerus-Sphaerirhynchia* Ecogroup can be recognized clearly at this time in Britain. In the Middle and Upper Llandovery *Pentamerus* becomes very common and dominates the *Pentamerus* Community, as does its descendant *Pentameroides*. *Pentameroides* disappears at the end of the Llandovery, and its habitat occupied by the *Sphaerirhynchia* Community in the late Silurian. Locally in the Ludlow the large pentameride *Kirkidium* occurs in great numbers and can be assigned to this ecogroup. The Wenlock *Sphaerirhynchia* Community is chiefly developed in limestones, but in the Ludlow it occurs in both limestones and clastic sediments. In Wenlock clastic sediments the ribbed spiriferid *Homoeospira* is present; it gives its name to the *Homoeospira* Community, but clastics containing this community are rare in Britain (the water was mostly too deep). We include the Reef Assemblage with this ecogroup since most Wenlock Limestone reefs are associated with brachiopods of the *Sphaerirhynchia* Community.

22 Pentamerus Community

This community was dominated by the large brachiopod *Pentamerus* (up to 10cms long), or its descendant *Pentameroides*. Although to seaward of the *Eocoelia* Community, the habitat of the *Pentamerus* Community was still fairly shallow water; many of the sands and shelly limestones show evidence of turbulence, such as cross-bedding and scour marks. Shells and shell debris are often much more abundant here than in land-derived sediment, so that limestones are produced.

The shell of *Pentamerus* is very thick at the umbones (posterior) and very thin at the other end (anterior). It rested on its posterior end (the umbones) and was much more stable than other shells of even thickness. In adults the pedicle became atrophied, and the pedicle opening blocked by shell material. In some beds with abundant *Pentamerus* in this upright growth position the shells are touching, and occasionally have interfered with one another's growth.

Pentamerus is so named after the five chambers inside the shell separated by plates which help to strengthen it. There is a Y-shaped plate in the larger (pedicle) valve, and two plates are present in the smaller (brachial) valve; these plates extend for half or more of the length of the shell. A section through both valves near the umbones is shown in the inset to Fig. 22. Halfway through the Upper Llandovery *Pentameroides* developed from *Pentamerus*. In *Pentamerus* the two plates in the brachial valve first diverge and then continue parallel to each other at some distance apart; in *Pentameroides* the two plates also diverge at first, but then converge again at about one-third of the way down the valve to become a single plate.

Both the *Pentamerus* Community and the slightly deeper *Stricklandia* Community contain *Atrypa,* solitary conical corals, and also compound corals, particularly *Favosites* and the chain coral *Halysites. Eocoelia* is also present in the *Pentamerus* Community, but in very much smaller numbers than in the *Eocoelia* Community. *Eocoelia* (like all brachiopods except spire bearers and pentamerids) is progressively smaller in size in the deeper water communities. This decrease in size can be linked with a decrease in food supply with depth. It is perhaps significant that the exceptions to this rule include the spire-bearing brachiopods, in which a much more efficient filter-feeding system existed. The *Pentamerus* Community occurs in the Llandovery of the Welsh Borderland and Carmarthenshire.

a *Pentamerus:* lateral view and horizontal cross-section

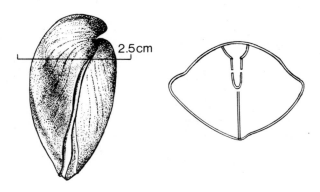

2.5cm

Fig. 22 Pentamerus Community
a *Pentamerus* (Brachiopoda: Pentamerida)
b *Halysites* (Coelenterata: Tabulata)
c streptelasmatid (Coelenterata: Rugosa)
d *Atrypa* (Brachiopoda: Spiriferida)
e *Hallopora* (Bryozoa: Ectoprocta)
f *Eocoelia* (Brachiopoda: Rhynchonellida)

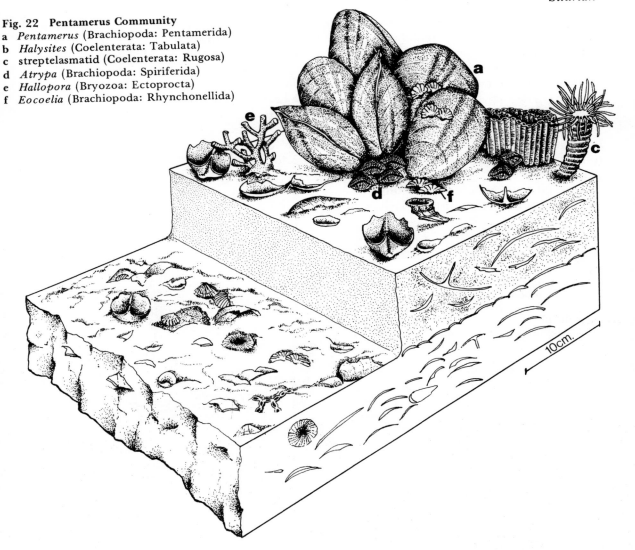

23 Sphaerirhynchia Community

Sphaerirhynchia is a globular rhynchonellid which is very rare in the Llandovery, but which becomes common, especially in limestones, in the Wenlock after *Pentameroides* becomes extinct. In the Ludlow of Britain the *Sphaerirhynchia* Community is equally common in limestones or clastic sediments. The *Sphaerirhynchia* specimens in the Wenlock Limestone are smaller and have a smoother valve margin than those in the Ludlow. The larger size and the more strongly deflected valve margins of the Ludlow *Sphaerirhynchia* may be adaptations to calmer water. Both these

109

features helped the brachiopod to separate the water sucked in from that which was expelled, having been extracted from it. This separation was not needed in the more turbulent water of the Wenlock Limestone environment.

The most abundant general in the *Sphaerirhynchia* Community are often *Atrypa* and *Howellella*, but these spire-bearing brachiopods are not confined to this one community. Other common brachiopods include strophomenides, spiriferides and *Protochonetes ludloviensis*. Bivalves, gastropods and trilobites can occur, but are not very common. The distribution of corals and bryozoans is patchy.

The *Sphaerirhynchia* Community occurs in the upper parts of the Wenlock limestone of the Welsh Borderland.

Fig. 23   Sphaerirhynchia Community
a *Leptostrophia* (Brachiopoda: Strophomenida)
b *Streptelasmatid* (Coelenterata: Rugosa)
c *Atrypa* (Brachiopoda: Spiriferida)
d *Sphaerirhynchia wilsoni* (Brachiopoda: Rhynchonellida)
e *Liospira* (Mollusca: Archaeogastropoda)
f *Protochonetes ludloviensis* Brachiopoda: Strophomenida)
g *Howellella* (Brachiopoda: Spiriferida)
h *Eospirifer* (Brachiopoda: Spiriferida)
i *Loxonema* (Mollusca: Mesogastropoda)

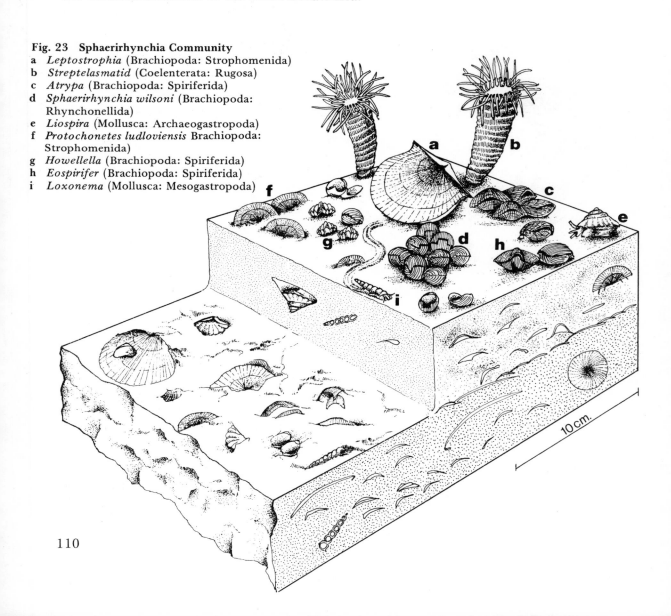

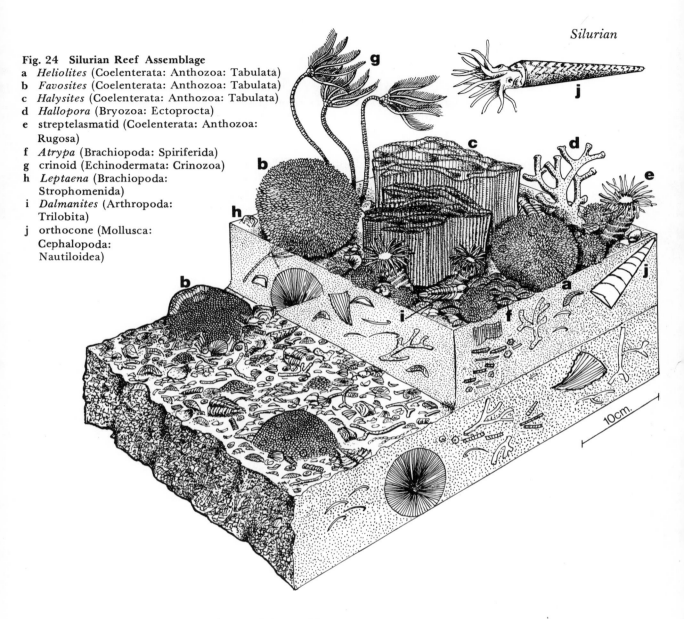

Fig. 24 Silurian Reef Assemblage

a *Heliolites* (Coelenterata: Anthozoa: Tabulata)
b *Favosites* (Coelenterata: Anthozoa: Tabulata)
c *Halysites* (Coelenterata: Anthozoa: Tabulata)
d *Hallopora* (Bryozoa: Ectoprocta)
e streptelasmatid (Coelenterata: Anthozoa: Rugosa)
f *Atrypa* (Brachiopoda: Spiriferida)
g crinoid (Echinodermata: Crinozoa)
h *Leptaena* (Brachiopoda: Strophomenida)
i *Dalmanites* (Arthropoda: Trilobita)
j orthocone (Mollusca: Cephalopoda: Nautiloidea)

24 Reef Assemblage

A reef limestone (or bioherm) is a rock composed of remains of organisms which are still in their life position. None of the debris of a true reef has been transported and therefore a reef limestone can be recognized by the absence of bedding. Reef limestones are, of course, subject to attack by wave action, and fragments of eroded reef can be laid as bedded sediments in the surrounding sea.

Reef limestones are only common in the Silurian of Britain during the late Wenlock Series, and the following description is

111

based on the assemblage in the Much Wenlock Limestone of Shropshire (Scoffin, 1971).

The frame builders were dominantly tabulate corals (*Heliolites, Favosites, Halysites*) and large, domed stromatoporoids (twenty or more genera have been recognized). The frame is bound together by calcareous algae and bryozoans, with subsidiary solitary rugose corals and flat stromatoporoids. Crinoids are also abundant.

In addition to the frame builders and binders, the Wenlock reef assemblages contain very varied shelly faunas which never occur in large numbers. These minor elements of the reef fauna include numerous trilobite species, brachiopods and molluscs. Most of these flourished adjacent to the reefs, rather than within them. Clusters of *Atrypa* are sometimes found in hollows within the reef.

THE STRICKLANDIA-ISORTHIS ECOGROUP (25-6)

The smooth pentamerid *Stricklandia* dominates the *Stricklandia* Community from the Lower Llandovery until high in the Upper Llandovery, when it evolves into the ribbed *Costistricklandia*. In the Lower Wenlock *Costistricklandia* gradually decreases in abundance and it is extinct by the Upper Wenlock. As *Costistricklandia* declines, the finely ribbed *Eospirifer* becomes more common, but by the Upper Wenlock the characteristic brachiopods of the ecogroup are the dalmanellid *Isorthis clivosa* and associated stropheodontid brachiopods. *Isorthis* continues in this habitat through the Ludlow, but in Wales and the Welsh Borderland there is no record of this ecogroup in the latter part of the Ludlow, for the water in Britain and much of North America was too shallow.

25 Stricklandia Community

In the Llandovery there are all gradations between the *Pentamerus,* Community and the deeper water *Stricklandia* Community. In borderline cases the assignment of an assemblage to one or other community depends simply upon whether *Stricklandia* or *Pentamerus* is more abundant. The sediments in which the *Stricklandia* Community occurs are usually fine sands, silts or clays. Though small scale cross-bedding may be present there is seldom any indication of severe turbulence or strong bottom currents. Like *Pentamerus, Stricklandia* rested on its beak, without a functional pedicle in adult life. Some beds are crowded with vertical individuals in their growth position. In addition to *Stricklandia* itself, the *Stricklandia* Community often contains the spire-bearing brachiopods *Atrypa* and *Eospirifer,* and other genera (including corals)

Fig. 25 Stricklandia Community

a crinoid (Echinodermata: Crinozoa
b *Costistricklandia* (Brachiopoda: Pentamerida)
c *Atrypa* (Brachiopoda: Spiriferida)
d *Clorinda* (Brachiopoda: Pentamerida)
e *Salopina* (Brachiopoda: Strophomenida)
f *Pholidostrophia* (Brachiopoda: Strophomenida)
g *Leptaena* (Brachiopoda: Strophomenida)
h *Eospirifer* (Brachiopoda: Spiriferida)
i *Eocoelia* (Brachiopoda: Rhynchonellida)
j *Whitfieldella* (Brachiopoda: Spiriferida)
k *Loxonema* (Mollusca: Mesogastropoda)
l *Euomphalopterus* (Mollusca:
    Archaeogastropoda)

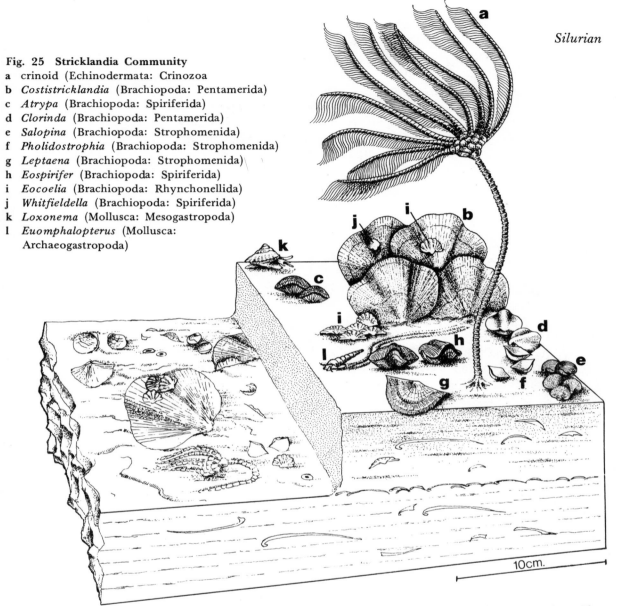

10cm.

which are typical of the shallower *Pentamerus* Community. The community may also contain *Clorinda*, *Leangella* and other brachiopods and straight nautiloids which were more typical of deeper water. Crinoids are common at some localities, but have a very sporadic distribution; solitary conical corals and compound corals may be present.

All the brachiopods, bryozoans, crinoids and corals were suspension feeders; these make up the vast bulk (perhaps 98 per cent) of the animals found fossil in this community. The gastropods and trilobites were deposit feeders and may also have been scavengers. It is most probable that little organic matter was left to decay on

113

the sea floor, because there is hardly any indication of it in the sediments, nor is the sea floor likely to have been stagnant, as this would not have suited bottom dwellers. The only fossil carnivores are echinoderms and nautiloids, but it is very likely that there were some soft-bodied carnivores, including worms, in the Silurian.

The *Stricklandia* Community occurs in the Upper Llandovery beds of Ayrshire, Wales and the Welsh Borderland, and is also very common in eastern North America.

26   Isorthis clivosa Community

In the Lower Wenlock, *Costistricklandia* gradually disappears and *Eospirifer* becomes more abundant in the *Stricklandia-Isorthis* Ecogroup. The diversity in this community continues to be greater than in those of shallower water habitats, so that no one genus is dominant. In the Upper Wenlock, the distribution of *Eospirifer* becomes more sporadic and the commonest brachiopods are the two dalmanellids *Dalejina* and *Isorthis* and the atrypoid *Atrypa*. In addition, there are numerous strophomenide genera, including *Protochonetes minimus, Leptostrophia* and *Leptaena*. Most strophomenides lacked a functional pedicle, and *Protochonetes* developed spines which acted as anchors; *Strophonella* and *Leptaena* both developed curved shells so that the opening between the valves (for feeding and respiration) was raised above the sea floor. In *Leptaena* this takes the form of a sharp bend in the shell. *Protochonetes minimus* is very much smaller than *P. ludloviensis*, which is most common in the shallow water *Salopina* Community (21), but may also occur in the *Sphaerirhynchia* Community (23).

Dalejina and *Atrypa* were often more abundant than *Isorthis clivosa* in this community, but they also occur in greater numbers in the shallower *Sphaerirhynchia* Community and in the deeper *Dicoelosia* Community. *Isorthis clivosa* was more restricted in its distribution and the community is therefore named after it, but other species of *Isorthis* were abundant in shallower environments.

Bivalves and gastropods are occasionally present in the *Isorthis* Community, but they are seldom as abundant as in shallower water Silurian Communities.

In this community the proportion of genera varies from one locality to another. This is a characteristic feature of all the deeper water Silurian Communities. It reflects the sporadic distribution of bottom-dwelling marine animals (benthos) on the sea floor at all times (from the Cambrian to the present day). Many brachiopod genera occur quite randomly on the sea floor, but articulated

Fig. 26 Isorthis clivosa Community
a *Isorthis clivosa* (Brachiopoda: Orthida)
b *Dalejina* (Brachiopoda: Orthida)
c *Protochonetes minimus* (Brachiopoda: Strophomenida)
d *Leptostrophia* (Brachiopoda: Strophomenida)
e *Mesopholidostrophia* (Brachiopoda: Strophomenida)
f *Leptaena* (Brachiopoda: Strophomenida)
g *Atrypa* (Brachiopoda: Spiriferida)
h *Acaste* (Arthropoda: Trilobita)

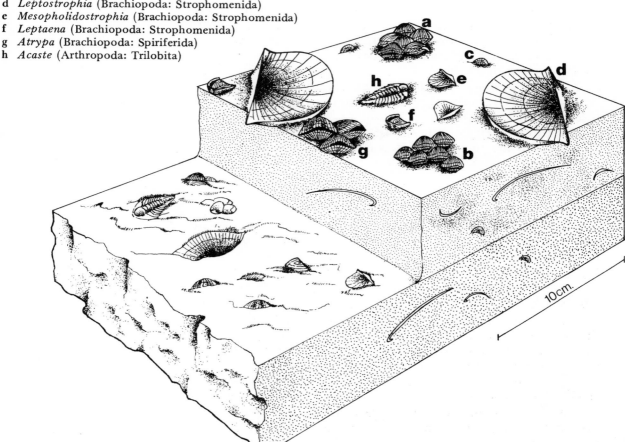

specimens of some genera are found in clusters. The clustering is particularly noticeable in smaller brachiopods with a relatively large pedicle opening. The pedicle was functional throughout the life of these animals which probably attached themselves to shell fragments or to shells of living animals. The clusters, which are usually all of one species, originate when a group of brachiopods anchored themselves on living members of their own species. *Isorthis, Dalejina* and *Atrypa* all had functional pedicles; although they usually lived as isolated individuals, they are occasionally found in clusters.

The *Isorthis clivosa* Community occurs in the Wenlock and Ludlow beds of Wales and the Welsh Borderland.

115

THE CLORINDA-DICOELOSIA ECOGROUP (27—29)

This ecogroup consists of the Silurian shelly benthos living in the deepest water habitats. In the Llandovery Series, this ecogroup is represented by the *Clorinda* Community, although locally *Eoplecodonta* and *Dicoelosia* were more common than *Clorinda*. In the Wenlock and Ludlow, *Dicoelosia* became generally more common while *Clorinda* was rare, and the ecogroup, at least in shales and sandstones (clastic sediments), is represented by the *Dicoelosia* Community. In limestones, *Dicoelosia* is rare, and *Eoplectondonta* is again the characteristic genus. In addition to the *Clorinda* and *Dicoelosia* Communities this ecogroup includes the *Visbyella trewerna* Community which consists of a few small brachiopods living near the maximum depth limits of bottom dwellers.

27   Clorinda Community

Assemblages from this community commonly contain over twenty genera of brachiopods (and sometimes as many as 35). In addition, there is a large diversity of other animal types present, but never a high density.

Trilobites are more abundant in the deep water habitats of this community than in any other Silurian habitat, but they are never as common as they were in the Cambrian and Ordovician. Crinoids, carpoids, ostracodes, starfish and bryozoans are also present. Bivalves and gastropods are rarer and less diverse than in shallower water assemblages. In addition to these bottom dwellers, there is a fauna of swimmers and floaters (nekton and plankton). These include both straight and coiled nautiloids, and sporadic graptolites.

The brachiopods are thinly spread over the bedding planes. Like modern deep water benthos, they do not occur in clusters and never reach the densities seen in the shallower communities. Although the *Clorinda* Community is known from coarse sandstones, the majority of occurrences are in fine silts or muds. This reflects the weakness of the currents, which were not fast enough to transport coarse sand, in the deep water environments.

The commonest brachiopods are the pentamerid *Clorinda,* the strophomenides *Eoplectodonta, Coolinia* and *Mesopholidostrophia,* the dalmanellids *Resserella* and *Skenidioides* and the spire bearers *Glassia, Atrypa* and *Eospirifer: Dicoelosia* is present, but less common.

As in modern times the deep sea environment contains considerably more species than nearer the shore, though it has lower densities. The *Clorinda* Community normally contains five or six times as many species as the *Lingula* Community. But towards the extreme depth limits of the bottom-dwelling brachiopods there is

Fig. 27   Clorinda Community
a *Clorinda* (Brachiopoda: Pentamerida)
b *Cyrtia* (Brachiopoda: Spiriferida)
c *Eoplectodonta* (Brachiopoda: Strophomenida)
d *Atrypa* (Brachiopoda: Spiriferida)
e *Glassia* (Brachiopoda: Spirferida)
f *Skenidioides* (Brachiopoda: Orthida)
g *Calymene* (Arthropoda: Trilobita)

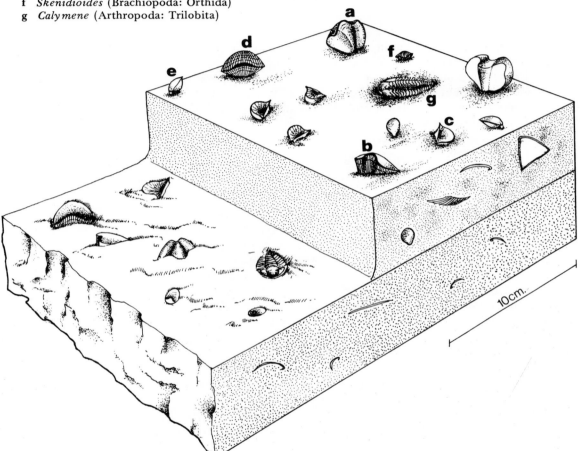

a reduction in diversity. Five or six species characteristic of the
Clorinda Community were able to live in water deeper than the
remainder. These species are known as the Marginal *Clorinda* Community and occupy the deepest habitats of any shelly fauna in the
Lower Silurian.

The *Clorinda* Community occurs in the Upper Llandovery of
Wales and the Welsh Borderland, the Lower and Middle Llandovery
of Scotland, and comparable assemblages in North America, Asia
and Australia.

117

28 Dicoelosia Community

The *Dicoelosia* Community occurs in the Upper Silurian (Wenlock and Ludlow Series). It contains many of the same genera seen in the Lower Silurian *Clorinda* Community, the main differences between the two being in the relative abundance of *Dicoelosia* and *Clorinda*.

The pentamerid *Clorinda* became rare in the Wenlock and Ludlow, while the dalmanellid *Dicoelosia* became abundant in clastic sediments. *Dicoelosia* is a small brachiopod (about ½cm long) with bilobed shell (inset to Fig. 28). Brachiopods are filter feeders which admit water to the shell, circulate it internally to extract oxygen and food particles, and then eject it. For small shells living in quiet water, it becomes important that the same water is not circulated repeatedly and many brachiopods develop characteristics which prevent this from happening. The deep groove in *Dicoelosia* was probably developed to separate the water going in (at the top of the two lobes) from the water coming out (in the groove).

Many brachiopods in the *Dicoelosia* Community are small and widely scattered. This may be a reflection of the reduced supply of plankton for food in the deep habitat of this community (though at least acritarchs, a common part of the plankton, are more abundant in deeper waters). For example, all three strophomenides (*Mesopholidostrophia, Leangella* and *Protochonetes minimus*) common in the community are small compared with their closest relatives living inshore.

Other common brachiopods include the dalmanellids (*Dalejina, Isorthis clivosa, Resserella*) and the spire-bearers *Eospirifer* and *Cyrtia* and *Glassia*.

Bivalves, gastropods and trilobites form a low proportion of the groups present, although they are often diverse. In Upper Silurian limestones, *Dicoelosia* became rare. The commonest brachiopods in the limestones deposited at these great depths include *Cyrtia, Eoplectodonta, Atrypa* and *Resserella*.

The *Dicoelosia* Community occurs in the Wenlock and early Ludlow beds of Wales and the Welsh Borderland.

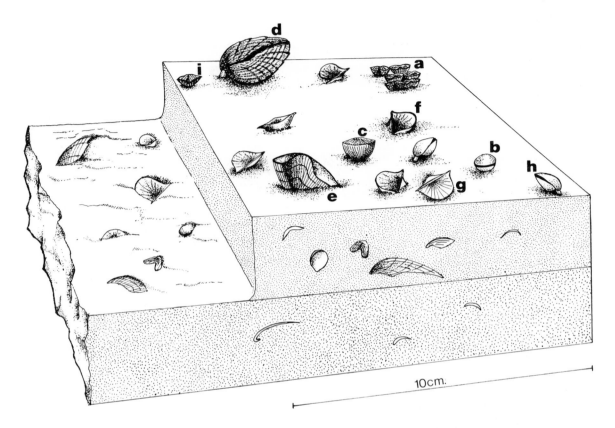

Fig. 28 Dicoelosia Community
a *Dicoelosia* (Brachiopoda: Orthida)
b *Nucleospira* (Brachiopoda: Spiriferida)
c *Resserella* (Brachiopoda: Orthida)
d *Atrypa* (Brachiopoda: Spiriferida)
e *Cyrtia* (Brachiopoda: Spiriferida)
f *Eoplectodonta* (Brachiopoda: Strophomenida)
g *Mesopholidostrophia* (Brachiopoda: Strophomenida)
h *Glassia* (Brachiopoda: Spiriferida)
i *Skenidioides* (Brachiopoda: Orthida)

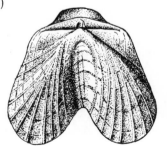

a

1cm.

119

Fig. 29   Visbyella trewerna Community
a   *Dalmanites* (Arthropoda: Trilobita)
b   orthoceratid (Mollusca: Nautiloidea)
c   *Protochonetes minimus* (Brachiopoda: Strophomenida)
d   *Visbyella trewerna* (Brachiopoda: Orthida)
e   *Bracteopleptaena* (Brachiopoda: Strophomenida)
f   *Cyrtograptus* (Hemichordata: Graptoloidea)

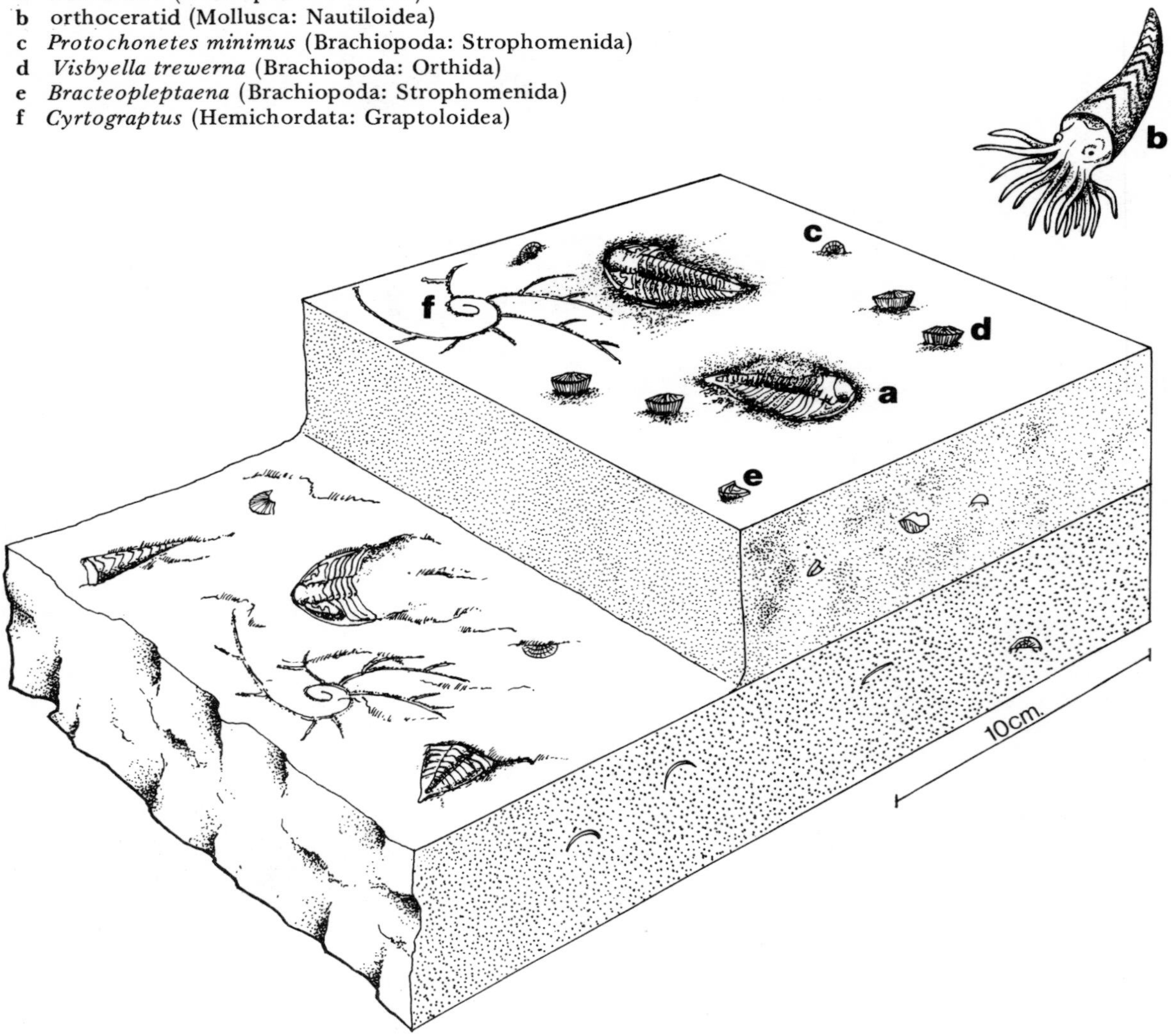

29   Visbyella trewerna Community

This community is a restricted variety of the *Dicoelosia* Community, but with many elements of the latter absent. It is dominated by *Visbyella trewerna,* similar very small dalmanellids, and the small spined strophomenide, *Protochonetes minimus.* It also includes three brachiopod species less common in the *Dicoelosia* Community: *"Clorinda" dormitzeri, Mesounia* and *Bracteoleptaena.* Trilobites, especially *Dalmanites,* are characteristic. *Dalmanites* is unusual in having a long tail spine, and large lenses in its compound eyes, which were raised above the level of its head-shield. So, like the Llandovery Marginal *Clorinda* Community, the *Visbyella trewerna* Community contains species more adapted to these deep conditions.

Today there are three depth related features seen in level bottom communities: diversity increases and size and abundance decrease with depth. Comparison between modern suspension feeders and Upper Silurian brachiopods with regard to these changes (Hancock et al. 1974) suggests that the *Visbyella trewerna* Community may have extended to depths greater than the average shelf. The decrease in diversity may also be related to oxygen deficiency in the deeper waters.

Graptolites and other pelagic forms occur preserved with the *Visbyella trewerna* Community. Among the latter, straight nautiloids such as *Michelinoceras*, and *Cardiola interrupta* are common. *Cardiola* is a bivalve not closely related to any modern form. Its mode of life is conjectural, but, as its ecological distribution resembles that of graptolites (commoner in very deep water deposits) it may have lived near the surface of the sea, perhaps attached to drifting algae.

The *Visbyella trewerna* Community occurs in deep water Upper Silurian sediments in North Wales.

GRAPTOLITIC ECOGROUPS (30)

Animals and plants floating on the surface of the ocean and those living at intermediate depths are not considered as components of the benthic ecogroups and communities. Although their remains sink to the sea floor on death, there was no important relationship between the benthos and these nektonic and pelagic forms.

30 Silurian graptolite Assemblage

The dominant Silurian animals in this category are the graptolites, of which there is a large variety. In the Lower Silurian, biserial diplograptids (which are common in the Ordovician) were present together with monograptids (which appeared in the Silurian). Most of the monograptids have a single branch, made up of connected thecae (each of which contained a separate animal), but the cyrtograptids show side branches coming off the convex sides of the main branch. The cyrtograptids occur mainly in the Wenlock. Other groups of Silurian graptolites are the retiolitids (which have a skeleton in the form of a network of connecting rods, and are thus not easy to find in the field) and the dendroids (which include forms ranging from the Cambrian to the Carboniferous).

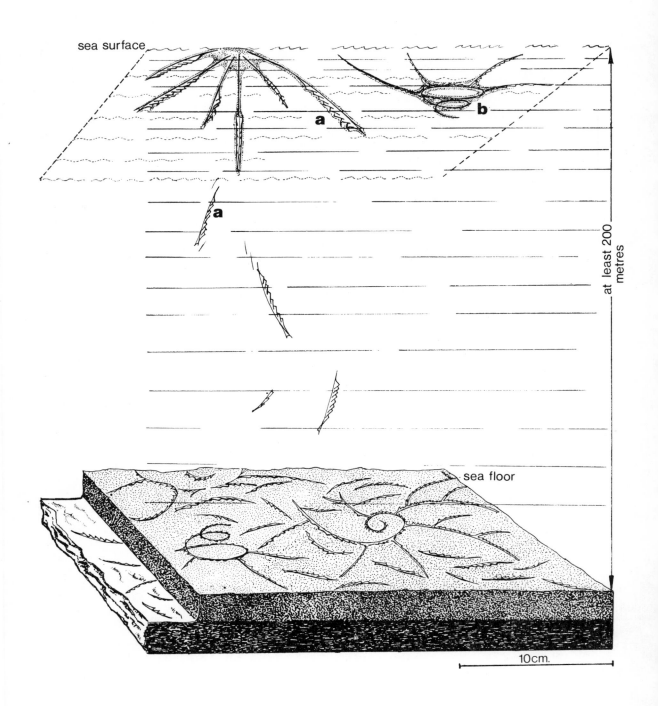

Fig. 30 Silurian graptolite Assemblage
a *Pristiograptus* (Hemichordata: Graptoloidea)
b *Cyrtograptus* (Hemichordata: Graptoloidea)

122

Besides the dendroids, monograptids alone survive through the up-permost Silurian and into the Lower Devonian. Our reconstruction shows a Wenlock *Cyrtograptus* and the monograptid *Pristiograptus*. This monograptid has several single-branched colonies, each con-nected by a protein thread (the nema) to a central float of soft tissue.

It has been suggested that different graptolites lived at differ-ent water depths. Graptolite faunas are most diverse in shales without benthic faunas. But in shallower water with benthic brachiopods, the number of graptolite genera decreases progres-sively towards the shallow water habitats. For example, cyrtograp-tids and monograptids with lobed thecae are confined to areas with a *Clorinda* Community or to deeper waters; biserial diplograp-tids occur only in water deeper than the *Pentamerus* Community, while only retiolitids, *Monograptus priodon* and *Monograptus tunicis* (and their close allies) are present in water as shallow as for the *Eocoelia* Community (Berry and Boucot, 1972).

Only a very limited number of animals floated (or were sus-pended) with the graptolites; these include straight nautiloids, the bivalve *Cardiola* and the horny brachiopod *Orbiculoidea*. Many of the varied deep water brachiopod communities occur in black shales, rich in carbonaceous matter. It is probable that the carbon is derived from floating seaweed to which the *Cardiola*, the *Orbicu-loidea*, and perhaps also some graptolites, were attached.

Graptolite communities occur in deep water shales in southern Scotland, the Lake District and Wales.

SILURIAN COMMUNITIES IN NORTH AMERICA AND EUROPE

During the Silurian, the same assemblages are present over North America, Europe and much of Asia.

Shelf seas covered much of North America and the Russian platform (including Gotland) and many shallow marine areas were at great distances from the nearest land, so that no land-derived sediment was deposited. Instead, limestones, consisting of shells, corals and algal debris, accumulated through biological action. In these environments, the proportion of brachiopods and other in-vertebrates may be locally rather different from those seen in similar habitats in areas of clastic (sand and shale) sedimentation.

In many areas of North America, the limestones have been altered to dolomite and the original organic structures are no longer present. Some rocks may show no obvious sign of fossils at all.

Changing salinity also affects marine communities. In the Late Silurian of parts of North America salt deposits indicate increased salinity in areas of higher evaporation and these more saline waters have a very reduced number of genera compared with normal marine habitats. A similar effect is seen in areas of reduced salinity.

In much of central Europe, the Silurian is entirely represented by platform mudstones. The absence of common shelf benthos has been explained by Berry and Boucot (1967) as a result of the water being too cold for them. However, we suggest that the absence of brachiopods in any particular place may be due either to too great a depth of water (i.e. greater than that under which the *Clorinda* Community lived) or to foul bottom conditions, which may be found at any depth. The latter may or may not have included low temperature conditions, but some brachiopods today thrive in the Arctic, and we reject cold water as a likely explanation of the structure of the central European Silurian.

Devonian

The most significant evolutionary development during the Devonian was the colonization of the land. Although the first land plants appeared in the Late Silurian, it was during the Devonian that they developed in abundance to provide the first true soils, and to support the first land animals. At first the plants would have existed without much assistance from the surface materials. In the Lower Devonian, the dominant plants were rootless and leafless psilophytes, but by the end of the period many different types had evolved: ferns, seed ferns, horsetails (e.g. calamites) and club mosses (lycopods), many of them reaching a very large size. Most of these more advanced plants developed roots, and thus obtained nutrients from the soil.

It is difficult to imagine the landscape before these soils developed. Although many areas would have had a high rainfall, there could be no soils without land plants. Much of the ground would be reddened by ferric iron oxides, as there would be an insufficient quantity of organic matter to reduce the iron to its green ferrous state. The lack of organic matter is the main reason why many Devonian (and earlier) river deposits are red in colour, like much of the Old Red Sandstone of north-west Europe.

The first land plants were followed almost immediately by land animals. Many of these were arthropods (spider-like mites and wingless insects) which had a covering of chitin. Although this covering had developed originally in their marine ancestors, it served these first land animals both by giving support to their bodies and by protecting the body tissues from drying up in the terrestrial environment. Many of these arthropods were sap-sucking and spore-eating forms similar to some living animals. Most of them were ground dwellers, but insect flight may have evolved during the Late Devonian (when tall plants first appeared), though the first fossil winged insects are not known until the Carboniferous.

In the Devonian there is a close correlation between the increase in complexity of spore structures and the diversity of terrestrial arthropods. Although some of these early land animals were predators, and others were probably detritus feeders (e.g. the myriapods), many of the early mites, chelicerates and wingless

125

insects may have fed on spores and dispersed them as a result; some of the complex ornament on Devonian spores may have been a direct adaptation to assist this dispersion (Kevan et al, 1975).

The Devonian also saw large evolutionary radiations in fish, culminating in the emergence of the first amphibians in the Late Devonian. The Early Devonian is renowned for the heavily armoured benthic ostracoderms. These had no jaws, but by the Middle Devonian the fish included groups of large predatory placoderms (arthrodires) which were the dominant carnivores in both marine and freshwater environments.

In north-west Europe and north-eastern North America, there are large areas of river and lake deposits: the Old Red Sandstone. This continental area extended continuously from northern Germany, England and eastern Canada to Scandinavia and north Greenland and much of the North-west Territories (Fig. g) after the close of the Iapetus Ocean sometime in the Early Devonian. Marine deposits were laid down around the margins of this continent in Germany, south-west England, the United States and much of western Canada.

Marine faunas in the basal Devonian are similar to those present in Silurian beds. Brachiopods and corals continued to be among the commonest benthos throughout the period; but there was a major change in pelagic faunas towards the end of the Lower Devonian, when the monograptids became extinct and several families of ammonoids developed. After Early Devonian times, the ammonoids become the best indicators of stratigraphic time zones. Because they had calcareous skeletons and not chitinous skeletons like the graptolites, cephalopod-rich limestones formed in areas receiving little terrigenous sediment instead of graptolitic shales (Tucker, 1974).

In shallow marine sandy and muddy areas, the brachiopods continued to be the most commonly preserved benthos. Spire-bearers (spiriferoids, atrypoids and athyroids) are commonest, but strophomenoids, chonetoids and rhynchonelloids also continue from the Silurian. Towards the end of the Devonian the productellids made their appearance; these brachiopods were anchored by spines on the surface of very convex pedicle valves, a new development that was to see its full expression in the productids of the Carboniferous. Bivalves and trilobites also continued from the Silurian; they show some new forms, especially those associated with reefs (bioherms). In general, however, the trilobites decreased in diversity during the Devonian, with only one family (the proetids) surviving into the Carboniferous.

In Devonian carbonate shelf environments (see House, 1975), the tabulate corals and the stromatoporoids reached their acme; rugose corals (which became more common in the Carboniferous) were also more common than during the Silurian. These groups

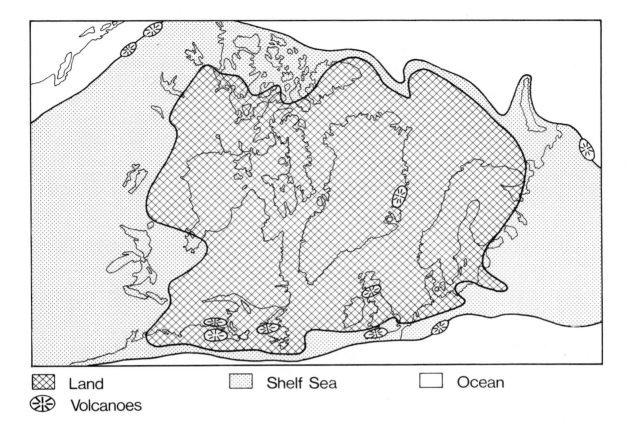

▨ Land	▨ Shelf Sea	☐ Ocean
✪ Volcanoes		

Fig. g. *Geography of North America and northern Europe during the Devonian (modified after House, 1967). The large land area is 'the Old Red Sandstone Continent' which formed after the closure of the Iapetus Ocean in the Late Silurian or Early Devonian. The Old Red Sandstone sediments were laid down in rivers and lakes, and yield freshwater fish and early land plants.*

were responsible for the construction of bioherms and other carbonate accumulations in many warm shallow-water environments during the Devonian, which provided new niches for other benthic marine organisms, notably brachiopods, trilobites, molluscs and ostracodes. The role of algae in Devonian bioherms was variable. Encrusting red algae (solenoporans) are conspicuous in some Australian and Canadian reefs, but are absent in German and Belgian reefs. Filamentous blue-green algae (such as *Girvanella* and *Sphaerododium*) are geographically more widespread.

The Devonian system is divided into seven stages named after localities in Belgium and Germany where good marine sequences occur. They are:

127

Upper Devonian	Famennian
	Frasnian
Middle Devonian	Givetian
	Eifelian
Lower Devonian	Emsian
	Siegenian
	Gedinnian

These stages can be recognized by graptolites (two zones at the base of the Gedinnian) ammonoids (goniatitids and clymeniids), pelagic ostracodes and conodonts. However, each of these groups was restricted to a certain habitat and so is only found in certain sediments. The thin-shelled pelagic ostracodes are only preserved in fine-grained sediments. The ammonites, like *Nautilus* today, avoided turbulent environments, and are most common in pelagic carbonates or offshore muds. Conodonts occur in distinct assemblages related to reef distributions (off-reef, reef, and back-reef communities seem to be distinct). Some benthic fossils are also of use in stratigraphic correlation, notably the spiriferides.

In river and lake environments fish are the principal means of correlation, though the significance of plants and spores is important because of their usefulness in correlation between marine and non-marine environments.

The illustrated environments have been based on marine and freshwater Devonian faunas in the British Isles. Although including marine sequences in the nominate area for the system in Devon in south-west England, the British Isles are perhaps best known for the Old Red Sandstone facies, which represent the first major development of Phanerozoic continental sedimentation. The marine facies, exposed mainly in south-west England, Devon and adjacent Cornwall, were much disturbed during the Hercynian orogeny and are still relatively poorly understood compared with the marine Devonian elsewhere in Europe, and in North America and Australia.

31 Lower Devonian Freshwater Communities

The Devonian was the first period in which there were extensive non-marine faunas and floras. The fluvial, lacustrine and estuarine environments supported the first major freshwater faunas, which may have initially invaded these environments during the Silurian.

The most common fish are ostracoderms (*Pteraspis, Anglaspis, Poraspis*) which probably looked something like large armoured tadpoles. Lacking lateral fins they had little more manoeuverabil-

Fig. 31 Lower Devonian Freshwater Community

a   *Turonia* (Vertebrata: Agnatha)
b   *Pteraspis* (Vertebrate: Agnatha)
c   eurypterid (Arthropoda: Chelicerata: Merostomata)
d   *Cephalaspis* (Vertebrata: Agnatha)
e   acanthodian spine (Vertebrata: Osteichthyes)
f   plant debris (Plant)
g   fish scales (Vertebrata: Agnatha)

129

ity, unless they were able to utilize their exhalent respiratory currents. Their mode of life is thought to have been benthic, mud or weed being sucked into their jawless mouths, though no marks have been found in the muds to support such an interpretation. Alternatively they might have fed mouth upward on floating vegetation and detritus like tadpoles. The remains are always of adult animals and it is likely that the larval stages were spent in coastal marine waters which would facilitate their migration. On metamorphosis, the fish might have migrated to the rivers and acquired their characteristic dermal plating.

Cephalaspis is another armoured fish with fixed lateral fins which probably gave slightly better stability. This fish is characterized by large sensory areas on the headshield and high-placed eyes. Our reconstruction shows a *Turonia* in the water above, its body covered with minute denticles. It is depicted as dead, drifting downstream, slightly buoyed by gases from its decomposition. The skeleton would soon rupture and the denticles scatter and be incorporated into the sand and mud. Acanthodian spines can also be found in this environment. The acanthodians are small spiny fish which, though they had jaws, were generally edentulous. Many were microphagous, swimming in deeper water and probably not competing with the agnathids.

Associated with the fish are fragments of eurypterid arthropods, which were scavengers and predators. Individual segments of the carapace can be recognized by the scale-like ornament, and trackways of these animals, generally preserved on the soles of fine sandstones, are known associated with overbank and lake sediments towards the tops of the cycles. A good deal of plant debris can also be found, but no plants have been found in their position of growth.

The Devonian of Wales and the Welsh Borderland is dominantly fluviatile though surprisingly few channels or channel associations are known compared with modern analogues. Sedimentation tends to occur in a cyclical manner with relatively coarse cross-bedded sandstones and conglomerates at the base of each cycle representing channel and lag deposits laid down under fairly high velocities. These pass upwards into muds, silts and often rippled sandstones representing quieter sedimentation on the flood plain or in pools and lakes. Soils with calcareous nodules (caliche), known locally as cornstones, are often present at the tops of the cycles indicating tropical soils formed in alternating wet and dry seasons. The section illustrated is constructed from the base of a cycle in the section of the Lower Old Red Sandstone (Dittonian) at Lydney, Gloucestershire, England. No determinations have been made of the plants at this locality, but rocks of about the same age in Wales and the Welsh Borderland yield simple vascular as well as non-vascular plants.

130

Fish are most common in the conglomerates associated with the bed load but the remains are generally disarticulated and fragmentary. Complete specimens were probably preserved by sudden overwhelming during floods. The more complete carcases at the top of the figure are shown resting on a rippled surface which would probably be reworked later, the debris being transported elsewhere downstream.

Thus the figure illustrates an assemblage of dead individuals swept together by flood waters, but exactly where the fish lived remains uncertain.

32 Lower Devonian Swamp Community

During the Lower Devonian land plants, which had first appeared in the Late Silurian, became common. Although the rootless and leafless psilophytes are seldom well preserved, at the world-famous locality of Rhynie in Aberdeenshire, Scotland the plants were silicified, sometimes in the position in which they grew on a delta marsh at the edge of a small lake, in a volcanic and fumarolic area. They are preserved as chalcedony, having been petrified by the siliceous waters often before much decay had taken place. An indication of how early the silification took place is the good preservation of cell structures. The plants were sometimes petrified in an upright growth position but more frequently delta streams cut through the vegetation and subsequently deposited their sand and plant load, much of the latter partly decomposed.

Rhynie is important not only because of the good preservation of these plants, but because the flora contains some of the earliest vascular forms. Three genera of vascular plants are known, *Rhynia* being the most common. Its simple, leafless, upright stems arose from rhizomes and the stems branched dichotomously, some terminating with an ovoid sporangium; the stem itself had a narrow woody conducting centre for water surrounded by a broad zone of soft tissue and an outer cuticle broken by stomata. There is much variation in the state of preservation and sometimes only the woody centre and cuticle are preserved. There are many indications of how decay was accomplished since preservation of fungal hyphae and reproductive structures are clear. *Horneophyton* is a rather similar plant.

Asteroxylon, a forerunner of the giant Carboniferous lycopods bore short leaves. The stem is much like that of *Rhynia*, but the central conducting zone (xylem) is star-shaped.

Fig. 32 Lower Devonian Swamp Community
a *Rhynia major* (Pteridophyta: Psilophytes)
b *Asteroxylon* (Pteridophyta: Lycopods)
c *Lepidocaris* (Crustacea: ? Branchiopoda)
d *Rhyniella* (Arthropoda: ? Hexapoda)
e *Protocarus* (Chelicerata: Arachnida)
f *Palaeocharinoides* (Chelicerata: Arachnida)

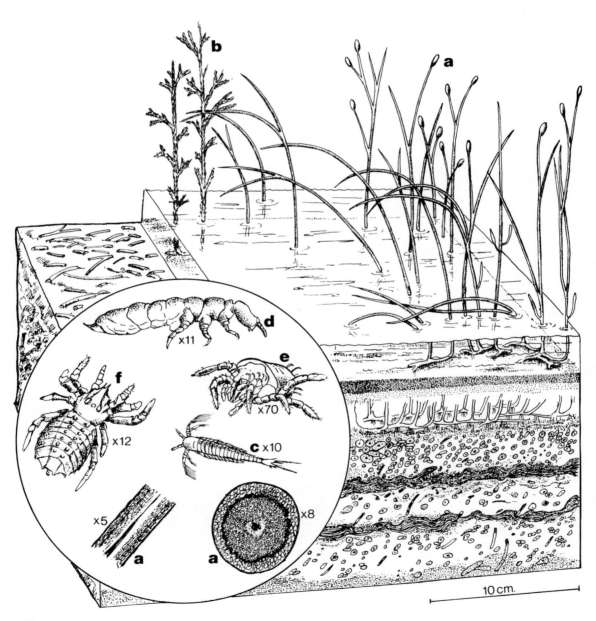

The Rhynie chert is also remarkable in that evidence of the earliest plant-animal associations have been preserved. There are a number of minute arthropods in the chert (preserved analogously to insects in Tertiary amber). The mite *Protocarus crani*, with a body length of 0.5mm or less, had strongly cutinized and pointed chelicerae suggesting that it may have pierced plant or animal tissue to imbibe liquid food. Stem injury is not uncommon, some of which could have been due to damage by arthropods. In one instance a group of arachnids have been located in sporangia where they may have been feeding on the spores. The possible spider *Plaeocteniza* and the spider-like *Palaeocharinoides,* both less than 3.0mm in length, were small predators. *Lepidocaris* is a small (3.0mm long) elongated branchiopod shrimp, probably plankton-eating. It must have been preyed on by larger forms of which no evidence has yet been found. *Rhyniella* is a primitive collembolan (silverfish, which used to be considered as a primitive wingless insect). It is 1—2mm long and was possibly a scavenger, though with its well-developed incisor areas it could have pierced plant cuticle. Like so many rich fossil localities, Rhynie shows strong preservational bias, in particular, the larger faunal elements are absent.

There is no actual exposure at Rhynie and the fossils have been, in part, obtained from loose blocks collected from fields and walls. A trench was dug early this century when the significance of the plant-bearing cherts was first realized.

33 Middle Devonian Lacustrine Community

Lacustrine environments are frequently found in the Devonian, especially in and adjacent to the Caledonides. However, the fauna indicates that there were connections with modern Russia and Canada, so we can conclude that sedimentation did not always take place in completely landlocked basins. One of the lakes encompassed Caithness and the Orkney and Shetland Islands. Water circulation in the lakes was at times incomplete so that the bottom waters became stagnant, and hence anaerobic, providing good preservation potential for the fauna living and dying in the surface waters.

The fauna is quite diverse and includes fish which occupied a variety of ecological niches. A shoal of small acanthodians (spiny sharks) is shown feeding on plankton in the surface waters. The large crossopterygian *Osteolepis* with lobe-like fins and a powerful tail was probably a scavenger, but neither it nor the predator *Cheirolepis*, a primitive actinopterygian fish, would have been able

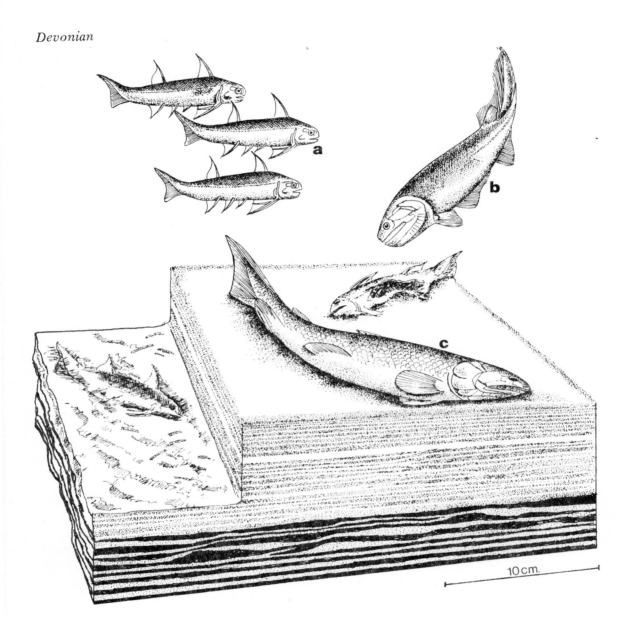

Fig. 33 Middle Devonian Lacustrine Community
 a *Diplacanthus* (Vertebrata: Osteichthyes: Acanthodii)
 b *Cheirolepis* (Vertebrata: Osteichthyes: Actinopterygii)
 c *Osteolepis* (Vertebrata: Osteichthyes: Crossopterygii)

to penetrate to the foul substrate. Bottom-dwelling armoured placoderms such as *Coccosteus* are not figured. They, and the lung fish *Dipterus*, must have lived closer to the shore.

Overall, the fauna shows great advances over that of the Lower Old Red Sandstone. The fish had developed powerful jaws, a variety of fins and large eyes; all necessary for effective predation.

The Achanarras Fish Beds of Caithness and the Sandwick Fish Beds of the Orkney Islands are a famous and well-known horizon of Middle Devonian age. The sediments are laminated impure limestones and comparisons may be made with sediments now being deposited on the floors of modern deep lakes. The fauna is well preserved though compressed. The bottom waters would have been considerably deeper than the waters actually lived in, but this distance is reduced in the figure.

34 Middle Devonian Muddy Shelf Community

The marine muddy shelf environment is one that shows relatively little change during Phanerozoic times, frequently being dominated by protobranch bivalves. The environment illustrated has a diverse fauna, which may reflect more permanent substrates.

Most of the faunas were suspension feeders. Brachiopods are common, particularly spiriferides. Some are winged, and have large spiral lophophore supports, and seem well adapted for life in water with relatively low plankton levels. Others are more globose such as athyrids and atrypids; these sometimes developed spines or frills to aid stabilization. Globular rhynchonellids are more typical of turbulent environments.

Also conspicuous here are small chonetids with rake-like cardinal spines aiding stability. Small turbinate or cylindrical solitary rugose corals and colonies of the tabulate *Heliolites* are quite common. The latter is one of the few Palaeozoic corals that developed intrapolypoidal skeletal integration. Crinoids must have been rooted in the mud or attached to shells. Scavengers are represented by large-eyed phacopid trilobites which were quite often buried in the mud and are preserved without disarticulation.

The substrate illustrated is very much the result of a reasoned guess. Clastic sedimentation decreased in south Devon during the Middle Devonian. Organic remains comprise a greater part of the sedimentary record than hitherto, but muds were still important and sedimentation was often interrupted by volcanic lavas and tuffs.

Few vital substrates have been preserved. The figure has been based on observations made along the south side of Hope's Nose, Torquay, in south-west England. Patches of shells in a good state of preservation are probably in or close to their life position.

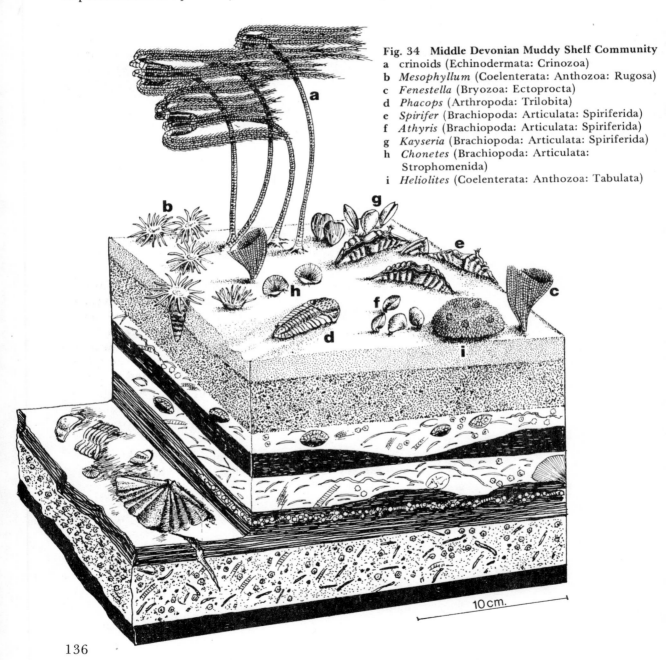

Fig. 34 Middle Devonian Muddy Shelf Community
a crinoids (Echinodermata: Crinozoa)
b *Mesophyllum* (Coelenterata: Anthozoa: Rugosa)
c *Fenestella* (Bryozoa: Ectoprocta)
d *Phacops* (Arthropoda: Trilobita)
e *Spirifer* (Brachiopoda: Articulata: Spiriferida)
f *Athyris* (Brachiopoda: Articulata: Spiriferida)
g *Kayseria* (Brachiopoda: Articulata: Spiriferida)
h *Chonetes* (Brachiopoda: Articulata: Strophomenida)
i *Heliolites* (Coelenterata: Anthozoa: Tabulata)

10 cm.

Fig. 35 Middle Devonian Off-reef Community
a nautiloid (Mollusca: Cephalopoda: Nautiloidea)
b *Thamnopora* (Coelenterata: Anthozoa: Tabulata)
c *Syringopora* polyps (Coelenterata: Anthozoa: Tabulata)
d coral (Coelenterata: Anthozoa: Tabulata)
e *Athyris* (Brachiopoda: Articulata: Spiriferida)
f *Syringopora* (Coelenterata: Anthozoa: Tabulata)
g stromatoporoid (Porifera: Stromatoporoid)

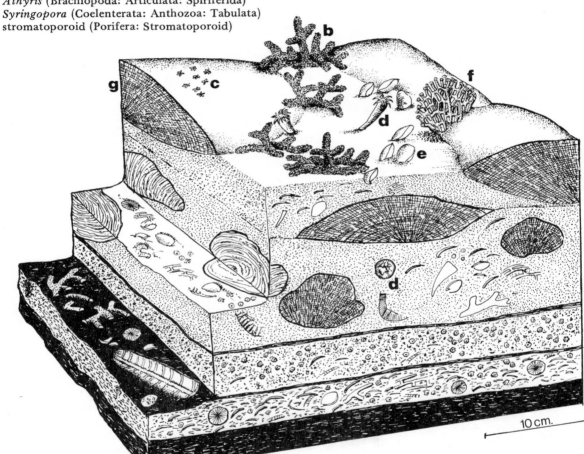

35 Middle Devonian Off-reef Community

The marine shelf faunas of the Devonian are similar in aspect to
those of the Silurian, whether in clastic or in calcareous facies. In
the latter the epifauna is dominated by spiriferoid and atrypoid
brachiopods and the fauna is in general more diverse than in the
Silurian. This diversity is exemplified particularly by the trilobites,

many of which have highly spinose and extravagantly sculptured skeletons.

The environment illustrated is that of a shallow shelf with a rich assemblage of corals and brachiopods. It was not a reef, for only a few beds show the organisms in their position of growth. The bedding surfaces represent the result of periodic events when the substrate was subjected to considerable turbulence halting growth, levelling off the beds and often depositing a layer of mud or coarser sediment that later had to be recolonized. In fact, most of the beds result from storm action, with broken masses of corals and disarticulated shells jumbled in a muddy matrix.

Stromatoporoids are plentiful, particularly massive forms. Exactly what stromatoporoids were is still not certain. The cellular skeleton did not house polyps (like corals), but they may have been the supports for hydrozoans or possibly algal tissue. Popular opinion at present is that it was the tissue of sclerosponges and that the star-shaped canals are some manifestation of the sponge exhalent system. Stromatoporoids often encrust rugose corals and are here depicted growing in association with the tabulate coral *Syringopora*. Shrub-like *Syringopora* also occur without the stromatoporoid association, exhibiting a growth form suitable for living in an environment subject to mud influxes whilst still maintaining intracolony contacts. Between the low stromatoporoid mounds are abundant but fragmentary *Thamnopora* (also a tabulate coral). These probably lived between the massive growths as illustrated. There are occasional solitary corals embedded in the calciluite and a variety of brachiopods of which the smooth-shelled athyrids are the most conspicuous. Crinoids and bryozoans were also important, but seldom survive in position of growth. All these animals were filter and tentacle feeders. Trilobites probably scavenged but their skeletons were easily disarticulated. Nautiloid cephalopods of various but mainly carnivorous feeding habits must have lived in the waters above. The water was almost certainly too deep for substantial algal growth.

The substrate figured has been based on some 20 metres of Middle Devonian (Givetian) at Triangle Point, at the west end of Meadfoot beach, Torquay, south Devon, England. The rich assemblage lived on a shallow shelf almost free of terrigenous influence. Nearby, in the slightly younger rocks at Lummaton, organic growth was sufficient for bioherms to form.

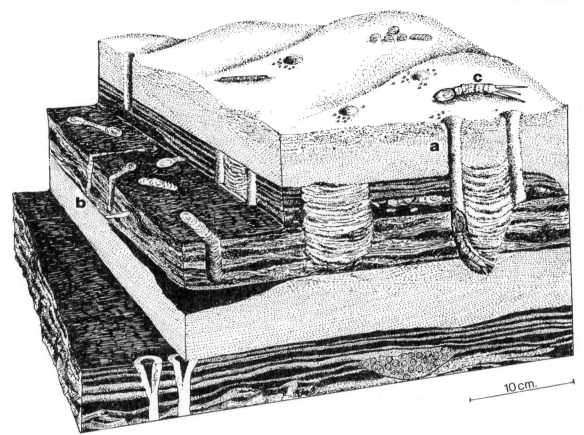

36 Upper Devonian Coastal Deltaic Community

Fig. 36 Upper Devonian Coastal Deltaic Community

a *Diplocraterion* (Trace fossil: ? crustacean)

b *Teichichnus* (Trace fossil: ?annelid)

c *Echinocaris* (Arthropoda: Crustacea: Malacostraca)

Coastal environments are not well represented in the geological record. The fauna is specialized and of low diversity.

Fig. 36 is a reconstruction from the Upper Devonian Baggy Formation of north Devon, England. The layered thin sands and muds were probably deposited just off a distributary and were subject to repeated scour by storms and floods. Burrowing offered protection for the fauna against turbulence and, probably more importantly, against changes in temperature and salinity. The burrows show both downward and, more commonly, upward migration suggesting that the animal responsible liked to maintain a constant depth below the substrate, yoyo-ing in response to the changing substrate level. Many modern animals make U-shaped burrows such as the amphipod *Corophium* on muddy tidal flats, but these are much smaller burrows, less permanent and easily renewed. It is, of course, by no means certain that the animal responsible for the trace fossil *Diplocraterion* was an arthropod;

various types of worms also make U-burrows. No hard parts have been preserved with the burrows, but associated with *Lingula* in the bay and coastal lagoon facies is the small phyllocarid arthropod *Echinocaris* preserved as phosphatic internal moulds. *Diplocraterion* is also present. Some of the phyllocarids may have burrowed but it is not very likely that the lobed and granular carapace of *Echinocaris* would have much aided burrowing. Small ovoid faecal pellets are distributed in the sediment around and within the burrows. A few other burrows are present, the most conspicuous being *Teichichnus* which is likely to have been made by a sediment-feeding polychaete worm. The steep wall-like burrow system is rather similar to that made by the living *Nereis*.

Scouring by currents has usually removed all traces of the fauna, apart from the burrows. We know little about the other benthic fauna nor what the nekton was; a few fish scales have been reported but are inadequate reasons for any informed guess to be made in the reconstruction.

The north Devon area in the Devonian was seldom far from the southern shoreline of the Old Red Sandstone continent, which moved across the area several times during the period. Oscillations were particularly marked towards the end of the period, just preceding the main transgression of the basal Carboniferous, when the terrestrial input via deltas and estuaries was only just compensated by subsidence.

The Upper Devonian Baggy Formation (from which Fig. 36 is constructed) is a suite of shallow water sediments deposited in a variety of nearshore environments but always under some degree of influence from a delta or distributary bringing in freshwater, sediment and plant debris. An offshore, benthic fauna, consisting of an assemblage similar to that in Fig. 37 occurs at scattered horizons, and thick-shelled mussels (*Dolabra*) are associated with the shallower sandstones. Bay environments are represented by muddy sediments where *Lingula* is often conspicuous, associated with bands of small bellerophontid gastropods swept together by wave action.

37   Upper Devonian Clastic Shelf Community

Devonian offshore shelf environments show little change in faunal or sedimentological aspects from those of the preceding Silurian. The sandy or muddy shelves were still colonized by a diverse fauna of brachiopods, bivalves, trilobites and echinoderms. The only major evolutionary innovation was the productid brachiopod in the Middle and, more typically, the Upper Devonian. However, the

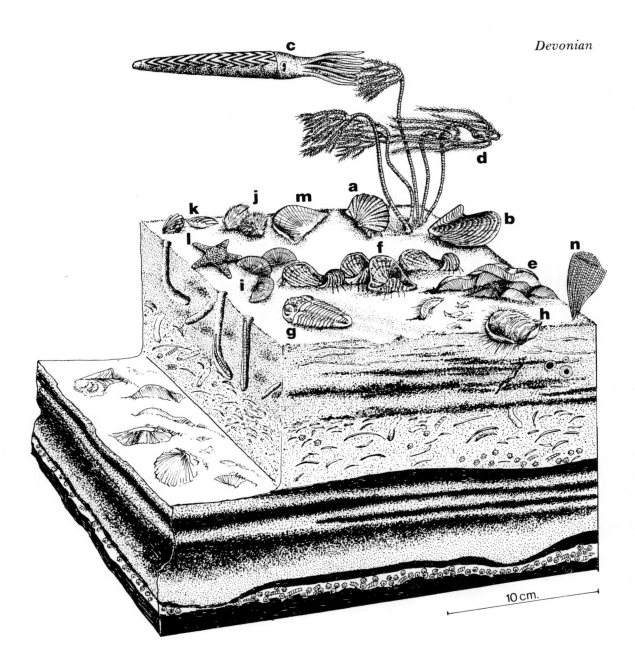

Fig. 37 Upper Devonian Clastic Shelf Community

a *Pterinopecten* (Mollusca: Bivalvia: Pterioida)
b *Ptychopteria* (Mollusca: Bivalvia: Pterioida)
c *Actinoceras* (Mollusca: Cephalopoda: Actinoceratoidea)
d crinoids (Echinodermata: Crinozoa)
e *Cyrtospirifer* (Brachiopoda: Articulata: Spiriferida)
f *Mesoplica* (Brachiopoda: Articulata: Strophomenida)
g *Phacops* (Arthropoda: Trilobita)
h *Productella* (Brachiopoda: Articulata: Strophomenida)

i *Chonetes* (Brachiopoda: Articulata: Strophomenida)
j *Athyris* (Brachiopoda: Articulata: Spiriferida)
k *Camarotoechia* (Brachiopoda: Articulata: Rhynchonellida)
l *Palaeaster* (Echinodermata: Asterozoa)
m *Schellwienella* (Brachiopoda: Articulata: Strophomenida)
n *Fenestella* (Bryozoa: Ectoprocta)

141

Upper Devonian environment illustrated is almost the last representative of the facies so typical of the Ordovocian, Silurian and Devonian. The post-Palaeozoic rise in infaunal bivalves and echinoderms and the decrease in epifaunal brachiopods meant a change in the overall aspect of the benthic faunas.

The depositional environment represented in Fig. 37 by the Pilton Formation of southern England varies a good deal in detail. Generally, turbulence was too frequent for any colonized substrates to be preserved. The general composition of the Pilton fauna is similar to Silurian examples, but of course the taxa present are quite distinct. Only a few taxa are illustrated. In the Upper Devonian, productellid and productid brachiopods are conspicuous. The productellids *(Whidbornella* and *Hamlingella)* are extremely spinose — the latter may even be described as hairy. The spines on the convex valve were for anchorage, while those on the upper brachial valve, would have protected the animal from settlement of pelagic larvae. The productid *Mesoplica* probably used its fewer and stouter spines as an anchoring device. It is quite likely that the productellids and productids colonized somewhat different substrates but no evidence for this is available. None have been found in life position.

Chonetes and the schellwienellid brachiopods probably lived flat on the substrate as adults, on their convex brachial valve, when they lacked a functional pedicle; *Chonetes* was perhaps kept stable by its cardinal spines. There are a number of winged spiriferides. The large *Cyrtospirifer verneuili* is conspicuous throughout the Devonian succession. It often developed a broad and flat cardinal area which must have aided its stability. In the Carboniferous part of the formation it is replaced by *Spirifer tornacensis.* There are a number of more globose brachiopods including athyrids and rhynchonellids, probably adapted to sandier substrates, though some may have been attached to crinoids. The water would have been too deep and muddy for algal growth.

Echinoderms are quite abundant. Elements of crinoids occur in almost every sandstone, but may well have been transported some distance. Starfish, which probably preyed on bivalves, also occur. Their not infrequent complete preservation in thick shelly sandstones testifies to the rapidity of their interment.

The bivalves are mainly epifaunal types, probably most attached by byssal threads to other shells and sand. The bryozoans were also suspension feeders. They formed delicate colonies easily fragmented by turbulence. *Fenestella* and stick bryozoans were probably attached to shell fragments.

There is only one species of trilobite in the Devonian part of the Pilton Formation, *Phacops accipitrinus,* which often reaches over 5cm in length. It probably scavenged and preyed on small soft-bodied elements. It is an interesting species, being the last phacopid, and has a broad geographical distribution across Europe, North Africa and into Asia. In the Carboniferous portion there are a few proetids. There is little evidence of substantial bioturbation by a soft-bodied infauna.

The Pilton Formation of north Devon in south-west England is some 700m thick and straddles the Devonian-Carboniferous boundary; there was ample time for considerable evolution and faunal change to have taken place. The section illustrated is mainly modelled on the shore section at Downend, Croyde, where the faunas are all transported. The actual substrate shown is modelled on similar facies in slightly older strata in New York State described by Bowen et al. (1974), where faunas were occasionally preserved in life position.

38   Late Devonian Basinal Environment

The pelagic animals in the Devonian were quite different from those of the Ordovician and Silurian. Goniatitic cephalopods (and in the Upper Devonian the clymenids) had replaced the graptolites which had become extinct during the Lower Devonian. The conodonts were probably more abundant and there must have been considerable numbers of thin-shelled planktonic ostracodes.

Fig. 38 illustrates an Upper Devonian basinal environment exemplified by Saltern Cove in south Devon. Saltern Cove, during the Upper Devonian, was at least 100km south of the shoreline and only mud reached the area from the Old Red Sandstone continent to the north. But the bottom was by no means level. The topography of submarine ridges (often volcanic), slopes and hollows favoured gravitational slumping so that the Saltern Cove area, which was on a slope, received slumped muds, disrupted blocks of limy sediment, tuffs, conglomerate (often with coralliferous blocks) and agglomerates. The bottom was probably often rocky. There are no examples of pelagic limestones at Saltern Cove, but an example is present near Chudleigh, a locality to the north, where the limestones (often nodular) may be packed with the remains of pelagic animals, particularly ammonoids and conodonts, indicating very slow sedimentation. The ammonoids at Saltern Cove are small, but this is because it is only the inner juvenile whorls that have been preserved as haematitic internal moulds.

Ostracodes are quite common but are always of pelagic type with rather thin shells bearing a pattern of fine 'fingerprint' lines. The occasional small bivalve *Buchiola* may occur with both valves together convex side down, suggesting that after death the open valves sank to the substrate where they rested undisturbed. *Buchiola* was probably attached to weed during life, together with soft-bodied organisms. Small blind trilobites (*Trimerocephalus*), which scavenged and burrowed for worms are also present. Blindness in trilobites is typical of the deep-water Devonian and Carboniferous environments and contrasts with the large eyes of shallow-water species of the same group. (See Figs. 34, 37).

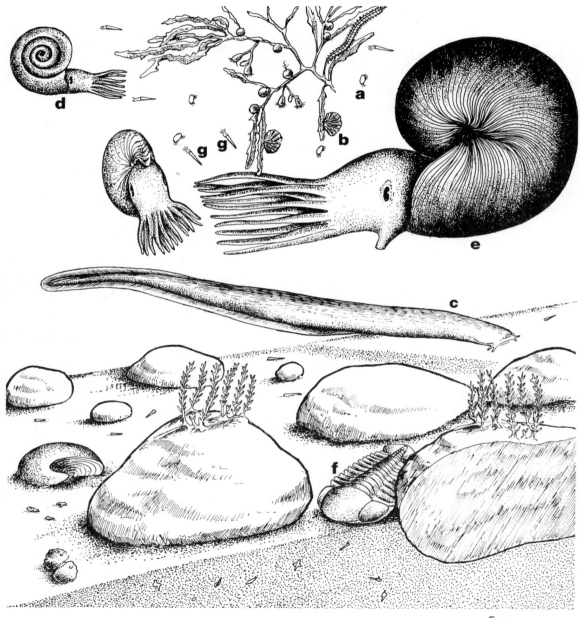

5cm.

144

Observers at the locality will probably be able to find scattered conodonts, millimetre-sized complex tooth-like structures, which can most easily be picked out from small greenish reduction centres, where red ferric iron has been reduced by organic matter to green ferrous iron. Conodonts are composed of calcium phosphate and are likely to have affinities with the chordates. But the nature of the organisms that housed them is still unknown. There are several current hypotheses: the one suggested in the figure is that they were part of the mouth structure (teeth supports) for a widespread, but mainly pelagic, hagfish-like predator and scavenger.

The ammonoids, conodonts and ostracodes show considerable stratigraphic variation and are important zone fossils.

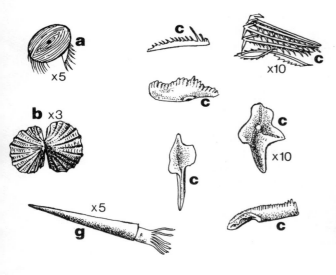

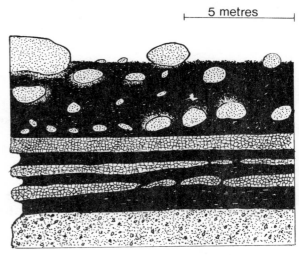

Fig. 38 Late Devonian Basinal Environment
a ostracodes (Arthropoda: Crustacea: Ostracoda)
b *Buchiola* (Mollusca: Bivalvia: Cryptodonta)
c conodont animal (Conodontophora)
d *Archoceras* (Mollusca: Cephalopoda: Ammonoidea)
e *Tornoceras* (Mollusca: Cephalopoda: Ammonoidea)
f *Trimerocephalus* (Arthropoda: Trilobita)
g *Styliolina* (Mollusca: Cricoconarida)

Vertical section through the sediment at Saltern Cove.

Carboniferous

In many parts of the world the Carboniferous deposits preserve the record of a major transgression of the seas onto land areas, and the subsequent retreat of the seas to give continental or near-continental deposits in the upper part of the system. Palaeomagnetic evidence indicates that much of the United States, Europe and parts of Asia lay in the equatorial belt (Fig. h) and there is evidence of hot climates, including evaporite deposits, in these regions.

Although the early Carboniferous transgression was general, it was not a continuous one; its progress was interrupted with pauses and temporary withdrawals of the sea. Except in areas in which subsidence was greater than elsewhere (and such areas are commonly referred to as basins) the hot, arid climate and shallow seas often gave rise to limestone deposition. With a few local exceptions the lands invaded by the sea were apparently not of high relief and land detritus generally occurs only near coastlines.

There were wide expanses of shallow or very shallow seas in which many different types of benthic communities flourished, each specially adapted to the prevailing local ecological conditions. Two factors appear to have been paramount in controlling the composition of these communities: depth and salinity. The nature of the enclosing sediment, be it mudstone or limestone, is also important but seems to have been of lesser significance. Sand in the Carboniferous was often too rapidly deposited for extensive faunal communities to develop.

The depth of water involved in most cases was probably rather small and abyssal deposits are very rarely seen. In the lower part of the Carboniferous a series of communities, linked with gross lithology was outlined by Ramsbottom (1973, 1974) (see Table V). Each type of limestone or mudstone contained its characteristic species of bottom-attached benthic fossils; these were dominantly brachiopods and corals, the mollusca being much less important. Swimming animals such as goniatites, fish and active bivalves are much rarer in the shallow water calcareous coastal deposits, but they were often the only types which flourished in the subsiding basinal areas in which mudstone was being deposited and in which the sea bottom may well have been de-oxygenated;

this prevented colonization by a bottom-dwelling fauna. The reasons for the probable de-oxygenation of the sea floor in basins are still a matter for debate.

At times seas of extremely shallow depth developed over wide areas. These conditions produced intense evaporation which gave rise to varying degrees of hypersalinity, and the fauna and floras which flourished contained only those euryhaline species which could tolerate such specialized conditions. Here stromatolitic algae became dominant, and grew on sheet or mound-like masses. It is

Fig. h. *The world during the Carboniferous. Positions of continents modified after Briden et al. 1974; land areas after Hodson and Ramsbottom, 1973.*

probable that the amount of relief on the stromatolite growths is directly proportional to the closeness of the shoreline and the disturbance of the water, for those with most relief appear to have grown furthest from the shorelines.

When hypersaline conditions occur in any locality, a faunal diversity gradient develops between those areas where salinity is normal and where it is extreme. In the Lower Carboniferous beds of the Northumberland Trough, in northern England, normal open sea conditions became more saline towards the east. There is more-over a definite order of disappearance of different fossil groups, reflecting their varying tolerance of hypersalinity (Table VI).

A similar type of faunal diversity gradient occurs when conditions become less saline than normal; these conditions are best studied in some of the Upper Carboniferous marine bands (Calver, 1968). In this case the freshwater end of the gradient contains 'non-marine' bivalves of a type comparable with living freshwater types of today.

A third type of diversity gradient reflects the transition from the diverse communities of the shelf to the fewer species of the subsiding basinal areas. Such gradients can be inferred from data (R. B. Wilson, 1974) from the Lower Carboniferous Macgregor Marine Beds of south-west Scotland, and also from the increase in diversity towards the old shorelines in the Namurian beds of South Wales.

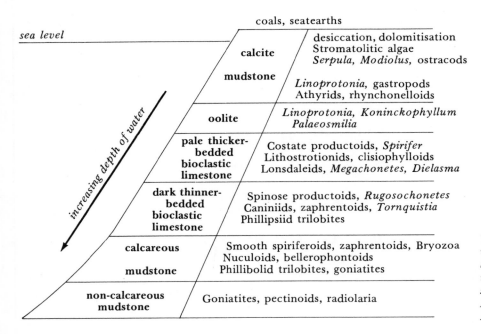

Table V *Hypothetical section, not to scale, to show the lithological and faunal changes which indicate deepening or shallowing of waters in the British Dinantian (from Ramsbottom 1973, fig. 1,).*

Foraminifera	
Rugose corals	Normal
Spiriferoids, chonetids, most productoids	Marine
Crinoids, Bryozoa	Limestones
Syringopora	
Antiquatonia teres, Composita ambigua	Porcellanous
'Camarotoechia' proava	Limestones
Gastropods (most	(slightly hypersaline)
Serpula, Spirorbis	
Sanguinolites roxburghensis, Schizodus	Dolomitic
Turreted gastropods	cementstones
Modiolus latus	(hypersaline)

increasing salinity ↓

Table VI *The order of disappearance of various groups of fossils in the Northumberland Trough (northern England) as the water increased in salinity.*

A common feature of all these diversity gradients is that there is usually a large number of individuals of few species in the less favourable environment, but not necessarily an actual decrease in the number of animals living there.

STRATIGRAPHICAL CLASSIFICATION

In Western Europe, the Carboniferous is divided (Table VII) into two sub-systems. At the base is the Dinantian, which contains the limestones of the major Carboniferous transgression, and at the top the Silesian, which contains much mudstone and sandstone (and includes the Coal Measures) and represents the retreat of the Carboniferous seas.

The Dinantian (also used as a Series name) is divided into stages which broadly coincide with the larger marine transgressions and regressions shown by the rocks; each stage is recognizable from the new faunal elements brought in by each transgression.

Although the actual species in each type of community vary throughout the Dinantian, the ecogroups as units survive throughout and even extend into the Upper Carboniferous.

The Silesian is divided into stages which are based for the most part on individual marine transgressive horizons, identified by widely distributed goniatites, but occurring through a very limited stratal thickness. The goniatites were evolving so fast in the Upper Carboniferous that each marine horizon generally contains its own distinctive goniatite species; thus there are many different assemblages but rather few communities.

In Russia and North America the Carboniferous rocks are classified in a different way, partly because of facies differences, but the broad correlation between Western Europe, Russia and North America is not a great problem.

Western Europe			U.S.S.R.		U.S.A.		
Sub-System	Series	Stages					
Upper Carboniferous (Silesian)	Stephanian	Stephanian C / Stephanian B / Stephanian A / Cantabrian	Gzhelian / Kazimovian	Upper Carboniferous	Virgilian / Missourian		Pennsylvanian
	Westphalian	Westphalian D / Westphalian C / Westphalian B / Westphalian A	Moscovian	Middle Carboniferous	Desmoinesian / Atokan		
	Namurian	Yeadonian / Marsdenian / Kinderscoutian	Bashkirian		Morrowan / ....?............		
		Alportian / Chokierian / Arnsbergian / Pendleian	Serpukhovian	Lower Carboniferous	Chesterian		Mississippian
Lower Carboniferous (Dinantian)	Dinantian	Brigantian / Asbian / Holkerian / Arundian / Chadian	Viséan		Meremecian / Osagian	Valmeyeran	
		Courceyan	Tournaisian		Kinderhookian		

Table VII *Stratigraphical Nomenclature and correlation in the Carboniferous of Western Europe, U.S.S.R. and U.S.A. The Series names for Western Europe are those recognized by the International Subcommission on Carboniferous Stratigraphy. The Stage names for the Dinantian are those recently proposed by George et al. (1976).*

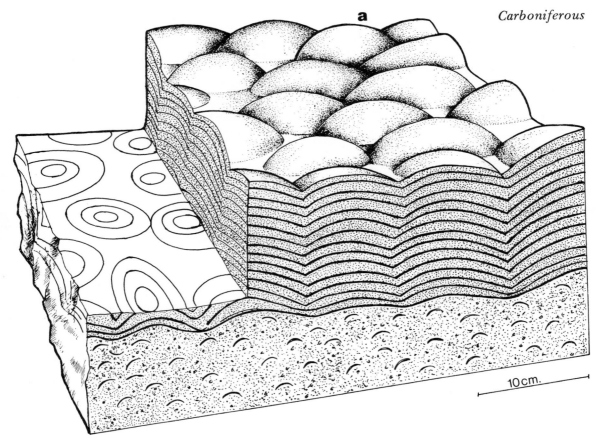

Figs. 39 and 40 are described by Dr. M. R. Leeder.

Fig. 39 Tidal-flat Community A
a stromatolite domes (Algae)

39   Tidal-flat Community A

Stromatolitic limestones are developed by the blue-green algae and
are believed to be for the most part accretions formed by the trapping of sediment. The algal filaments themselves are usually not
calcified. These algae are prolific in extremely shallow coastal
waters and are tolerant of hyper-salinity, and here they often form
the only colonisers. The structural details of the fossil algal domes
may be obscured by dolomitisation, which was partly contemporary, and at times by silification as well. The size and shape of the
particular growth forms seems to have depended largely on the
degree of turbulence of the waters and upon the growth of the
algae over initial irregularities of the substrate. Dome-like growths
are commonest in low energy environments, and they may be
single isolated domes or laterally-linked domes forming a continuous substrate cover.

The laterally linked stromatolite domes illustrated probably
accreted in an intertidal, or sheltered shallow subtidal environment.

40   Tidal-flat Community B

Many nearshore facies in the Lower Carboniferous contain well developed algal stromatolites. Those illustrated here are based on examples from the Lower Border Group (Lower Viséan) in southern Scotland, and represent actual in situ assemblages (Leeder, 1975). The isolation of pillars was probably the result of tidal action where scouring by currents prevented all but temporary lateral linkage between adjacent pillars. The pillars grew together upon a carbonate sand deposit made up of molluscan fragments *(Polidevcia attenuata)*, pellets, ooids and eroded algae. The crinkled algal mat crept over the pillars because of the regression of the shoreline. The mat probably accreted in the high intertidal zone. Spirorbid worms sometimes colonised the partly lithified protruding pillars and crinkles.

In areas of slightly deeper water the blue-green algae occurred as rounded nodules with the form of the genus *Osagia*.

Fig. 40 Tidal-flat Community B

a   stromatolite pillar (Algae)
b   *Polidevcia attenuata* (Mollusca: Bivalvia: Palaeotaxodonta)
c   *Spirorbis* (Annelida: Serpulid)
d   temporary bridging laminae between pillars
e   carbonate sand
f   sediment-rich laminae (in inset)
g   algal-rich laminae (in inset)

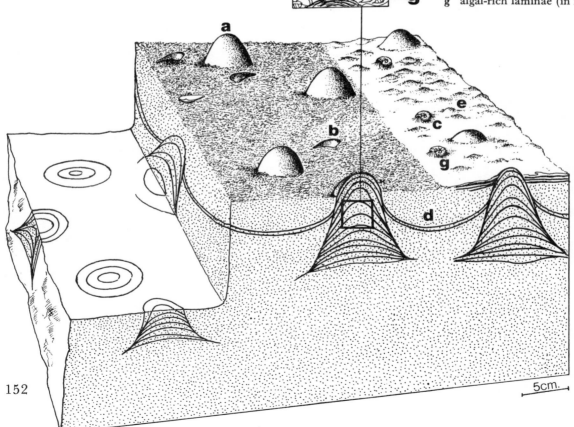

5cm.

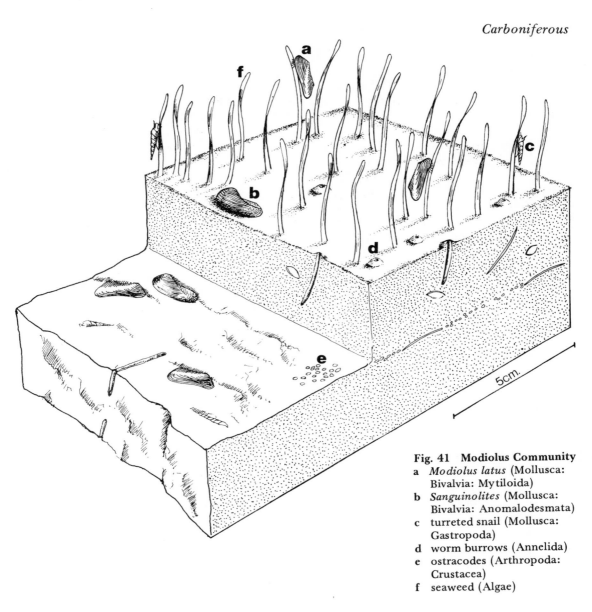

Fig. 41 Modiolus Community
a *Modiolus latus* (Mollusca:
   Bivalvia: Mytiloida)
b *Sanguinolites* (Mollusca:
   Bivalvia: Anomalodesmata)
c turreted snail (Mollusca:
   Gastropoda)
d worm burrows (Annelida)
e ostracodes (Arthropoda:
   Crustacea)
f seaweed (Algae)

41 Modiolus Community

Dolomitic calcite mudstones are often present with very restricted
faunas. These beds are believed to have been hypersaline because
of the extreme shallowness of the water and its high temperature,
accounting for the chemical deposition of calcium carbonate and
its contemporaneous dolomitization (Ramsbottom, 1973). Under
such conditions life must have been harsh, and it is not surprising
that the fauna is highly specialized and by no means abundant.
The community living in these areas has been called the *"Modiola"*

Phase by Dixon and Vaughan (1911). The bivalve *Modiolus* is the most characteristic member and was perhaps a burrower, though it may also have been byssally attached to seaweed (as shown). Another, less common burrowing bivalve is *Sanguinolites*. A few small turreted gastropods also occur, and there are traces of burrows believed to have been made by worms, though some of them may be of crustacean origin. Some bedding planes are covered with abundant smooth-shelled ostracodes, though only two or three species are represented, and the low diversity indicates the hypersalinity deduced from other evidence. There are a few rare traces of stromatolitic algae in these beds.

Beds of this type occur in the shallowest parts of the regressive phases (Ramsbottom, 1973) of the British Carboniferous.

42 Composita Community

This community is well developed in calcite mudstones, which typically occur in cycles (1 to 4m thick) showing alternations of fine and slightly coarser-grained limestones. The finer beds of calcite mudstones are perhaps of chemical origin, the calcium carbonate being deposited from the warm water in the same manner as is the "fur" in a kettle. In each cycle the grain size decreases upwards into calcite mudstones often with small rounded algal nodules; sometimes thin stromatolitic beds are present at the top of each cycle.

Fossils occur in the middle part of these cycles; the community consisting almost entirely of the smooth-shelled spiriferide *Composita* (usually *C. ficoides*) together with a few gastropods. The *Composita* often form shell beds, with the shells crowded together; both valves are usually preserved together in place, indicating that conditions were quiet. Occasionally a few rhynchonellids ("*Camarotoechia*") and *Linoprotonia* are found, but corals are virtually absent. Such a restricted and specialised fauna is taken to indicate a slightly hypersaline environment, with shallow, warm water.

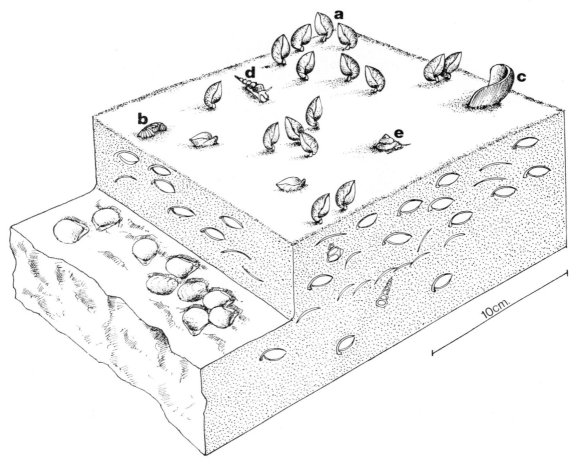

Fig. 42 Composita Community
a *Composita* (Brachiopoda: Articulata: Spiriferida)
b *'Camarotoechia'* (Brachiopoda: Articulata: Rhynchonellida)
c *Linoprotonia* (Brachiopoda: Articulata: Strophomenida)
d turreted gastropod (Mollusca: Gastropoda)
e *Straparollus* (Mollusca: Gastropoda: Archaeogastropoda)

43 Oolitic Limestone Community

The fauna of the oolites in the Carboniferous tends to be rather sparse. The repeated shifting of the bottom material, through current and possibly tidal action in water probably no more than a metre or two deep, seems to have prevented all but the larger species becoming established: most of these forms are fairly massive, presumably to give them a better hold on the mobile sea floor.

155

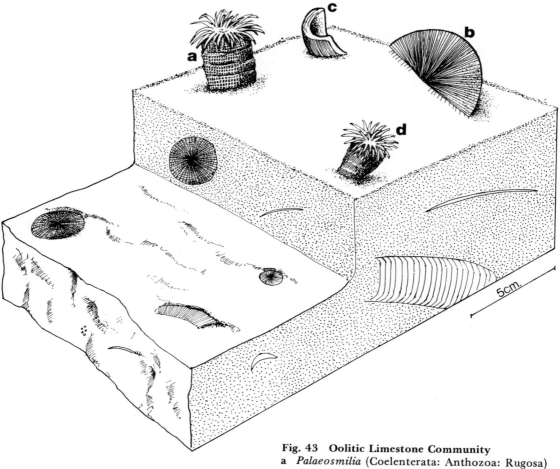

Fig. 43 Oolitic Limestone Community
a *Palaeosmilia* (Coelenterata: Anthozoa: Rugosa)
b *Megachonetes* (Brachiopoda: Articulata: Strophomenida)
c *Linoprotonia* (Brachiopoda: Articulata: Strophomenida)
d *Koninckophyllum* (Coelenterata: Anthozoa: Rugosa)

Large simple corals (*Palaeosmilia, Koninckophyllum*) and large brachiopods (*Megachonetes, Linoprotonia*) are the commonest fossils, and many of these appear to have been carried by currents and damaged to some extent. Microfossils are scarce, though small foraminiferida often form the central core round which oolite grains have developed. The community shown here is typical of that found in the Gully Oolite of Somerset and Avon, (the equivalent of the *Caninia* Oolite of south Wales); this oolite is found in the regressive period near the close of Major Cycle 2 (Ramsbottom, 1973), and is of Chadian age. It can be traced as far east as Aachen in West Germany. Because of their very shallow water origin, oolite limestones are useful in delineating palaeogeography.

156

REEF COMMUNITIES

The following account of Reef Communities (Nos. 44–48) of mid-Viséan (Asbian) age has been contributed by Mr. D. Mundy.

Mid-Viséan reef limestones are generally associated with a marginal situation between a stable shallow water 'shelf' environment and a deeper subsiding 'basin' environment, and are usually characterized by a marked contemporary topographic relief. This can take the form of discrete 'knolls' (as in the 'Craven reef belt') or a continuous 'apron' (as in north Derbyshire). In both cases the reef slopes now present represent palaeoslopes; during the later history of the reef these slopes could span a considerable depth range (122m near Castleton in Derbyshire, England).

The reef limestones are characteristically poorly bedded and the typical matrix between the fossils is a light grey calcilutite (the 'clotted' calcilutite of Schwarzacher, 1961). Bioclastic limestones of various grain sizes are also present. Fabrics once known as 'reef tufa' involving fibrous calcite (the 'radiaxial fibrous calcite' of Bathurst, 1959) are common, and have been interpreted as modified early diagenetic cements filling cavities.

The reef communities are well known for their diversity of taxa and abundance of individuals. However, the richly fossiliferous areas on the reef are often separated by limestones with a poor macrofossil content. Considered as a whole the fauna is dominantly benthic, with a great diversity of brachiopods, bivalves, gastropods and cephalopods, the occasional rostroconch and rare scaphophods. Bryozoans are common, particularly fenestrate, pinnate, and encrusting forms. Corals, including solitary and colonial rugose, tabulate and heterocorals are fairly well represented. Echinoderms, particularly crinoid remains, are abundant, but calices are rarely preserved; blastoids and echinoids are also present. Arthropods are common, particularly trilobites and ostracodes. Other less conspicuous components include sponges, annelids and rare fish remains. The microfauna is dominated by foraminiferida but conodonts are occasionally present. The flora includes stromatolitic, dasycladacean and other algae.

The relative percentages of shelly fauna in a large sample of shells from the Cracoe reefs of Yorkshire is:

Articulate brachiopods	82.32%
Bivalves	9.93%
Cephalopods	3.25%
Gastropods	3.13%
Rostroconchia	1.32%
Inarticulate brachiopods	0.04%
Scaphopods	0.03%

The reef communities illustrated are based on large samples of assemblages from the Cracoe reefs of Yorkshire, and are typical of other communities in reef limestones in other areas. They include an Algal Reef Community (No. 44) and several reef slope communities (reef slope is here used in the sense of the fore reef of Wolfenden, 1958). The term 'shallow water' is not defined in absolute depth. Such a community generally occupies the upper (topographic) portions of the reef and is associated with some evidence of algae. The deeper water communities occupy low levels on the palaeoslopes and are not associated with algae. It should be noted that many elements of the reef phase fauna are found at all depths.

44   Algal Reef Community

Autochthonous algal frameworks producing wall-like structures were recognized in the mid-Viséan reefs of northern England by Wolfenden (1958). These frameworks represent the shallowest water deposits of the reef communities. They were probably subject to strong wave and current action. The community illustrated is based on an algal reef exposed on the top of Stebden Hill, Cracoe.

The primary frame-builder was an encrusting algal stromatolite (presumably a blue-green alga) which produced a finely laminated limestone clearly indicating its growth pattern; it normally has cavities within its structure. In section there is little evidence of the algae responsible apart from the occasional development of *Aphralysia* cells. The stromatolitic framework is often greatly modified by the assembled encrusting fauna, particularly the bryozoan *Fistulipora* and *Tabulipora*, the lithistid sponges *Microspongia* and *Radiatospongia*, and the tabulate corals *Michelinia* and *Emmonsia*. Groves of small solitary rugose corals grew on parts of the framework. These are generally enveloped in the stromatolite and it appears that the latter not only provided the corals with a firm substrate, but also supported them during growth. The tabulate coral *Cladochonus* grew in the same way. A conspicuous small solitary rugose coral is *Cyathaxonia cornu*, which is generally associated with deeper water muddy environments. The associated shelly fauna includes the brachiopods *Leptagonia, Streptorhynchus, 'Reticularia' elliptica* and *Stenoscisma*, as well as small rare productoids (not shown here, but including *Stipulina, Rugicostella* and *Undaria*). The bivalve *Pachypteria*, normally a rare genus, was attached, oyster-like, to the substrate. The large ostracode *Entomoconchus* is common and is frequently found in large numbers in the 'internal sediment' of growth cavities. Colonies of fasiculate *Lithostrotion* are frequently associated with algal reef.

158

Fig. 44 Algal Reef Community
a *Michelinia* (Coelenterata: Anthozoa: Tabulata)
b *Emmonsia parasitica* (Coelenterata: Anthozoa: Tabulata)
c *Cladochonus* (Coelenterata: Anthozoa: Tabulata)
d *Cyathaxonia* (Coelenterata: Anthozoa: Rugosa)
e *Fistulipora* (Bryozoa: Ectoprocta)
f *Fenestella* (Bryozoa: Ectoprocta)
g lithistid sponge (Porifera: Demospongea)
h *Pachypteria* (Mollusca: Bivalvia: Pterioida)
i *'Reticularia'* (Brachiopod: Articulata: Spiriferida)
j *Stenoscisma* (Brachiopod: Articulata: Rhynchonellida)
k *Streptorhynchus* (Brachiopod: Articulata: Strophomenida)
l *Leptagonia* (Brachiopod: Articulata: Strophomenida)
m *Entomoconchus* (Arthropoda: Crustacea: Ostracoda)

159

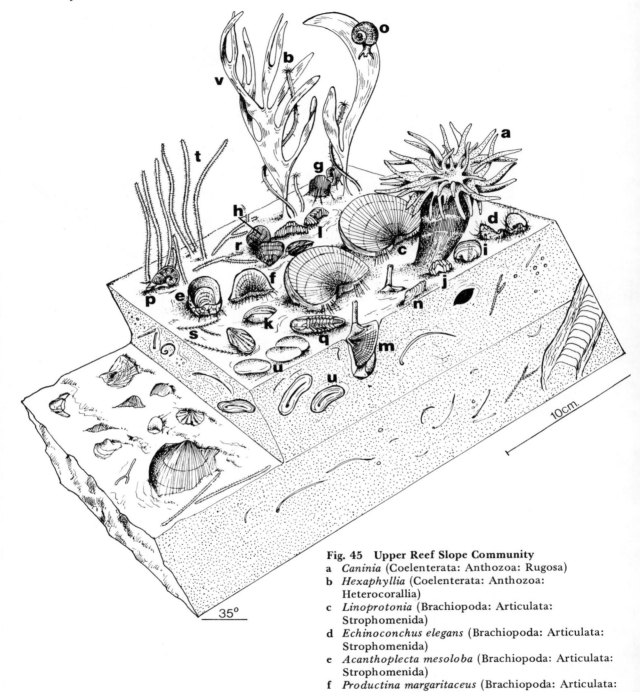

Fig. 45 **Upper Reef Slope Community**
a *Caninia* (Coelenterata: Anthozoa: Rugosa)
b *Hexaphyllia* (Coelenterata: Anthozoa: Heterocorallia)
c *Linoprotonia* (Brachiopoda: Articulata: Strophomenida)
d *Echinoconchus elegans* (Brachiopoda: Articulata: Strophomenida)
e *Acanthoplecta mesoloba* (Brachiopoda: Articulata: Strophomenida)
f *Productina margaritaceus* (Brachiopoda: Articulata: Strophomenida)
g *Alitaria* (Brachiopoda: Articulata: Strophomenida)
h *Rhipidomella* (Brachiopoda: Articulata: Orthida)

45 Upper Reef Slope Community

This community is probably typical of shallow water reef palaeo-slopes. It represents a slightly deeper water than the algal reef and occurs just below the latter on the reef slope. The matrix is a light grey calcilutite which has a clotted texture, and contains little recognizable organic detritus. Irregular patches of fibrous calcite are common. Algae, including *Koninckopora, Girvanella,* and oncolitic encrustation of shell fragments, give an indication of shallow water independent of topographical position. The shelly fauna is dominated by brachiopods, which include small and medium sized productoids and a variety of pedunculate genera. Small rhynchonellids such as *Pleuropugnoides pleurodon* and the terebratulid *Dielasma hastatum* are particularly common and are often found in localized clusters. The molluscan fauna includes the unusual semi-infaunal rostroconch *Conocardium alaeforme,* and local concentrations of gastropods; the latter could be indicative of the presence of non-preserved vegetation (Black, 1954). Occasionally large solitary rugose corals, particularly *Caninia,* occur, and the heterocoral *Hexaphyllia* is locally abundant (our tentative reconstruction shows them supported by conjectured vegetation). Bryozoans are largely of the 'stick' type with *Penniretepora* and *Rhombocladia* being characteristic. Trilobite remains are common throughout, particularly the genus *Cummingella.*

This shallow water reef environment was probably well aerated and the sea turbulent, and it is possible that marine vegetation was very significant.

i *Aulacophoria keyserlingiana* (Brachiopoda: Articulata: Orthida)
j *Spiriferellina insculpta* (Brachiopoda: Articulata: Spiriferida)
k *Dielasma* (Brachiopoda: Articulata: Terebratulida)
l *Pleuropugnoides pleurodon* (Brachiopoda: Articulata: Rhynchonellida)
m *Conocardium* (Mollusca: Rostroconchia)
n *Parallelodon* (Mollusca: Bivalvia: Arcoida)
o *Straparollus* (Mollusca: Gastropoda: Archaeogastropoda)
p *Bellerophon* (Mollusca: Monoplacophora)
q *Cummingella* (Arthropoda: Trilobita)
r *Rhombocladia* (Bryozoa: Ectoprocta)
s *Penniretepora* (Bryozoa: Ectoprocta)
t *Koninckopora* (Algae)
u algal encrusted shells
v seaweed (Algae)

46   Lower Reef Slope Mollusc Community

In deeper water reef environments there is often a great diversity of taxa, and communities are often very variable. Much of the fauna appear to have lived also in shallow water environments. There is no evidence in this community for the presence of algae.

Crinoid remains are more abundant here than in the shallower environments and are frequently associated with the coprophagous gastropod *Platyceras*. The rugose coral *Amplexus coralloides* flourished in the deeper water and occurred locally in dense groves. *Amplexus* is regarded by most authors as the coral most typical of the reef phase, but it is particularly abundant in deeper water reef communities. The presence of large delicate fronds of the bryozoans *Fenestella* and *Polypora* are indicative of the more tranquil conditions of this depth. The community illustrated shows molluscs, including a variety of bivalves and cephalopods, dominating the shelly fauna. Goniatites tend to be common in the deeper water environments, but elsewhere on the reef they are poorly represented, apart from localized concentrations (such as that described by Ford, 1965). The bivalves include shallow burrowers and epibyssate forms. It is noticeable that among the reef faunas infaunal detritus-feeding bivalves are markedly absent. Brachiopods are less abundant in this particular community, but can be more dominant elsewhere in the deeper water environments. Trilobites are slightly less well represented here than in shallow water environments, apart from *Brachymetopus* which is common. Large worm tubes of the genus *Serpula* are often present, and their exposed sections are frequently encrusted by the bryozoan *Tabulipora*.

The lithologies are variable, ranging from calcilutites to calcarenites; coarser calcirudites composed of crinoid debris are also present. The cavity spaces between fenestellid fronds are often filled with successive layers of fibrous calcite.

Fig. 46   Lower Reef Slope Mollusc Community
a *Plicatifera plicatilia* (Brachiopoda: Articulata: Strophomenida)
b *Alitaria* (Brachiopoda: Articulata: Strophomenida)
c *Pugnax pugnus* (Brachiopoda: Articulata: Rhynchonellida)
d *Tylothyris subconica* (Brachiopoda: Articulata: Spiriferida)
e *Fusella triangularis* (Brachiopoda: Articulata: Spiriferida)
f *Posidoniella vetusta* (Mollusca: Bivalvia: Pterioida)
g *Aviculopecten* (Mollusca: Bivalvia: Pterioida)

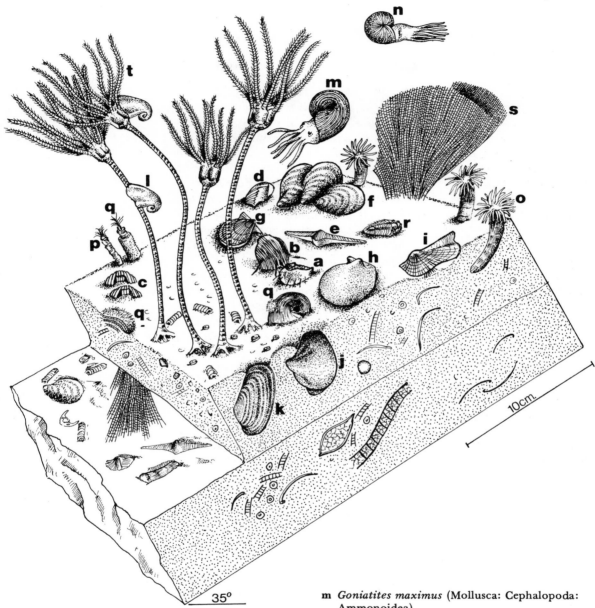

h   *Streblopteria hemisphaericus* (Mollusca: Bivalvia:
    Pterioida)
i   *Parallelodon reticulatus* (Mollusca: Bivalvia:
    Arcoida)
j   *Schidozus* (Mollusca: Bivalvia: Trigonioida)
k   *Edmondia* (Mollusca: Bivalvia: Anomalodesmata)
l   *Platyceras* (Mollusca: Gastropoda:
    Archaeogastropoda)

m   *Goniatites maximus* (Mollusca: Cephalopoda:
    Ammonoidea)
n   *Bollandoceras* (Mollusca: Cephalopoda:
    Ammonoidea)
o   *Amplexus coralloides* (Coelenterata: Anthozoa:
    Rugosa)
p   *Serpula* (Annelida)
q   *Fistulipora* (Bryozoa: Ectoprocta)
r   *Brachymetopus* (Arthropoda: Trilobita)
s   *Polypora* and *Fenestella* (Bryozoa: Ectoprocta)
t   crinoids (Echinodermata: Crinozoa)

163

47   Lower Reef Slope Brachiopod Community

In this deeper water reef community many features are similar to those of the previous community with which it merges. The lithology is a medium grey lime mud which probably was originally a fairly soft substrate. The dominant brachiopods include various productoids, spiriferoids, rhynchonellides and an orthide. The productoids are semi-infaunal forms supported in the soft sediment by spines extending from the convex pedicle valve. Some genera such as *Plicatifera* had a juvenile stage during which they clung to a host (such as a fenestrate bryozoan) by clasping spines (Brunton, 1966). The other brachiopods were largely epifaunal pedunculate forms. The large orthide *Schizophoria resupinata* and the rhynchonellide *Pugnax acuminatus* probably lived on the substrate, and the latter species may have been able to stand a slight amount of sediment covering by holding its high anterior margin clear of the sediment. The bivalves in this community are represented mainly by species of *Leiopteria* which often occur in large numbers. Other bivalves such as the epibyssate *Parallelodon bistriatus* are also present. Cephalopods are well represented, as they were in the previous community. Larger orthocone nautiloids are found, which have a tendency to come to rest facing down the slope. In addition the community includes crinoids, fenestrate bryozoans, trilobites and the coral *Amplexus* (not figured here).

The fine sediment is easily deposited in the empty shell cavities. At the time of deposition the surface of this sediment would have been horizontal; by comparing this surface with the adjacent beds the original palaeoslope can be determined.

Fig. 47   Lower Reef Slope Brachiopod Community
a *Acanthoplecta mesoloba* (Brachiopoda: Articulata: Strophomenida)
b *Plicatifera plicatilis* (Brachiopoda: Articulata: Strophomenida)
c *Krotovia spinulosa* (Brachiopoda: Articulata: Strophomenida)
d *Antiquatonia* (Brachiopoda: Articulata: Strophomenida)
e *Alitaria* (Brachiopoda: Articulata: Strophomenida)
f *Echinoconchus punctatus* (Brachiopoda: Articulata: Strophomenida)

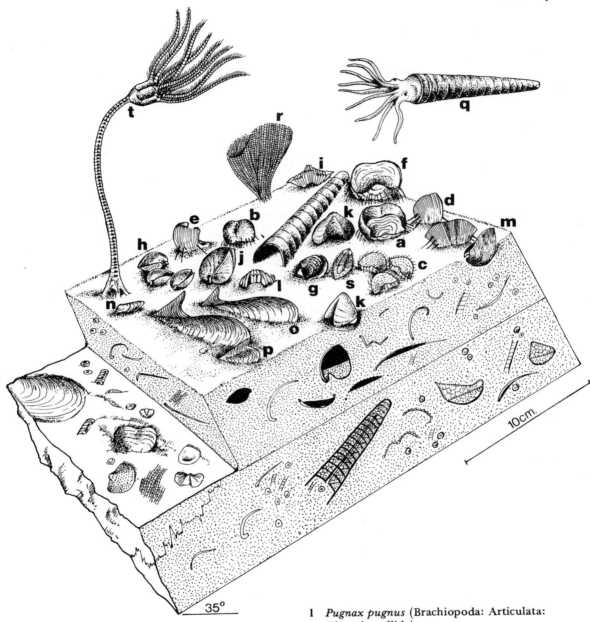

35°

10cm.

g *Overtonia* (Brachiopoda: Articulata: Strophomenida)
h *Phricodothyris* (Brachiopoda: Articulata: Spiriferida)
i *Spirifer bisulcatus* (Brachiopoda: Articulata: Spiriferida)
j *Martinia* (Brachiopoda: Articulata: Spiriferida)
k *Pugnax acuminatus* (Brachiopoda: Articulata: Rhynchonellida)

l *Pugnax pugnus* (Brachiopoda: Articulata: Rhynchonellida)
m *Schizodus* (Mollusca: Bivalvia: Trigonioida)
n *Parallelodon* (Mollusca: Bivalvia: Arcoida)
o *Leiopteria grandis* (Mollusca: Bivalvia: Pterioida)
p *Leiopteria laminosa* (Mollusca: Bivalvia: Pterioida)
q orthocone (Mollusca: Cephalopoda: Nautiloidea)
r *Fenestella* (Bryozoa: Ectoprocta)
s *Brachymetopus* (Arthropoda: Trilobita)
t crinoids (Echinodermata: Crinozoa)

165

48   Productus-Buxtonia Community

This brachiopod-dominated reef slope community is characterized by the productoids *Productus productus, Buxtonia* and *Echinoconchus subelegans,* the davidsoniacean *Schellwienella crenistria,* the orthide *Schizophoria connivens* and the spiriferide *Martinia,* These six species are remarkable for their consistent association, and the community is probably one of the most well defined in the reef phase. Occasionally the productoid *Striatifera striata* occurs in this community. Bivalves are present but are much less numerous than the brachiopods; they include *Leiopteria, Sulcatopinna* and *Myalina.* The surface of the shells are often encrusted by the bryozoans *Fistulipora* and *Tabulipora. Cyclus radialis,* a crustacean, is common and appears to have replaced the trilobites in this environment. The semi-infauna and epifauna of this community tend to occur in shell beds with large numbers of individuals, the resulting rock often being a coquina with the spaces between the shells filled with fibrous calcite. Drifted assemblages contain a high percentage of epifaunal shells. It is probable that *Cyclus* lived between the shells in the original shell beds.

In northern England, this community is well developed on the middle and lower reef slopes of P_{1_a} age and is thus younger than the communities Nos. 44—47, which are of B_2 age. It is however closely associated with encrusting algal stromatolites which suggests that the community probably lived in fairly shallow water; it may thus have developed low on the reef during periods when part of the reef was above water.

166

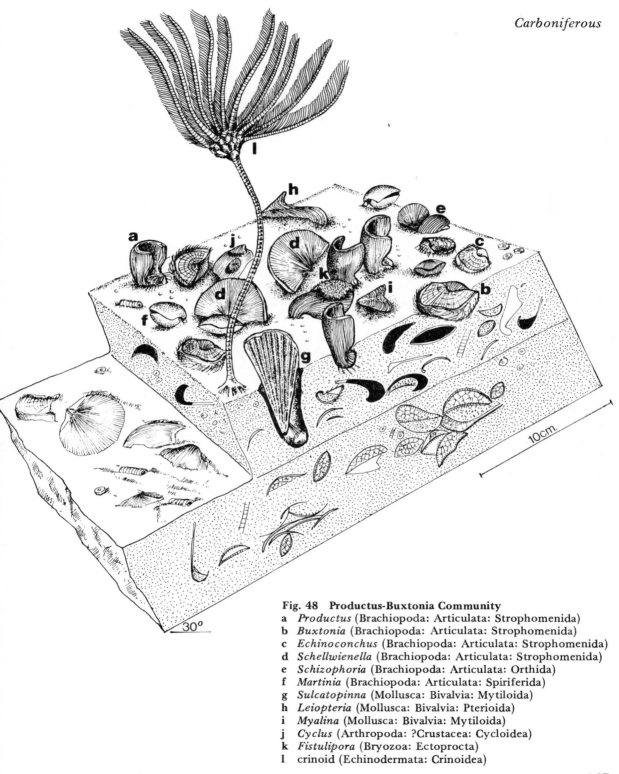

Fig. 48 Productus-Buxtonia Community
a *Productus* (Brachiopoda: Articulata: Strophomenida)
b *Buxtonia* (Brachiopoda: Articulata: Strophomenida)
c *Echinoconchus* (Brachiopoda: Articulata: Strophomenida)
d *Schellwienella* (Brachiopoda: Articulata: Strophomenida)
e *Schizophoria* (Brachiopoda: Articulata: Orthida)
f *Martinia* (Brachiopoda: Articulata: Spiriferida)
g *Sulcatopinna* (Mollusca: Bivalvia: Mytiloida)
h *Leiopteria* (Mollusca: Bivalvia: Pterioida)
i *Myalina* (Mollusca: Bivalvia: Mytiloida)
j *Cyclus* (Arthropoda: ?Crustacea: Cycloidea)
k *Fistulipora* (Bryozoa: Ectoprocta)
l crinoid (Echinodermata: Crinoidea)

167

49   Coral-calcarenite Community

This community, mainly characterized by the presence of large and massive compound corals and an assortment of brachiopods, is developed in pale crinoidal bioclastic limestones. The species found in this clear water environment include the varied macrofauna illustrated, and also benthic foraminiferida (too small to be shown in the figure). The bulk of the rock is formed of broken shell fragments and shell debris, apparently for the most part transported into the area.

The particular community figured here is found in beds of Asbian (D_1) age in the north of England. The large coral colonies can reach up to a metre across. Somewhat similar communities are found in most of the transgressive (deeper water) phases of cycles of deposition (Ramsbottom, 1973). One of the most characteristic species in such communities of mid-Viséan age is the alga *Koninckopora inflata*, which can be found in practically every thin section of these limestones, often more or less broken, and not commonly preserved complete. The large and massive compound coral colonies sometimes occur upside down, indicating a somewhat turbulent environment. These large colonies probably provided a number of niches in their shadows where small brachiopods, especially smooth spiriferides, could flourish. Trilobites with tumid form, which crawled around on the sea floor, possibly as scavengers, are not uncommon in this community.

Fig. 49   Coral-calcarenite Community
a   *Lithostrotion arachnoideum* (Coelenterata: Anthozoa: Rugosa)
b   *Lithostrotion portlocki* (Coelenterata: Anthozoa: Rugosa)
c   *Paladin* (Arthropoda: Trilobita)
d   *Koninckopora* (Algae)
e   *Antiquatonia* (Brachiopoda: Articulata: Strophomenida)
f   *Spirifer* (Brachiopoda: Articulata: Spiriferida)
g   *Syringopora* (Coelenterata: Anthozoa: Tabulata)
h   *Composita* (Brachiopoda: Articulata: Spiriferida)
i   *Rugosochonetes* (Brachiopoda: Articulata: Strophomenida)
j   orthocone (Mollusca: Cephalopoda: Nautiloidea)
k   *Schellwienella* (Brachiopoda: Articulata: Strophomenida)
l   *Linoprotonia* (Brachiopoda: Articulata: Strophomenida)
m   *Koninckophyllum* (Coelenterata: Anthozoa: Rugosa)
n   rhynchonellid (Brachiopoda: Articulata: Rhynchonellida)

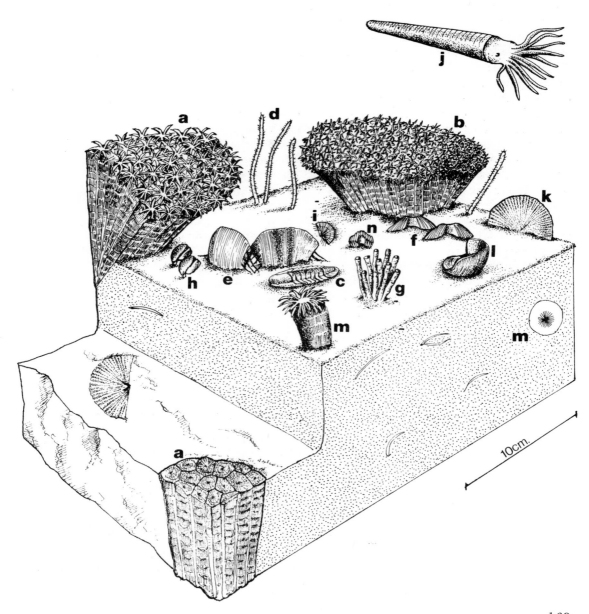

50 Brachiopod Calcarenite Community

This community occurs in dark relatively thin-bedded bioclastic limestones which differ from the paler limestones of similar type mainly in the amount of terrigenous matter present. This factor, however, is enough to cause considerable differences in the faunal associations between the clearer water pale limestones (community 49) and the darker limestones formed in more turbid waters. Large brachiopods, mainly with a costate shell ornamentation, such as *Gigantoproductus, Semiplanus* and *Daviesiella*, are characteristic, while the commonest of the smaller brachiopods are the spiriferoids. The smaller productoids include many spinose forms such as *Pustula* and *Buxtonia*. The corals are not only the simple forms like *Dibunophyllum* and '*Caninia*', but also compound forms, both massive and fasciculate which may form large colonies more than half a metre across. Foraminiferida are abundant in these communities, but are too small to be shown in the figure. The large coral colonies may still sometimes be upside down, but the environment as a whole was less turbulent than that of the previous community (49).

Communities of this type are found in several transgressive phases of the major Dinantian sedimentary cycles (Ramsbottom, 1973) and are characteristic of many of the limestones of Brigantian (D_2) age in northern England and elsewhere.

Fig. 50 Brachiopod Calcarenite Community
a *Dibunophyllum* (Coelenterata: Anthozoa: Rugosa)
b *Eomarginifera* (Brachiopoda: Articulata: Strophomenida)
c *Gigantoproductus giganteus* (Brachiopoda: Articulata: Strophomenida)
d *Semiplanus latissimus* (Brachiopoda: Articulata: Strophomenida)
e *Lonsdaleia* (Coelenterata: Anthozoa: Rugosa)
f *Aviculopecten* (Mollusca: Bivalvia: Pterioida)
g small gastropod (Mollusca: Gastropoda)
h *Composita* (Brachiopoda: Articulata: Spiriferida)
i *Lithostrotion junceum* (Coelenterata: Anthozoa: Rugosa)
j *Rugosochonetes* (Brachiopoda: Articulata: Strophomenida)
k *Schellwienella* (Brachiopoda: Articulata: Strophomenida)
l *Buxtonia* (Brachiopoda: Articulata: Strophomenida)
m orthocone (Mollusca: Cephalopoda: Nautiloidea)
n zaphrentoid (Coelenterata: Anthozoa: Rugosa)
o *Chaetetes* (?Porifera)

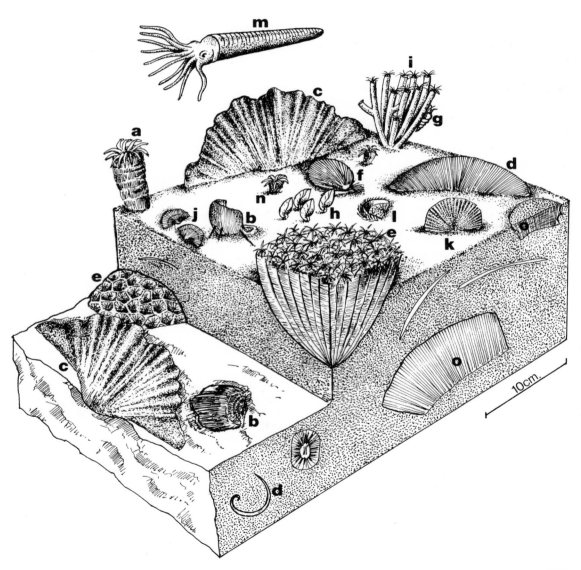

51 Mud Community

These faunas occur in muds which are considered to have been deposited in slightly deeper water than the dark thin-bedded limestones (Community 50), but there are many elements in common. Usually, however, the large productoids such as *Gigantoproductus* are missing from the Mud Communities.

Small corals (such as *Cyathoxonia* and *zaphrentoids*), bryozoans (especially *Fenestella* and *Penniretepora*), crinoids, some echinoids (usually found dissociated), smooth spiriferides such as *Martinia* and *Crurithyris*, small chonetoids, a few productoids including *Productus* s.s., phillibolid trilobites (as opposed to the phillipsiid types more characteristic of shallower water shelf limestones), bivalve pectinoids and occasional goniatites comprise this community. Among the smaller fossils are ostracodes of many types, holothurians (only found as dissociated spicules), arenaceous and calcareous foraminifera, conodonts and occasional radiolaria. These groups are sometimes of exceptional diversity, as many as 50 or 60 or even more species being present.

Communities of this type are best developed in north-west Europe in high Dinantian beds, but are also found in Namurian shelf or near-coastal environments, and, rarely, in the Westphalian marine beds.

Fig. 51 Mud Community
a *Fenestella* (Bryozoa: Ectoprocta)
b crinoid (Echinodermata: Crinoidea)
c *Cyathoxonia* (Coelenterata: Anthozoa: Rugosa)
d *Rotiphyllum* (Coelenterata: Anthozoa: Rugosa)
e orthocone (Mollusca: Cephalopoda: Nautiloidea)
f *Sudeticeras* (Mollusca: Cephalopoda: Ammonoidea)
g *Productus* (Brachiopoda: Articulata: Strophomenida)
h *Crurithyris* (Brachiopoda: Articulata: Spiriferida)
i holothurian (Echinodermata: Echinozoa)
j rhynchonellid (Brachiopoda: Articulata: Rhynchonellida)
k bryozoan (Bryozoa)
l *Rugosochonetes* (Brachiopoda: Articulata: Strophomenida)
m *Aviculopecten* (Mollusca: Bivalvia: Pterioida)
n *Pinna* (Mollusca: Bivalvia: Mytiloida)
o *Edmondia* (Mollusca: Bivalvia: Anomalodesmata)
p *Archegonus* (Arthropoda: Trilobita)
q ostracodes (Arthropoda: Crustacea)
r *Penniretepora* (Bryozoa: Ectoprocta)
s gastropod (Mollusca: Gastropoda)

52   Namurian Goniatite Community

In the Namurian the basinal areas of relatively deep water, in which non-calcareous mudstones were being deposited, commonly contain no benthic fauna during the marine episodes. But there is an abundance of swimming forms, especially of goniatites and pectinoids. Both these groups were free-swimming, though they may have come to rest on the bottom at times. Bivalves of pterioid form, such as *Caneyella* and more rarely *Posidoniella*, have the same distribution as the pectinoids and either had the same mode of life or were attached to plants of which there is now no trace.

The lack of benthos has been attributed to the lack of oxygen on the sea floor, though it was possibly due to the lack of a definitive division between the mud and the water, which would have made colonization of an unstable sea floor difficult (Holdsworth, 1966).

The community illustrated is of Marsdenian (R_{2a}) age and is typical of many of the Namurian marine horizons of north-west Europe. The goniatites *Reticuloceras* and *Hudsonoceras* are the common forms though nautiloids, both orthocones and coiled forms, do occur rarely. *Posidonia, Caneyella* and the large pectinoid *Dunbarella,* comprise the bivalves. Also present, though not shown in the figure, are radiolaria and conodonts (presumed to be derived from a free-swimming animal), and, rarely, fish, usually as fragments.

In some of the Namurian marine horizons there are large numbers of very small molluscan shells (spat), both gastropods and bivalves. They are of a size (< 1.0mm) at which the early planktonic shell sinks to the bottom and settles for its adult life, but in this case the animals died at once, presumably because of the unsuitable, possibly de-oxygenated, nature of the sea floor (Ramsbottom et al. 1962).

Goniatite communities of this type are also found in some of the Westphalian marine bands.

53   Namurian Basinal Benthic Community

An example of one of the rare benthic communities of the European Namurian basins is shown here. The goniatites and pectinoids occur as in the non-benthic Namurian faunas, but associated is a limited and restricted benthos which includes productoids, chonetoids, rare smooth spiriferides (not illustrated), crinoids and trilobites. Apart from the trilobites, which presumably acted as scavengers on the sea floor, the remaining benthic groups were probably all

Fig. 52 Namurian Goniatite Community
a *Hudsonoceras* (Mollusca: Cephalopoda: Ammonoidea)
b *Reticuloceras* (Mollusca: Cephalopoda: Ammonoidea)
c *Posidonia* (Mollusca: Bivalvia: Pterioida)
d *Dunbarella* (Mollusca: Bivalvia: Pterioida)

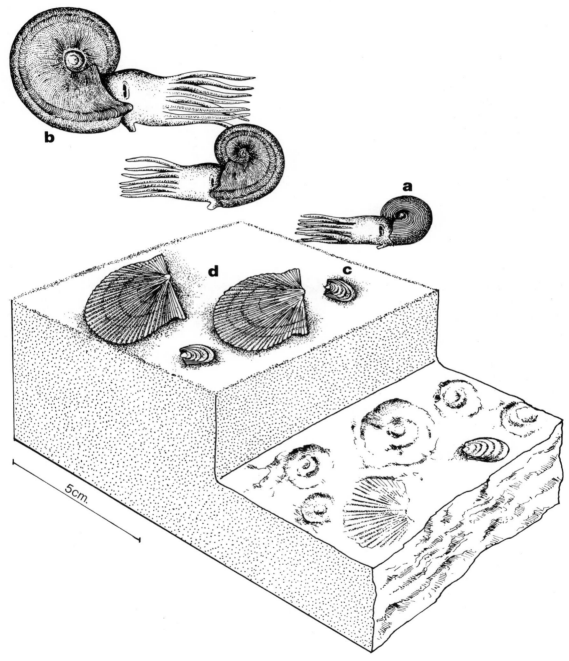

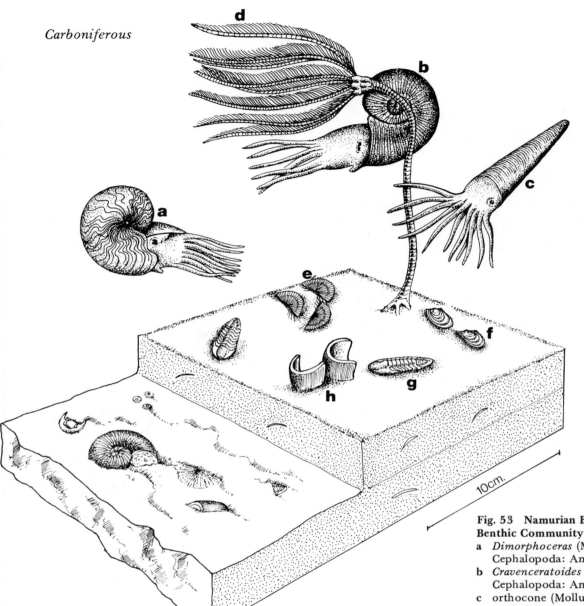

Carboniferous

Fig. 53 Namurian Basinal Benthic Community

a *Dimorphoceras* (Mollusca: Cephalopoda: Ammonoidea)
b *Cravenceratoides* (Mollusca: Cephalopoda: Ammonoidea)
c orthocone (Mollusca: Cephalopoda: Nautiloidea)
d crinoid (Echinodermata: Crinoidea)
e *Rugosochonetes* (Brachiopoda: Articulata: Strophomenida)
f *Posidonia* (Mollusca: Bivalvia: Pterioida)
g *Paladin* (Arthropoda: Trilobita)
h *Productus hibernicus* (Brachiopod: Articulata: Strophomenida)

filter feeders, and burrowing forms were not present. This fauna is widespread at one horizon (in E_{2b}) in the Arnsbergian stage in north-west Europe. It presumably indicates either a shallowing of the water or the development of bottom currents so as to allow colonization of the sea floor by benthic animals. The mudstone containing this fauna is slightly calcareous.

176

54 Westphalian Non-marine Bivalve Community

Fig. 54 Westphalian Non-marine Bivalve Community
a *Anthraconaia* (Mollusca: Bivalvia: Pterioida)
b *Carbonicola* (Mollusca: Bivalvia: Unionida)
c *Naiadites* (Mollusca: Bivalvia: Pterioida)
d *Spirorbis* (Annelida: Serpulid)
e *Geisina* (Arthropoda: Crustacea: Ostracoda)
f *Pelecypodichnus* (Trace fossils of Bivalve)

Non-marine bivalves inhabited the shallow waters which spread over extensive areas after the flooding of the coal forests. There are no true marine fossils present, though some of the bivalve genera, particularly *Naiadites* and *Curvirimula*, were apparently able to tolerate salinity conditions which might have been brackish.

Naiadites appear to have been byssally attached, possibly to drifting vegetation, which would explain the wide distribution of this genus. The small adherent worm tubes of *Spirorbis* are commonly found attached to the shells of *Naiadites*, and the relationship appears to have been a commensal one, for the spirorbids are

177

concentrated on the shell close to the inhalent siphonal current, which would have provided them with a source of food.

Carbonicola and the much rarer *Anthraconaia* were burrowers, and work by Eagar suggests that the variations in the shape of the shell in the different species indicate slightly differing habitats. The burrows of these bivalves, especially those of *Anthraconaia*, which made a deep vertical burrow, are found as the trace fossil called *Pelecypodichnus*. These are often aligned in one general direction, probably that of the current, and were oval or semi-oval depressions on the mud floor.

The fauna illustrated is thought to be typical of that found in the lower part of the Westphalian in north-west Europe. At this horizon a common benthic ostracode is *Geisina arcuata,* which is believed to have lived amongst the weeds presumed to have been either growing in the bottom muds or floating freely in shallow water.

Fig. 55 is described by Dr. R. Goldring.

55 Sea Level Lake Community

The Upper Carboniferous swamps and inland seas are often monotonously unfossiliferous except for plant fossils. The occasional marine transgressions are, of course, frequently marked by sediments rich in marine fossils: goniatites and productoid brachiopods. The example shown is taken from the Bude Formation of the Upper Carboniferous of south-west England. It is a rather special situation where bedding surfaces have frequently preserved the impressions of limulids moving over the substrate. Some surfaces are densely tracked. Limulids are one of a very conservative group of animals. Their morphology, indicated by their tracks, shows little evolutionary change from the Devonian to the present. In the illustration the trackways show in detail the marks of the telson, the legs, especially the pushers, each with its hand-like extremity and, occasionally, traces of the genal spines. These tracks were probably made by *Euproops*, but as is so frequent in the fossil record, no specimen has been found associated with the tracks. The tracked horizons were unstable substrates and therefore not the most favourable for carapace fossilization. The traces vary in sharpness and in the amount of detail preserved. The pushing appendages penetrated slightly deeper than the anterior appendages. The sediment has not been much disturbed by the animals' preambulations; they were certainly not digging extensively for food as do modern limulids living on sandy nearshore substrates. The individuals can be judged to have been of small size. There are also

Fig. 55 Sea Level Lake Community
a *Rhabdoderma* (Vertebrata: Osteichthyes)
b *Elonichthys* (Vertebrata: Osteichthyes)
c *Euproops* (Arthropoda: Merostomata)

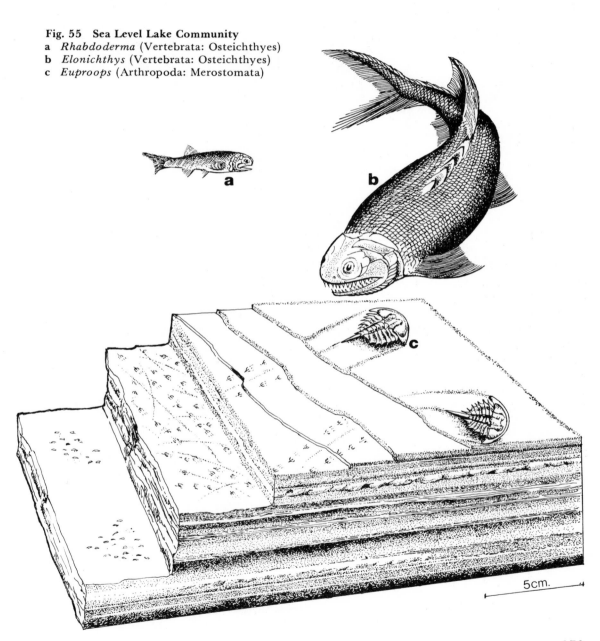

5cm.

Fig. 56 Westphalian Forest

a *Stigmaria* (Pteridophyta: Lycopod)
b *Lepidodendron* (Pteridophyta: Lycopod)

c tree fern (Pteridophyta: Fern)
d calamitid (Pteridophyta: Calamites)
e dragonfly (Arthropoda: Hexapoda)

some curious telson tracks in the form of a repetitive double curve (not figured). The curves are slightly out of phase and one curve is of rather greater amplitude than the others. The large curve follows and cuts the smaller and both must have been made by a unit which Dr. A. F. King envisaged as a pair of crabs mating with the male riding on the buckler of the female, just as in modern limulids. The tracked silts were not disturbed by any other organism, though in the black mudstones associated with the Bude Formation fish can be found in concretions, where the skeletons have resisted compaction. Two palaeoniscoids (primitive actinopterigians) are figured; both were probably carnivorous with sharp conical teeth separated by wide gaps. The smaller is *Cornuboniscus budensis* and the larger *Elonichthys aitkeni*. A small acanthodian (spiny shark), *Acanthodes wardi*, and the coelocanthid *Rhabdoderma elegans* also occur. The crustacean *Crangopsis* has also been found, and the black shales often contain the trace fossil *Planolites*.

56 Westphalian Forest

In the Coal Measures of Europe, which are mainly of Westphalian age, and in comparable beds in North America, vast areas (Fig. i) became intermittently covered with forest swamps of the type shown in this figure. The tallest trees, such as *Lepidodendron (Sigillaria* is the name for its stems, and *Stigmaria* for its roots), with their spreading branches formed the rather open canopy of the forest. Trees of rather lesser size included calamitids, especially *Calamites* itself, related to the horsetails of today and marked by prominent horizontal nodal lines. All these probably grew actually in or at the edge of shallow marshy pools. In among the trees grew the smaller tree-ferns. It is probable too that abundant representatives of the fungi, algae, mosses and liverworts also flourished in these forests, but they have left few fossil remains. The latter part of the Carboniferous saw the development of primitive amphibia and also of the early insects. One of these, a large dragonfly with a wing span of 50 or 60cm, is shown in the figure.

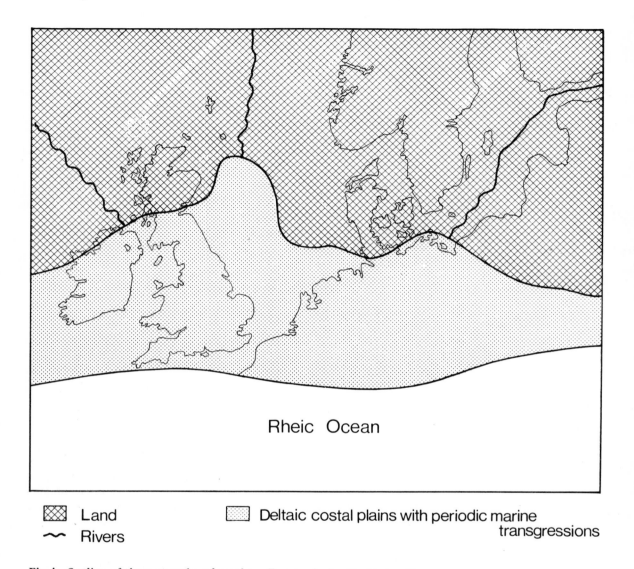

Rheic Ocean

☒ Land
〰 Rivers
▦ Deltaic costal plains with periodic marine transgressions

Fig. i. *Outline of the geography of northern Europe during the Carboniferous (early Westphalian). Some islands may have existed in the deltaic area. The exact positions of the rivers flowing southwards is quite uncertain. The Rheic Ocean may have closed in the late Westphalian (McKerrow and Ziegler, 1972).*

CARBONIFEROUS COMMUNITIES OF OTHER AREAS

Communities comparable to those figured, together with many others not dealt with here, are found in the limestones which occur in all the regions of the Carboniferous equatorial belt. However, the taxa involved at any age vary in the different regions of the world.

Broadly there are three main faunal realms:

1 *Eurasia* — stretching from China to Western Europe and including North Africa, the Middle East, and possibly Japan and Australia. Here the faunas of the Viséan and Namurian are characterized by the large productoids such as *Gigantoproductus* and *Semiplanus*. In the Viséan the large and stratigraphically very informative chonetoids *Delepinea* and *Daviesiella* are widespread. Among corals *Lithostrotion* and *Lonsdaleia* are typical, and in the Upper Viséan and Namurian *Dibunophyllum*.

2 *The American Cordilleran* region. This area contains many of the faunal elements of the Eurasian area, but often in a slightly modified form. For example the large productoids are represented by *Titanaria*, a rare genus which differs significantly from *Gigantoproductus*, but is assigned to the same family. Among the corals the presence of *Lithostrotionella*, also common in China and Japan, is notable. There are also, as in all areas, a good many endemic species. The large chonetoids did not reach this region.

3 *The mid-western United States,* especially the Mississippi valley area. The faunas here contain a larger proportion of local forms than is usual in the Carboniferous. Large productoids and chonetoids are lacking and the coral faunas are to a large extent distinctive. The curious screw-like bryozoan *Archimedes* is almost restricted to this area.

The reasons for this faunal distribution in the Carboniferous are connected with the contemporary geography (Fig. h, p. 147), which, together with winds, influenced the main oceanic currents. Thus the north-easterly currents in the Ural region caused migrations across the Arctic area to western America, and the generally easterly currents (and easterly winds, of which there is evidence from the distribution of volcanic ashes) caused migrations from China and Russia into western Europe. The mid-west region of the United States was apparently at the end of this chain of migration, and this would account for the high number of local species in that area, and for the later appearance there than elsewhere of the archaediscid foraminiferida.

Permian

During the Permian and especially the early part of the period the seas withdrew from the continental masses, to a greater or less extent, in many parts of the world (Fig. j). But some areas, including parts of Australia and the United States, contain marine sequences in which, in broad terms and with different taxa, many of the types of community seen in the Carboniferous persist.

Permian marine onshore faunas, like those of the Carboniferous, are dominated by the brachiopods, which occupy ecological niches later taken over by bivalves. Bryozoans become more evident than in the Carboniferous and are frequently the sediment-trapping animals in reefs, but the other animal groups remain much as before. One new development is represented by the large rock-forming fusulinid foraminiferida which began to develop in the later Carboniferous but reached their acme in the Permian. On land there was a great development of the reptiles, forerunners of the later dinosaurs.

Palaeomagnetic evidence for the Permian indicates that north-west Europe and the United States were in the equatorial belt, and this is confirmed by the distribution of evaporites and limestones. North-west Europe was a land area for most of the early part of the period and only in the late Permian (Zechstein) did the seas return as a shallow epicontinental gulf which covered an area from north-east England to the present borders of Poland and the U.S.S.R. (Fig. k). The climate was hot and arid. The Permian succession in the area is a cyclic one with four or five main influxes of oceanic water coming from the north. Each cycle was completed by the precipitation of salts from the evaporating sea-water. The resultant evaporite beds are the source of some of the salt, gypsum and anhydrite, and of all the potash resources in the area.

The fauna of the north-west European epicontinental seas was rather restricted, probably because of the development of hyper-salinity to varying degrees due to the hot, dry climate. There were no fusulinids and few cephalopods. The richest faunas occurred in the first of the Zechstein cycles (often designated Z1), especially in the barrier reefs formed at the margins of the Zechstein Sea. These reefs are comparable in many features, but not in scale, to

184

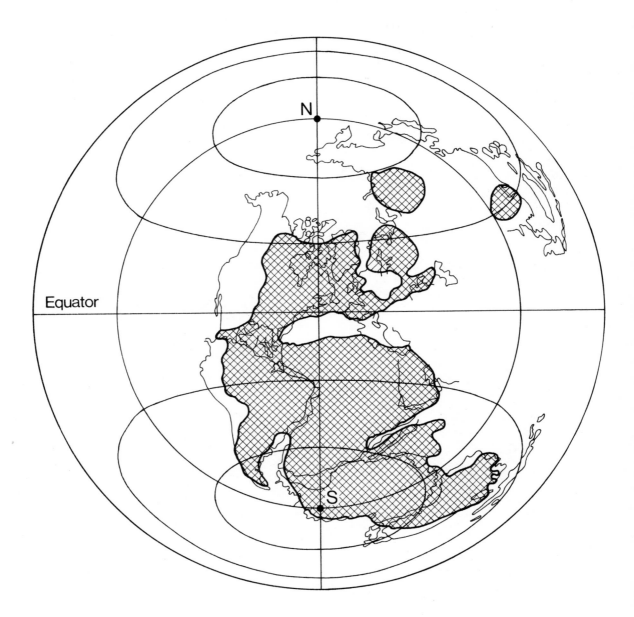

Fig. j. *The world during the Permian. After Early Permian time, it is probable that all the continents were united to form Pangea, the Permian supercontinent. Positions of the continents after Briden et al. 1974.*

185

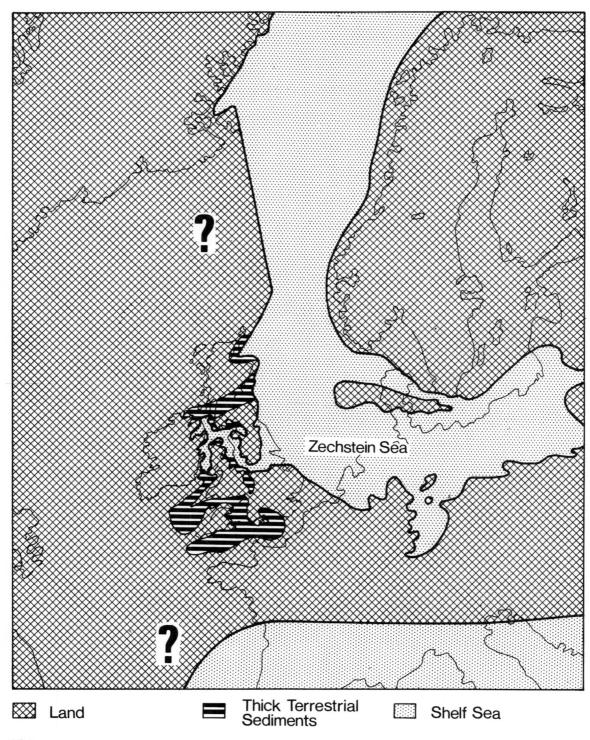

Land | Thick Terrestrial Sediments | Shelf Sea

the better known reef in the Guadaloupe Mountains of New Mexico, U.S.A.

One of the Zechstein 1 reefs was near what is now the coast of north-east England. It was about 1.6km wide and about 30km long. To the west was a shallow water and probably hypersaline lagoon, at least 15km wide in places, which supported a meagre, largely molluscan fauna. The fauna on the reef itself contained more molluscs on the lagoonal margin than on the basinal edge. A vertical section of the reef shows a progressive upwards impoverishment of the faunas which Trechmann (1913) attributed to increasing salinity during the reef's existence. Several of the more stenohaline brachiopods such as *Horridonia* and *Pterospirifer*, so prominent in the earlier part of the reef, were the first to die out, as was the bryozoan *Synocladia*. The role of the retiform bryozoa as frame-builders was increasingly taken over by algal stromatolites.

Mr. J. Pattison has contributed the following account of four Permian communities.

57   Barrier Reef Top Community

This community occupied the reef top near its seaward or basinal edge. The large ovoid bodies of the reef contain ramose bryozoa (*Acanthocladia* and *Thamniscus*) and small pedunculate brachiopods (*Dielasma*). The preserved bodies show fine laminations and are probably algal in origin. On and between them most of the benthos consisted of bryozoans and brachiopods. The retiform bryozoans *Fenestella* and *Synocladia* grew upwards and outwards in fan and funnel-shaped colonies and acted as reef frame-builders. The brachiopods included the small pedunculate rhynchonelloid *Stenoscisma* and larger, quite spectacular forms: the coarsely spinose productoid *Horridonia* and the very wide, alate *Pterospirifer*. These and most of the bivalves present were sessile suspension feeders. The commonest active bottom-dwellers were gastropods, of which at least some appear to have been carnivorous as the bivalves are commonly found with neatly bored holes in their beaks. The most abundant free-swimming organism was the small pectinid *Streblochondria? pusilla*.

The distribution of a particular kind of community within the reef was irregular. Forms not shown in the picture, although common in the same area at that time, are crinoids, the brachiopods *Cleiothyridina*, *Spiriferellina* and *Strophalosia* and the bivalve *Pseudomonotis*. Also left out are the micro-organisms found in the early part of the reef including fischerinid foraminiferida and bairdiid ostracodes. This kind of reef fauna can be seen at several localities in the Sunderland area of north-east England.

Fig. k. *Geography of the North Atlantic region during the Late Permian. The Zechstein Sea, east of Britain, was highly saline; its connection with the open oceans may have been to the north between Norway and Greenland, or there may have been a connection to the east. (After Pattison et al. 1973).*

Permian

Fig. 57 Barrier Reef Top Community

a *Acanthocladia anceps* (Bryozoa: Ectoprocta)
b *Fenestella retiformis* (Bryozoa: Ectoprocta)
c *Synocladia virgulacea* (Bryozoa: Ectoprocta)
d *Thamniscus dubius* (Bryozoa: Ectoprocta)
e *Dielasma elongatum* (Brachiopoda: Articulata: Terebratulida)
f *Horridonia horrida* (Brachiopoda: Articulata: Strophomenida)
g *Pterospirifer alatus* (Brachiopoda: Articulata: Spiriferida)
h *Stenoscisma humbletonensis* (Brachiopoda: Articulata: Rhynchonellida)
i *Mourlonia ? linkiana* (Mollusca: Gastropoda: Archaeogastropoda)
j *Bakevellia binneyi* (Mollusca: Bivalvia: Pterioida)
k *Parallelodon striatus* (Mollusca: Bivalvia: Arcoida)
l *Streblochondria? pusilla* (Mollusca: Bivalvia: Pterioida)

188

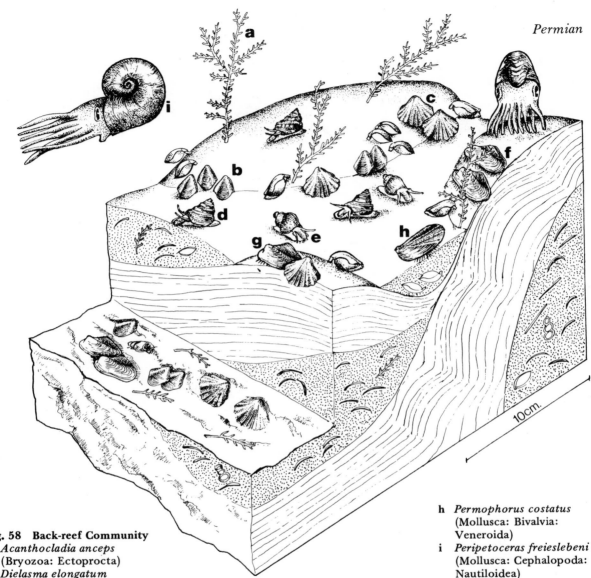

Fig. 58 Back-reef Community

a *Acanthocladia anceps*
(Bryozoa: Ectoprocta)
b *Dielasma elongatum*
(Brachiopoda: Articulata:
Terebratulida)
c *Stenoscisma schlotheimi*
(Brachiopoda: Articulata:
Rhynchonellida)
d *Yunnania* (Mollusca:
Gastropoda: Archaeo-
gastropoda)
e *Naticopsis minima*
(Mollusca: Gastropoda:
Archaeogastropoda)
f *Bakevellia binneyi* (Mollusca:
Bivalvia: Pterioida)
g *B. ceratophaga* (Mollusca:
Bivalvia: Pterioida)

h *Permophorus costatus*
(Mollusca: Bivalvia:
Veneroida)
i *Peripetoceras freieslebeni*
(Mollusca: Cephalopoda:
Nautiloidea)

58 Back-reef Community

The picture shows part of the reef top near its lagoonal margin
during the first Zechstein cycle at a later time than that shown in
the previous picture. Laminated algal sheets bound and encompassed pockets of shell and bryozoan debris. On and between the
upper surfaces of these sheets the benthos still included many
brachiopods, but more bivalves and still more gastropods. All the
species represented were survivors from the earlier reef but some
of them had increased in numbers because rivals, less tolerant of
the increasing salinity, had been eliminated. The stenohaline

Streblochondria? pusilla had disappeared and the only common nektonic species was the nautiloid *Peripetoceras*. In addition to the sessile benthic organisms shown, the strophalosiid brachiopod *Orthothrix* and the bivalve *Pseudomonotis* were also common in the reef at this time. Among the micro-organisms, foraminifera were more rare than in the earlier reef but ostracodes were abundant.

This kind of reef community can be found preserved near Sunderland.

59 Hypersaline Landlocked Basin Community

A number of basins west of the Zechstein sea were also inundated during the first Zechstein cycle but they were almost land-locked and the water in them was generally hypersaline. They supported faunas which were largely molluscan and included very few species. Of these species much the most common was *Bakevellia binneyi*, indeed to such an extent that the sea in these western basins has been called the *Bakevellia* Sea (Smith, 1970).

In one of the basins in north-west England was deposited the Manchester Marl, the lower part of which contains marine fossils. The substratum in and on which they lived was soft, consisting of mixed silt and mud. The fauna in this area is notable for the presence of the infaunal bivalve *Schizodus obscurus* which is large in size and abundant. The small *Bakevellia binneyi* lived in clumps wherever there were suitable points of attachment; these were possibly algal fronds which are no longer preserved, or empty *Schizodus* valves. Other bivalve species such as *Permophorus* were much more rare. The small gastropods probably grazed on algae. They are much more common in the thin limestones which occur in the Manchester Marl than in the intervening mudstones and siltstones, and the limestones were probably at least partly algal in origin. No attempt has been made here to illustrate the algae, as their form is unknown.

Micro-organisms in this hypersaline basin include foraminiferida and ostracodes, but no bryozoans or articulate brachiopods have been recorded. Their apparent absence was probably due to the hypersalinity.

Fig. 59 Hypersaline Landlocked Basin Community
a *Coelostylina? obtusa* (Mollusca: Gastropoda: Mesogastropoda)
b *Naticopsis minima* (Mollusca: Gastropoda: Archaeogastropoda)
c *Bakevellia binneyi* (Mollusca: Bivalvia: Pterioida)
d *Permophorus costatus* (Mollusca: Bivalvia: Veneroida)
e *Schizodus obscurus* (Mollusca: Bivalvia: Trigonioida)

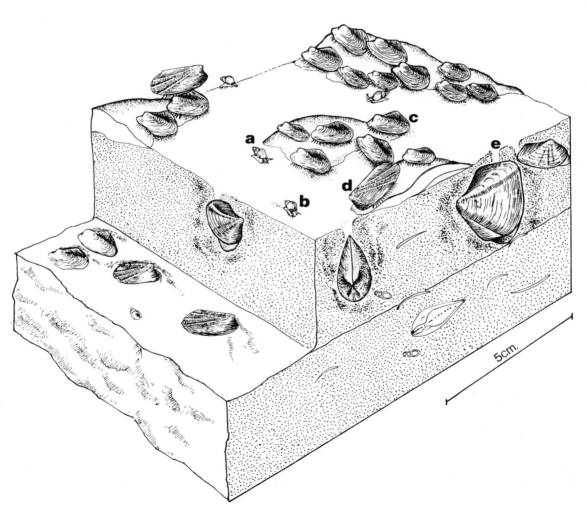

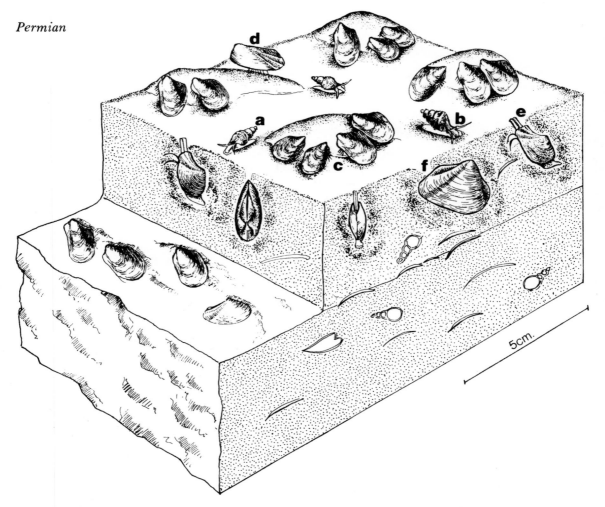

Fig. 60 Hypersaline Sea Community

a *Coelostylina? permiana* (Mollusca: Gastropoda: Mesogastropoda)
b *Yunnania helicina* (Mollusca: Gastropoda: Archaeogastropoda)
c *Liebea squamosa* (Mollusca: Bivalvia: Pterioida)
d *Permophorus costatus* (Mollusca: Bivalvia: Veneroida)
e *Phestia speluncaria* (Mollusca: Bivalvia: Palaeotaxodonta)
f *Schizodus obscurus* (Mollusca: Bivalvia: Trigonioida)

60 Hypersaline Sea Community

Most of the species which disappeared from north-west Europe
when the Zechstein 1 sea dried up did not return when the
Zechstein 2 sea invaded the same areas. The fauna of the second
Zechstein cycle included few species, of which molluscs were the
most conspicuous. Bryozoans and brachiopods were rare and in
many respects the Zechstein 2 sea biotas were comparable with

those of the *Bakevellia* Sea during Zechstein 1 time. It is probable that hypersalinity was again the main limiting factor. However, one notable difference is the apparent absence of *Bakevellia* in English Zechstein 2 deposits.

Although there were few species in the Zechstein 2 sea, at some times and in some areas benthic molluscan individuals were very common. The picture shows the probable nature of a shelf sea bottom (during deposition of the Concretionary Limestone). Smith (1972), has suggested that the shelly bands in the Concretionary Limestone are turbidites, so the fauna shown probably flourished some way to the west of the present coastal outcrops where the bands are exposed. *Liebea squamosa* is the most common species. Its distribution in Zechstein 1 rocks and its predominance in Zechstein 3 biotas both suggest that it was the most euryhaline of Zechstein bivalves. The relative abundance of infaunal forms (*Schizodus* and *Phestia*) indicate a soft substratum. The micro-organisms present included abundant ostracodes and nodosariid foraminifera.

Apparently the Zechstein 2 sea did not reach as far as that of Zechstein 1 and faunal remains like those shown here can be seen at outcrop only near Sunderland in north-east England.

FAUNAL AND FLORAL PROVINCES IN THE PERMIAN

Work on the distribution of Permian faunas and floras has suggested to several palaeontologists (Stehli, 1971; Chaloner and Lacey, 1973; Yancey, 1975) that the variations found in the biotas from various parts of the world can be best explained by the existence of a marked thermal gradient between equatorial and polar regions. There were also geographical variations within the equatorial belt, as there were in the Carboniferous, enabling several faunal provinces to be distinguished. These are shown especially well by the fusulinid foraminiferida (Ross, 1967). The Eurasian/Boreal province, covering most of Asia, eastern Europe and the Arctic area (including northern Canada) is characterised by foraminifera such as *Quasifusulina* and *Schwagerina*, and brachiopods including *Horridonia* and *Licharewia*. There was also a Cordilleran Province and a mid-Continent/South American province, each with characteristic faunas. Although there was a cosmopolitan element in most Permian faunas (e.g. *Pseudoschwagerina*) most areas also contain their own characteristic species. The explanation for the distribution of these faunas, and the communities which provide the data for such conclusions, appears to be the combination of the above-mentioned thermal gradient, the winds, and the ocean currents, as in the earlier Carboniferous.

Triassic

The widespread extinction of many groups of organisms towards the end of the Permian is one of the great enigmas of the fossil record. This wave of mass extinction affected a variety of marine organisms including: foraminiferida, tabulate and rugose corals, the trilobites, eurypterids and certain other arthropods, the goniatite cephalopods, productid and orthide brachiopods, and many groups of stalked echinoderms. On land the pelycosaur reptiles (paramammals) also became extinct. The demise of so many taxa requires some explanation.

This major phase of extinction was succeeded in the early Triassic by the appearance of many new forms. Some of these may have arisen to fill vacated ecological niches, whereas others carved out new niches for themselves in response to the newly developed trophic resources of the Mesozoic Earth. In the early Triassic, there is a great diversification in the Bivalvia (clams) with the appearance of the unionids, carditids, myacea and oysters. Other molluscan taxa were also undergoing diversification with a number of new gastropod families appearing upon the scene including the patellids (limpets), trochids (top shells), littorinids (periwinkles), cerithids (ceriths) and naticids (moon shells). A new ammonoid family also arose, (the phylloceratids). A whole suite of new reptile groups developed, namely the ichthyosaurs, rhynchosaurs, squamata (snakes) and archosaurs (including a series of orders like the crocodiles, dinosaurs, and pterosaurs). The first mammals also appeared before the end of the period while their ancestors, the paramammals, became extinct.

Extinction hypotheses have included cosmic radiation, pollution by volcanic emanations, world-wide salinity changes in the oceans and catastrophic climatic (especially temperature) fluctuations. The 'poisoning' hypotheses can be viewed with scepticism because of the selective nature of the extinctions, and the same criticism applies to the theory of catastrophes on a cosmic scale. World-wide salinity changes in the oceans would also have had a much more general effect, and would possibly have left some independent evidence in the form of changing proportions of evaporite minerals in accumulating salt deposits, but these are not to be

194

found. Finally, temperature or climatic deteriorations (that correspond at the present time with low diversities in high latitudes) do not seem to have caused mass extinctions during the Pleistocene and so cannot be regarded as the sole reason for mass extinctions at the end of the Permian.

The Permo-Carboniferous phase of earth movements resulted from the collision of previously separated continental areas and culminated, during the Permian, in the welding together of the huge supercontinent of Pangea (Fig. j). This collisional phase caused a great reduction in the area of shelf seas and caused those that persisted to be influenced much more by continental conditions. Marine environments that are subject to the least continental influences and remain stable through time tend to support the most diverse populations. Those that are most influenced by continental areas and suffer from marked seasonal changes in higher latitudes undergo higher environmental stresses and have lower faunal diversities. The creation of Pangea produced a supercontinent with restricted shelves. The vast spreads of late Palaeozoic epicontinental seaways were eliminated and the limited shelves became areas of high environmental stress allowing only the more eurytopic (tolerant) groups to survive. The most probable reason for the extinction of so many marine faunas is thus an increase in competition within a reduced area of habitable environments, coupled with the absence of any isolated refuges around Pangea (Schopf, 1974).

On the continents too, the joining together of once-separated areas brought terrestrial organisms into direct competition, while the continental environments themselves would have been extreme as a result of the rise of newly generated mountain chains (the Hercynides) and the establishment of a totally new global climate. All these stresses would have led to the survival of the most highly adaptable groups and the extinction of those most specialized to the pre-existing regimes. Extinctions at the bases of food chains would have had snowballing repercussions throughout the higher taxa (Valentine, 1973).

Following the Carboniferous-Permian (Hercynian) earth movements, northern Europe lay in the grip of predominantly continental conditions throughout much of the succeeding Triassic Period. In Britain and in the area now occupied by the North Sea, sediments of Triassic age are dominated by clastic sands and clays that are often a striking red colour.

The linear fold mountain belt of the Hercynian had been attacked by erosion during the Permian, and had also been broken up into a series of fault-controlled blocks that continued to supply coarse sediments locally at least during the early part of the Triassic. These local massifs included the Meseta in Spain, the Cornubian-Armorican Massifs of Britain and northern France, the

Massif Central, and further to the east, the Ardennes and Bohemia.

At the beginning of the Triassic, which also marks the beginning of the Mesozoic Era, the continental areas of the world are still united in the supercontinent of Pangea (Fig. j) that was later divided in its sub-tropics by the encroaching Tethyan Ocean and in the boreal region by an ancestral Arctic Ocean. Periodically the sea spread over the continental interiors to produce broad, but usually restricted, embayments (Fig. l). In the Permian, the advance of the Arctic seaway into northern Europe led to the formation of thick evaporite (Zechstein) salts under the present North Sea and northern parts of Germany. During the Triassic, the Tethyan transgression led to the development of the Muschelkalk dolomitic limestones which are also associated with evaporitic minerals.

Within the continental areas, deposition was limited to the subsiding basinal areas where Triassic deposits often succeed those of Permian without any noticeable break. The presence of evaporite minerals in many Triassic sequences suggests that the climate was predominantly arid. However, many of the sandier successions show evidence of having been water-laid. They contain fragments too large to have been moved by wind action and the finer grained beds often reveal filled, polygonal cracks that were produced during phases of desiccation. Thus, we have a picture of predominantly arid environments with infrequent but catastrophic flash floods spreading sheets of detritus from the massif areas over broad, subsiding and almost featureless sedimentary basins.

As we might expect animal remains are not very common under these conditions. However, there were times when rivers became less ephemeral and some quite rich reptile faunas became established accompanied by floras of horse-tails (Equisetales) and conifers. For example, the Keuper Sandstone fauna of the English Midlands contains amphibia, lungfish, shelled crustaceans and scorpions in an assemblage that became established during a damp phase when semi-permanent rivers drained the Armorican highlands and flowed northward into the Midlands Basin (Warrington, 1971). Later, in the same area, a marginal marine fauna was established during one of the few Triassic marine transgressive phases.

Fig. 1. *Geography of the North Atlantic region during the Late Triassic. Much of the areas shown as 'Desert playa lakes' was subject to invasion by the sea at various times, and contains local salt deposits and marls (the Keuper Marl) with restricted faunas. (After Hallam and Sellwood, 1976).*

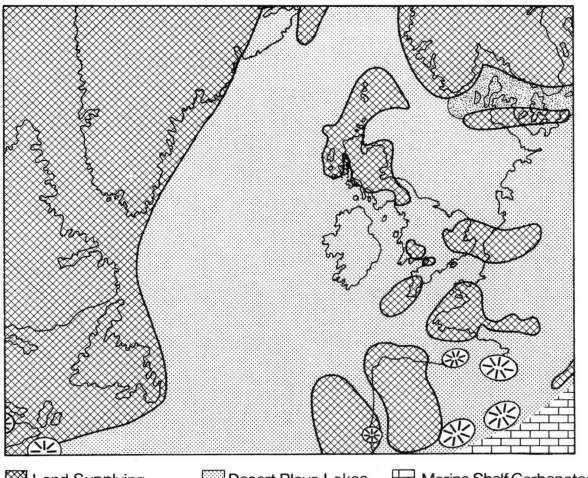

Land Supplying Sediment

Desert Playa Lakes

Alluvial Fans

Marine Shelf Carbonates

Volcanoes

61 Triassic Lagoon Scene

The reconstruction of this scene is based on material deposited in Warwickshire during the 'Mid-Triassic transgression'. Marine influences are indicated by the presence of *Lingula* and rare bivalves including some similar to *Modiolus* and *Pholadomya* (Rose and Kent, 1955). More recently the sediments associated with this fauna have been found to contain marine microplankton. However, the presence of the moulds of salt crystals and of desiccation cracks in some of these same sediments suggests very arid conditions with perhaps a Persian Gulf type of climate on the margins

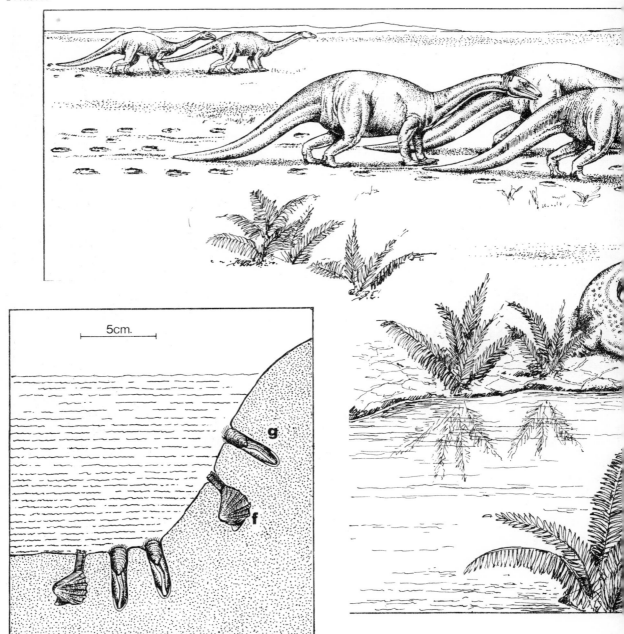

5cm.

g

f

Section to show infauna in lagoon

Fig. 61 Triassic Lagoon Scene
a rhynchosaur (Vertebrata: Reptilia: Lepidosauria)
b *Macrocnemus* (Vertebrata: Reptilia: Lepidosauria)
c nothosaur (Vertebrata: Reptilia: Euryapsida)
d *Mastodonsaurus* (Vertebrata: Amphibia)
e *Plateosaurus* (Vertebrata: Reptilia: Archasauria)
f *Pholadomya* (Mollusca: Bivalvia: Anomalodesmacea)
g *Lingula* (Brachiopoda: Inarticulata)

of the extended Muschelkalk sea. Of the vertebrate fauna, the fish were represented by animals such as the palaeoniscid *Gyrolepis* which certainly had marine affinities since it is found in the Muschelkalk of Germany. This genus actually returned to the British scene during the late Triassic marine transgression. Amphibian carnivores were represented by labyrinthodonts and some of these reached enormous sizes with their skulls alone reaching 1m in length. Feeding around the margin of the brackish shores were rhynchosaurs, a group of reptiles adapted to eating shellfish with their beaked mouths and flattened crushing teeth (Halstead, 1975). Another group of very widespread reptiles lived as semi-aquatic carnivores; these were the nothosaurs. Their pointed teeth, flattened tail and webbed feet suggest that they were active swimmers, with fish as their main diet. During the temporarily humid phases promoted by the transgressions, herds of thecodontosaurid dinosaurs grazed and browsed along the water courses and left abundant footprints and the occasional skeleton. Carnivorous lizards like *Macrocnemus* preyed upon the young of many of the other groups and may also have eaten the odd insect or two. The vertebrate fauna in our reconstruction is based upon the collection of Walker (1969) and on the reconstruction of Halstead (1975).

A surprising feature of this fauna is the widespread distribution of many of the individual components. Nothosaurs are found all over Europe, in China, Japan, Tunisia, Jordan and India. Rhynchosaurs are also found in Europe, India, Africa, North and South America. The lizard *Macrocnemus* is also found in other parts of Europe and in Texas while the terrestrial thecodontosaurid dinosaur *Poposaurus* (similar to *Plateosaurus* in Fig. 61) found in Warwickshire also occurs in Wyoming. This brief view of the geographical distributions of some of these animals indicates the homogeneity of the continental areas by mid- to late-Triassic times and reinforces the theory of continuous land connections between these now dispersed areas. As the Triassic progressed, the massif areas were worn down to mere stumps but a series of large rivers, perhaps far to the north, collected a vast amount of fine background mud, probably derived from a deeply weathered continental area. In east Greenland coals were forming periodically, so that either the climate was becoming more humid towards the north or some large Tigris-Euphrates type of river system was entering the generally arid European basin. Towards the end of the Trias this basin had the appearance of a vast and almost totally flat plain, parts of which were probably submerged by shallow hypersaline waters. Salt brines were received from two sources, first from a tenuous connection with the southern Tethys Ocean and second from Permian salt plugs that punched their way through the Triassic sediment column.

Over southern Europe the influence of the Tethys Ocean was

stronger and in Germany, southern France and in central and southern Spain the typical Triassic development can be seen in the characteristic threefold divisions from which the Trias takes its name: the sandy Bunter facies, followed by a dominantly carbonate Muschelkalk that is capped in turn by the more argillaceous Keuper facies. While the Bunter and Keuper facies are mostly similar to sequences developed further north, the Muschelkalk is not. The Muschelkalk fauna in central Germany and Spain shows typical high environmental stress associations. Faunas, although generally scarce, are normally dominated on individual bedding planes by a single species but individuals may be present in enormous numbers; bivalves (for instance *Gervillia, Myophoria* and *Hoernesia*) are the most important although some bedding planes may contain terebratulides while others may reveal ceratite ammonites.

Further to the south in Europe we enter the tectonically complex Alpine belt. Here carbonate facies are dominant, particularly in the late Trias, sometimes with the development of major 'reef' sequences like those of the famous Dachstein Riffkalk (Fischer, 1964). Some of the 'reef' complexes were composed of corals but some were constructed by hydrozoans and others contained abundant megalodont bivalves. Certainly, it is in the sequences nearer to Tethys that one finds the more diverse assemblages of Triassic marine faunas.

A large-scale reconstruction for the European Triassic might show an open oceanic belt running east to west in the south, flanked on its northern margin by a major 'reef' belt. This protected a huge 'lagoonal' area that extended northwards over thousands of square kilometres. The lagoonal belt was (as in the Muschelkalk) more influenced by the sea at some times than at others, but mostly remained a vast hypersaline region undergoing considerable evaporation. Slight elevations of sea level spread more marine conditions over this lagoonal area but the main phase of marine transgression did not occur until very late Trias-Early Jurassic times.

62 Late Triassic Tidal-flats

Towards the end of the Triassic, the first phases of marine transgression began and this first established sabkha shorelines with evaporating lagoons and eventually produced a set of restricted carbonate-shelf environments over parts of central and southern Britain. These latest Triassic (Rhaetic) deposits characteristically

Fig. 62 Late Triassic Tidal-flats
a algal mounds

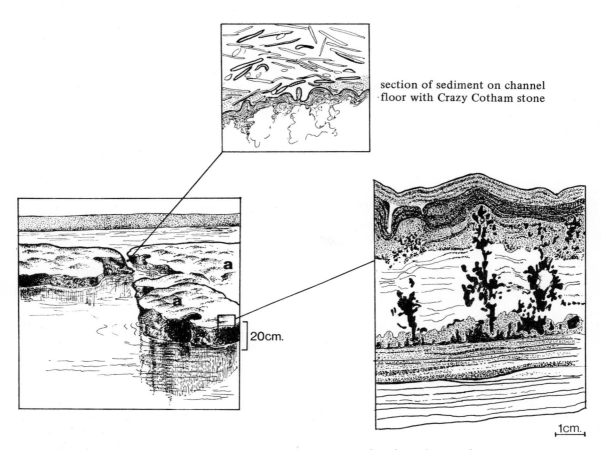

section of sediment on channel
·floor with Crazy Cotham stone

20cm.

1cm.

section through part of growing algal mound

contain low diversity-high density communities like those of the
Muschelkalk and, like these, also reflect life under high environmental stress conditions.

The illustration shows the probable mode of formation of one
of the more famous of the British Rhaetic beds, namely the
Cotham (or Landscape) 'Marble'. This consists of irregular 'bun-
shaped' mounds of banded muddy and shelly limestone with well
developed internal laminations that probably resulted from algal
growth (Hamilton, 1961).

Recent analogues of such structures have been found in modern hypersaline bays such as Shark Bay, Western Australia. In

Shark Bay the stromatolitic structures are composed of sand-sized shell debris that has been trapped and bound by filamentous algae. The Cotham Marble stromatolites are associated with lenses of tiny gastropods that were presumably grazing upon the algal surfaces.

Not all of the Cotham Marble presents the typical laminated and arborescent structures. In places, the rock has a brecciated appearance and this 'Crazy Cotham' was formed by the sediment deposited within channels that existed between the growing mounds. Individual flakes and fragments within the breccias were themselves derived from the erosion of stromatolitic material.

Both the Cotham and Shark Bay mounds are penetrated by desiccation cracks and in contemporary Shark Bay the height of the mounds approximates to the present day tidal range. It is possible to extend this analogy and to infer that the height of the Cotham mounds (200mm) gives an approximation of the palaeo-tidal range.

Rhaetic Beds outcrop on the coast in southern Devon, Avon and northwestern Scotland. Inland exposures occur in south-western England, Glamorgan and the Midlands. Poorer outcrops exist in Yorkshire. The Cotham Limestone facies is confined to the more southerly areas.

Jurassic

In the Jurassic many families of marine organisms that had evolved during the Triassic were diversifying rapidly; and on land the para-mammals had been replaced by the dinosaurs as the dominant group. The Triassic nothosaurs gave rise to the fish-eating plesio-saurs and pliosaurs, while in the air, the pterosaurs were in full flight. Late in the Jurassic the dinosaurs themselves gave rise to the birds.

The beginning of the Jurassic was a time of marine trans-gression on a world-wide scale. Broad seaways advanced on to the heartlands of the existing continents (Fig. m). Throughout the period (65 million years) climates seem to have been equable, with an absence of ice-caps even in areas that are thought, on palaeomagnetic grounds, to have been over the poles. The fact that both the north and south poles lay in oceanic areas may explain why ice-caps were unable to become established, as marine circu-lation would tend to equilibrate temperature differences between areas in the marine realm. This absence of ice-caps contrasted strongly with the extensive glaciations of the Palaeozoic on Gondwanaland. The absence of polar ice-caps probably caused a much less vigorous atmospheric exchange than at the present time.

The Jurassic also saw the beginning of the dispersal of the Pangea supercontinent. This resulted from the westerly extension of the Tethys Ocean and the beginning of rifting in the Central Atlantic (Fig. n). The separation of various parts of the continent provided greater scope for the development of provincialism in faunas as a result of their geographical isolation but these effects did not become very pronounced until later on in the period.

Through the Jurassic there was a progressive rise in sea levels around the world. This was probably caused by the growth of submarine oceanic ridges that accompanied the widening of the Jurassic oceans. This secular eustatic rise reached its peak in late Jurassic times when vast areas of the continent were flooded. The trend was reversed at the end of the Jurassic when there was a temporary, but widespread, withdrawal of the sea.

Fig. m. *The world during the Jurassic. Positions of the continents after Briden et al. 1974.*

LOWER JURASSIC

In the Lower Jurassic, the marine benthos consisted mostly of invertebrates; bivalves, gastropods and crustaceans are the commonest groups. These three groups all show major evolutionary advances when compared with their ancestors of the Palaeozoic. More bivalve taxa had developed siphons which enabled them

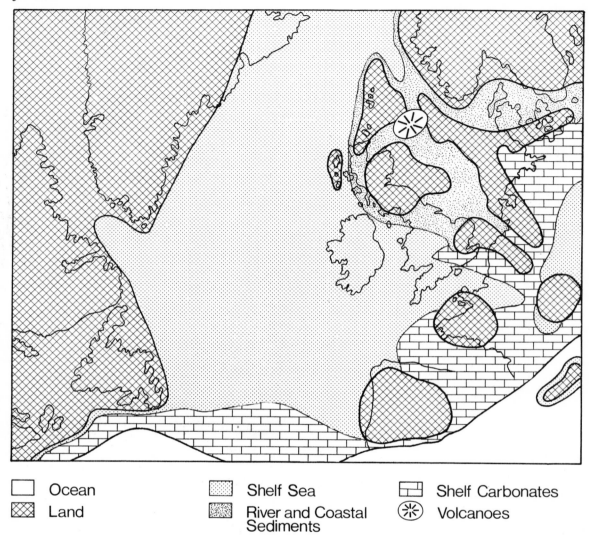

☐	Ocean	▦	Shelf Sea	⊞	Shelf Carbonates
▩	Land	▨	River and Coastal Sediments	✳	Volcanoes

Fig. n. *Geography of the North Atlantic region during the Middle Jurassic*

to burrow more efficiently; this was either for protection or for more efficient deposit feeding. Other bivalves, like the pectinids, had become adept at clapping their valves so they could either swim through the water by jet-propulsion or rest on the sea floor. The bivalve oysters, which appeared in the Triassic, were either cemented by one valve on a hard substrate, or (like the coiled *Gryphaea*) lay free upon muddy or sandy sea floors. Most gastropods retained their roles as grazers, detritus feeders and scavengers. The crustaceans did not always have thick enough skeletons to be well preserved, but the minute bivalved ostracodes are sometimes abundant. Nonetheless, many types of crustacean burrows are common; Jurassic marine sediments contain burrows comparable in many respects to those produced by modern crustaceans.

Many Jurassic marine communities are connected closely with a certain type of sediment; this is known as substrate dependence. It is notable that from the Mesozoic onwards many more communities were related to the substrate than appears to have been the case in the Palaeozoic, and many of the Jurassic communities described here are therefore named after the sediment in which they occurred.

Not only the type of sediment but also the rate of deposition controls the benthos in modern environments; many animals are unable to adapt to rapid sedimentation rates, especially when these conditions are linked with the presence of unstable substrates. The communities described in the following sections also show changes probably caused by a lower food supply, which, in modern environments, generally decreases with distance from the shore. Communities are also affected by the degree of oxygenation of the sea water near the sea floor; poorly aerated bottom conditions severely limit the variety of benthos and help to preserve organic matter in the sediment leading perhaps to the formation of bituminous and laminated shales.

Genera of nekton (the swimming animals) and plankton (the floating animals and plants) are in general more widely distributed in the Jurassic than are benthic genera. They were not affected by the nature of the sea floor, and wide oceans did not form a barrier to their migration; for example, some individual ammonite genera in the early Jurassic have almost a world-wide distribution. This is a reflection of the greater homogeneity of pelagic environments in contrast to the more heterogeneous mosaic-like environments found on the sea floors of shelf areas. The ammonites, in particular, have now been studied for well over a century, and because of their wide geographical distribution and generally robust shells they are pre-eminently suitable for stratigraphical correlation over large areas and between different marine environments. The Jurassic ended around 135 million years ago (Lambert, 1971) and the rocks deposited during this period have been divided into 64 zones based on the detailed study of ammonite evolution and the appearance of new genera and species. Thus, on average, each zone represents about 1.2 million years.

In the Lower Jurassic, the ammonite zones appear to be of two types; they are either characterized by successive species of the same genus or family (when the time divisions are thus directly related to rates of evolution), or by the sudden appearance of genera whose detailed ancestry is poorly known (in these latter cases, correlation is based on the assumption that migration has been rapid). Details of the ammonite species in each Lower Jurassic zone have been recorded by Dean et al. 1961.

Before the ammonites had been studied in detail, the early nineteenth century geologists divided up the beds entirely by the type of rock present — this was the approach adopted by William Smith, the canal engineer (1769—1839), who first recognized that strata could be identified by their fossils. The Lower Jurassic of Britain was named the Lias after the west country term for flat stones (Arkell, 1933), the Lias being distinguished from the beds above on the grounds of lithology; the clays and sands of the Lias are clearly distinct from the succeeding shelly oolitic limestones of the Middle Jurassic of England.

The Lias of England consists mostly of clay and muddy limestone, but sands and ironstones are also present at some levels, with oolitic ironstone formations being particularly characteristic of the Middle Lias. The Lower, Middle and Upper Lias are not equal divisions; in northern Europe they have been divided respectively into twelve, two and six ammonite zones.

The ammonites have one major failing as fossils for correlation: they were very sensitive to changes in salinity, and as a result they are absent or very rare in those areas that were partially isolated from the open sea. Many other planktonic and nektonic invertebrates were also stenohaline; benthic invertebrates included many animals which were stenohaline, but it seems from analogy with their living descendants that some benthos were euryhaline, notably the oysters, which often provided the only faunal indication of marine influence in saline or fresh water lagoon environments. In Britain, the broadly lagoonal environments of the late Triassic were followed by general marine conditions which extended across much of northern Europe (Fig. n), and ammonites are common in most Jurassic beds.

The Jurassic nekton included fish and reptiles but they are seldom a dominant element of the total fauna. Thus, in most of the succeeding reconstructions, the abundant invertebrate organisms have been given pride of place.

63 Bituminous Mud Community

This is a reconstruction from a bituminous shale. Probably, few of the animals lived on the sea floor and the assemblage is dominated by nekton and plankton: cephalopods (ammonites, nautiloids and belemnites), and marine vertebrates (ichthyosaurs, pliosaurs and fish) which were scavengers and carnivores. The fauna also includes filter feeders (bivalves, crinoids, serpulids and barnacles) that were attached to floating vegetation, wood and debris projecting above the anaerobic sea floor.

The sediment was rich in organic matter derived from the partially decomposed plant and animal remains trapped in the fine mud accumulating on the oxygen-deficient sea floor. Bituminous shales formed in two distinct environments during the Lias. The first was related to very shallow, but calm, water in 'lagoonal' conditions where the sea floor was anaerobic owing to its partial isolation from the open sea. The second type formed in the deepest parts of marine basins where normal nekton and plankton were present in the main body of sea water, but where limited water circulation allowed stagnant conditions to develop on the sea floor. The fossils in the first type of shale consist only of limited numbers of euryhaline benthos, like those seen in the basal (pre-ammonite) beds of the Lower Lias. The example shown is the second type. The cephalopods present were stenohaline, and probably lived in normal sea water well above the stagnant bottom. The sea was probably around 200 to 300m deep; comparable but deeper water, oxygen-poor environments occur today in the Black Sea, the Gulf of California (Calvert, 1964), and the Mediterranean.

Although benthic animals are usually absent in the deeper type of bituminous shale, some bedding planes may be crowded with bivalves, crinoids, or other animals that are normally benthic. A clue to this anomaly is provided by large logs of fossil wood, which are often found encrusted with large numbers of attached organisms (Seilacher et al, 1968; Hauff, 1953). On any one log, the attached animals were usually all of one species, but in our reconstruction we have included several kinds in order to show some of the varied genera which can occur on logs in the Lias. The mode of life of *Bositra [Posidonia]*, is based upon the interpretation of Jefferies and Minton (1965) but there is considerable doubt about the plausibility of swimming bivalves of this type, although this theory would help to explain the wide distribution of *Bositra* in sediments that often lack any other benthos.

The preservation of calcite and aragonite shells is often unusually good in bituminous shales. The impermeable nature of the sediment has often prevented solution or recrystallization of the calcium carbonate, but precipitation of pyrite (iron sulphide) has often occurred in open cavities and chambers within the shells. Sometimes the pyrite has prevented the shells from being flattened when the sediment was compacted by the weight of younger strata. In the stagnant water of this environment, neither action of currents nor of burrowers could dismember the reptile skeletons. Some ichthyosaurs have been found with their stomach contents preserved in place; these may include fish remains and the chitinous

Jurassic

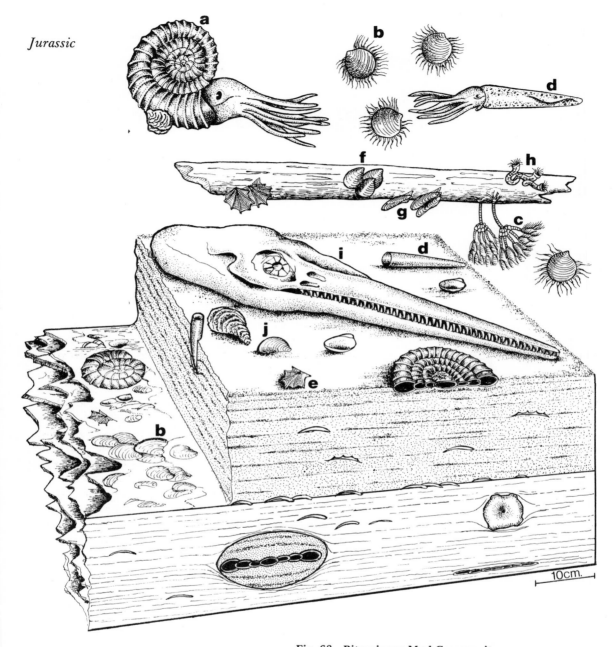

Fig. 63 Bituminous Mud Community
a *Dactylioceras* (Mollusca: Ammonoidea: Cephalopoda)
b *Bositra* (Mollusca: Bivalvia: Pterioida)
c *Pentacrinus* (Echinodermata: Crinozoa)
d belemnite (Mollusca: Cephalopoda: Coleoidea)
e *Oxytoma* (Mollusca: Bivalvia: Pterioida)
f *Inoceramus* (Mollusca: Bivalvia: Pterioida)
g *Gervillia* (Mollusca: Bivalvia: Pterioida)
h serpulids (Annelida – polychaete)
i ichthyosaur (Vertebrata: Reptilia: Euryapsida)
j *Liostrea* (Mollusca: Bivalvia: Pterioida – oyster)

210

hooks from belemnite arms. These same reptiles occasionally contain juveniles, presumably their young, which suggests that they were viviparous. The reptiles' skins have often been preserved, the hides having been tanned by chemicals in the anaerobic substrate.

The bituminous shale community illustrated is based upon the fauna from the Upper Lias of Yorkshire; both deep and shallow types of assemblage are found in the Lias of Yorkshire, central England and the Glamorgan coast and in the Blue Lias, 'Shales with Beef', Black Ven and Belemnite Marls of the coast of Dorset, England.

64 Restricted Clay Community

Since the sea floor was not stagnant, this community contains many burrowing animals in addition to the nektonic and planktonic cephalopods, reptiles and fish shown previously (Fig. 63). The fossil benthos is dominated by detritus feeders: protobranch bivalves (*Nucula* and *Nuculana* — the latter is distinguished by its pointed posterior), lucinoid bivalves (*Lucina, Mactromya*), and procerithid gastropods. These animals were all able to cope with the soft substrate which prevented many other burrowers from colonizing the area. Crustacea and worms also burrowed in this environment; they are recognizable from their burrows. One branching worm-burrow (*Chondrites*) is visible when it has been filled with slightly different mud from the surrounding sediment, and some simpler tubular burrows may be preserved in pyrite.

The burrowing activity obliterated the bedding laminae seen in the more bituminous muds, so many metres of sediment accumulated with no marked bedding being preserved. The sediments lack ripples and scour structures, and probably accumulated in depths greater than 30m, below the limits of wave motion and strong current activity.

The burrowing action of the protobranchs and the worms appears to have made the sea floor too soft and unstable for the successful development of a more diverse benthos. A careful search of the sediment can reveal an abundance of minute bivalves which were the shells of juveniles of other genera which are typical of less muddy shallower water environments. In the restricted clay environment the spat of these genera settled on unfavourable muds and failed to achieve adulthood.

The reasons why this habitat was successfully colonized by protobranchs and lucinoids in the Jurassic can be discerned from studies of their present day descendants. Modern protobranchs are shallow burrowers, living just below the sediment/water interface as mobile or semi-sessile detritus feeders. The lucinoids construct

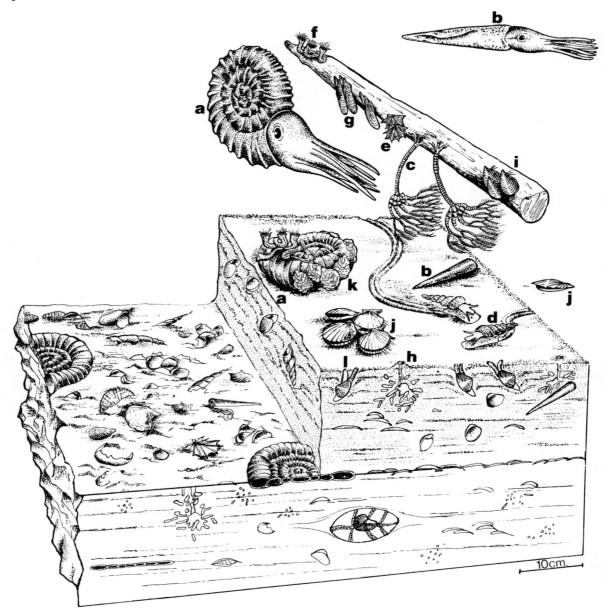

Fig. 64 Restricted Clay Community

a *Dactylioceras* (Mollusca: Cephalopoda: Ammonoidea)
b belemnite (Mollusca: Cephalopoda: Coleoidea)
c crinoid (Echinodermata: Crinozoa)
d *Procerithium* (Mollusca: Gastropoda: Mesogastropoda)
e *Oxytoma* (Mollusca: Bivalvia: Pterioida)
f serpulids (Annelida — polychaete)

g *Gervillia* (Mollusca: Bivalvia: Pterioida)
h *Chondrites* (Trace fossil: annelid burrow)
i *Inoceramus* (Mollusca: Bivalvia: Pterioida)
j pectinid (Mollusca: Bivalvia: Pterioida)
k *Liostrea* (Mollusca: Bivalvia: Pterioida — oyster)
l *Nuculana* (Mollusca: Bivalvia: Palaeotaxodonta — nuculoid)

mucus-lined tubes in the soft sea bed, down which they suck detritus. Allen (1958) noted that modern lucinoids are able to live in environments with a low food content and with restricted supplies of oxygen, but that they cannot compete with other burrowing bivalves in more normal environments.

The commonest suspension-feeding epifaunal elements are pectinids, bivalves which could keep their mantles free of mud by valve-clapping. The thin-shelled pectinids could create jet-propulsion by the same activity, and were therefore capable of limited swimming. *Inoceramus* was sedentary and byssally attached to shell fragments or to floating wood.

Ammonites and belemnites were the commonest nekton and plankton; some ammonites were encrusted by oysters and serpulid worms. If the whole shell was encrusted on both sides, the oysters must have grown while the ammonite was alive and swimming above the sea floor, the oysters taking advantage of the abundant food material where the ammonite was living. More often, however, encrustation only occurred on one side. Some free-swimming bivalves (*Bositra*) are also present, their separated valves occurring rarely in the sediment. This bivalve has been considered a swimmer (as depicted in Fig. 63) but may well have been an extremely tolerant benthic form.

Like the bituminous shales, these impermeable clays usually show excellent preservation of calcite and aragonite shells, and many cavities within the shells are filled with pyrite. Some shells are preserved in calcite nodules which formed before compaction when the sediment lost water.

The Restricted Clay Community can be seen in England in parts of the Black Ven Marls (Lower Lias) of the Dorset coast, and parts of the Alum Shales (Upper Lias) of Yorkshire.

65 Silty Clay Community

Both the fauna and the trace fossils (burrows) were more diverse in silty clays than in clays without silt particles, partly because the substrates were more stable, and partly because more abundant suspended food was available. In addition to the benthic detritus-feeding organisms (including protobranchs, lucinoids and the worms which made the *Chondrites* burrows), there were many benthic filter feeders; some of these lived on the sea floor (epifauna) and others burrowed at various depths within the substrate (infauna).

The epifaunal filter feeders included *Gryphaea*, pectinids and crinoids. Most modern oysters live cemented by the left valve to some hard substrate, but *Gryphaea* was free-living for most of its life; the spat settled on a hard substrate (often a shell fragment), but, as its shell grew, the area of attachment did not increase and the left valve became curved; eventually the shell tipped over to rest on the convex left valve, and the animal lay free on the sea floor. The pectinids were widespread in silty clays, particularly genera with thick, robust shells. Rippled silts and storm scours filled with cross-bedded sand suggest the influence of waves. In similar modern environments the shell thickness in bivalves often increases with wave turbulence, and the thick shells of the pectinids of the Lower Jurassic suggest a similar environmental response. The suspension-feeding crinoids lived in clusters that were destroyed by storms and most of their remains occur as isolated ossicles and plates. Brachiopods were rare in these silty clays, the most common form being the spire-bearer *Spiriferina*; this had a more efficient feeding mechanism than the other groups of Mesozoic brachiopods.

With the presence of silt, the most notable increase in diversity was in the infaunal and semi-infaunal bivalves (*Pholadomya, Pleuromya, Astarte, Cardinia* and *Pinna*). Modern representatives of these genera feed on matter suspended in the seawater. *Pinna* had a large shell; in life this stood vertical, the lower part being buried in the sediment and attached to particles in the sediment by byssal threads (seldom preserved fossil). The upper part of the shell is often encrusted with serpulids and small bivalves (*Liostrea*), which took advantage of the food-carrying inhalent currents of the *Pinna*. Many *Pinna* are totally encrusted; these were shells exhumed by storm action and often laid down pointing in the direction of the depositing current. The infauna also included crustacea and worms, seldom preserved as fossils, but represented by their burrows. In particular, the *Thalassinoides* and *Rhizocorallium* crustacean burrows were characteristic of this environment.

The nekton and plankton included the belemnites, ammonites and vertebrates seen in the clay communities, but, in addition, the silty clays contain larger ammonites with tubercles (e.g. *Eoderoceras*), which probably scavenged near the sea floor.

The diverse Silty Clay Community described above is present in the Lower Lias beds of Robin Hood's Bay on the Yorkshire coast in England (Sellwood, 1972). A sparser fauna is seen in some silty Middle Lias clays of Dorset, where sedimentation rates were higher. Intermediate communities occur in the Lias of east and west Scotland. In other silty clays the infaunal suspension feeders, which were very sensitive to turbidity, may be rare or absent.

The permeability of silty clays is higher than that of pure muds. Water moving through the sediment may have dissolved the aragonite shells of many bivalves and ammonites, leaving only the moulds preserved. Calcite shells (like oysters and belemnites) are less soluble and were normally preserved with little alteration.

Fig. 65 Silty Clay Community
a   *Eoderoceras* (Mollusca: Cephalopoda: Ammonoidea)
b   belemnite (Mollusca: Cephalopoda: Coleoidea)
c   crinoids (Echinodermata: Crinozoa)
d   *Spiriferina* (Brachiopoda: Articulata: Spiriferida)
e   *Pholadomya* (Mollusca: Bivalvia: Anomalodesmata)
f   *Pleuromya* (Mollusca: Bivalvia: Anomalodesmata)
g   *Gryphaea* (Mollusca: Bivalvia: Pterioida — oyster)
h   *Astarte* (Mollusca: Bivalvia: Veneroida)
i   *Cardinia* (Mollusca: Bivalvia: Veneroida)
j   *Pinna* (Mollusca: Bivalvia: Mytiloida)
k   serpulids (Annelida — polychaete)
l   *Liostrea* (Mollusca: Bivalvia: Pterioida — oyster)
m   *Chondrites* (trace-fossil — Annelida)
n   *Thalassinoides* (trace-fossil — Crustacea)
o   pectinids (Mollusca: Bivalvia: Pterioida)

215

66 Muddy Sand Community

There was a complete gradation between the communities that lived in mud and muddy sand. As the proportion of silt and sand increased, so there was an increase in diversity of both epifauna and infauna. The sand provided a more stable substrate than silt for those animals living on the sea floor, and the greater turbulence in the shallower water caused an increase in the amount of organic material in suspension and provided a richer food supply for suspension feeders and deposit feeders. The Muddy Sand Community includes many infaunal suspension-feeding bivalves such as the siphonate pholadomyoids (*Pholadomya* and *Pleuromya*) and the veneroids (*Cardinia, Astarte, Hippopodium* and *Protocardia*). The semi-infaunal *Pinna* was also present, and the fissure-dwelling pectinid *Camptonectes*, which was normally epifaunal, may sometimes be found attached to *Pinna* shells partially below the surface. Many of the infaunal and semi-infaunal bivalves occur with the valves vertical, in their positions of growth; other valves lie parallel with the bedding, the shells having been drifted by periodic fast currents in this shallow marine environment.

The infauna was also represented by many burrowers including *Chondrites* while suspension-feeding crustaceans formed the trace fossils *Thalassinoides, Rhizocorallium* and *Diplocraterion*. The 'U' burrows of the latter two forms sometimes show an upward migration of the burrow system in response to the periodically high rates of sedimentation (Goldring, 1964; Sellwood, 1970).

The epifauna was also more diverse in these more sandy beds. In addition to *Gryphaea* and thicker-shelled pectinids (*Pseudopecten, Camptonectes* and *Chlamys*), brachiopods (rhynchonellids and terebratulids) are locally common. They often occur in 'nests' representing colonies; close inspection of many of the brachiopods may show the pedicular scars of other brachiopods (Bromley and Surlyk, 1973). Brachiopods evidently anchored themselves on each other. Bivalves with a similar mode of life to the brachiopods include the byssally attached pterioids (*Oxytoma* and *Gervillia*), which probably swung free over the sea floor (Kauffman, 1969) in the turbulent water.

Swimming and floating scavengers were represented by ammonites, belemnites, reptiles (ichthyosaurs and plesiosaurs) and fish. Sea floor scavengers included gastropods (*Procerithium* and the thicker-shelled *Amberleya*) and large spinose eoderoceratid ammonites.

The diverse faunas occur in the Lower Lias of Yorkshire, intermediate diversity faunas occur in Scotland and fossils are sparser in many of the more rapidly deposited beds of Dorset and central England. The Dorset muddy sands also show some reduction in the numbers of bivalves in most beds.

Fig. 66 Muddy Sand Community

a *Dactylioceras* (Mollusca: Cephalopoda: Ammonoidea)
b belemnite (Mollusca: Cephalopoda: Coleoidea)
c *Isocrinus* (Echinodermata: Crinozoa)
d pectinids (Mollusca: Bivalvia: Pterioida)
e *Oxytoma* (Mollusca: Bivalvia: Pterioida)
f *Amberleya* (Mollusca: Gastropoda: Mesogastropoda)
g *Gryphaea* (Mollusca: Bivalvia: Pterioida — oyster)
h *Astarte* (Mollusca: Bivalvia: Veneroida)
i *Cardinia* (Mollusca: Bivalvia: Veneroida)

j *Pholadomya* (Mollusca: Bivalvia: Anomalodesmata)
k *Hippopodium* (Mollusca: Bivalvia: Veneroida)
l *Thalassinoides* (trace-fossil — Crustacea)
m *Diplocraterion* (trace-fossil — Crustacea or Annelida)
n *Pinna* (Mollusca: Bivalvia: Mytiloida)
o *Liostrea* (Mollusca: Bivalvia: Pterioida — oyster)
p *Camptonectes* (Mollusca: Bivalvia: Pterioida — pectinid)
q serpulids (Annelida — polychaete)
r *Chondrites* (trace-fossil — Annelida)
s *Eoderoceras* (Mollusca: Cephalopoda: Ammonoidea)

217

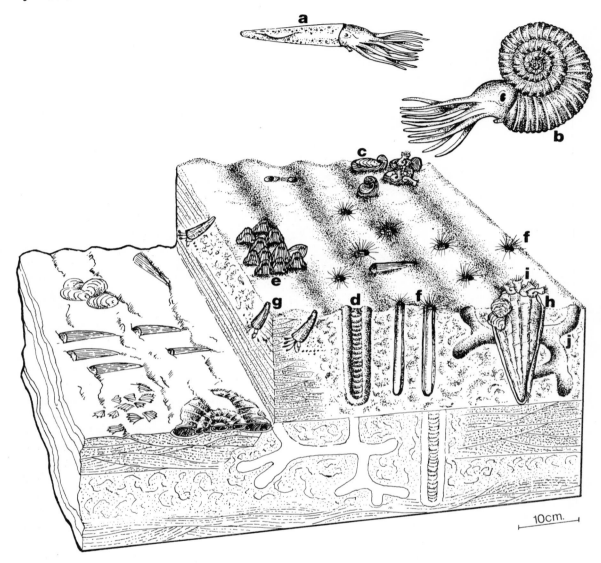

Fig. 67 Sand Community
a belemnite (Mollusca: Cephalopoda: Coleoidea)
b ammonite (Mollusca: Cephalopoda: Ammonoidea)
c *Gryphaea* (Mollusca: Bivalvia: Pterioida — oysters)
d *Diplocraterion* (trace-fossil)
e *Tetrarhynchia* (Brachiopoda: Articulata: Rhynchonellida)
f *Skolithos* (trace-fossil — Annelida)
g *Dentalium* (Mollusca: Scaphopoda)
h *Pinna* (Mollusca: Bivalvia: Mytiloida)
i serpulids (Annelida — polychaete)
j *Thalassinoides* (trace-fossil — Crustacea)

67 Sand Community

The marine Lower Jurassic sands contain oysters (*Gryphaea* and *Liostrea*), thick-shelled pectinids, the scaphopod *Dentalium,* and serpulids. Occasional 'nests' of brachiopods are present, and the semi-infaunal bivalve *Pinna* may occur. The substrate appears to have been too unstable and, at frequent intervals, deposition too rapid to have supported large populations of siphonate infaunal suspension-feeding bivalves. But these sands are extremely porous and many shells, especially those of aragonite, may have been dissolved (most of the fossils listed above are of calcite).

Burrows within the sands made by crustaceans and worms are similar to those seen in muddy sands (Fig. 66). *Diplocraterion, Rhizocorallium, Thalassinoides* and *Chondrites* were all present, and in addition there were vertical tubes (*Tigillites, Skolithos* or *Monocraterion*). We have based our interpretation of these assemlages on the work of Seilacher (1967) who suggested that a general dominance of simple vertical burrows in marine sediments indicates shallow turbulent water with a high concentration of suspended food. The addition of 'U' burrows and *Thalassinoides* indicates slightly less turbulent water, and meandering 'U' burrows and the complex spiral burrows of detritus feeders reflects quiet water with low concentrations of suspended food.

The oysters and serpulids lived attached to hard materials, usually other shells, but after death the oyster shells were normally redeposited by currents as disarticulated valves. *Dentalium* belongs to a small group of molluscs whose shell is in the form of an open cone. It moved through the sediment with its head downwards and its posterior extending into the water above. *Dentalium* can sort out the finer grained material in the sediment, digest the organic matter, and eject the rest up into the seawater. A bed worked over in this way by *Dentalium* has a low content of fine grained particles. *Dentalium* was often transported after death, and the shells aligned in accordance with the prevailing current direction.

Nektonic and planktonic organisms (ammonites and belemnites) were present in this environment, but the aragonite shells of the ammonites were often dissolved, except where the sands were calcified shortly after sedimentation. Calcified sandstone bands are therefore the most fruitful beds in which to search for fossils.

Sand Communities occurred in the Scalpa Sandstone of west Scotland, and in England in the Lower to Middle Lias Sands of the Yorkshire coast; in the Cotswold Sands of the Midlands and the Bridport, Yeovil, Thorncombe and Downcliff Sands of Dorset and Somerset.

68 Condensed Limestone Community

Areas of Jurassic shelf often show sequences commencing with mud and passing upward through sandy mud and muddy sand and finishing with sand (Communities 64 to 68). If the water became very shallow at a distance from land (i.e upon a shoal or 'swell') the currents carrying sediments were usually diverted to deeper parts of the sea. Little material was deposited on the sea floor in these shallow 'swell' areas except the remains of the animals and plants living there. Condensed limestones are therefore frequently found above sandy beds and the areas of slow subsidence (swells) can be related to the underlying structure in the pre-Jurassic rocks (Sellwood and Jenkyns, 1975). The main addition to the fragmented shell material which formed the sediments in these environments is carbonate mud, sometimes trapped by blue-green algae which lived on the sea floor. These algae were covered by mucus to which the mud became attached in fine stromatolitic laminae with the laminations picked out by limonite (the rusty hydrated oxide of iron) and by manganese minerals which gave a brown or pink colour to the stromatolites.

The Middle-Upper Lias Junction Bed of Dorset is an excellent example of a condensed limestone. In this bed ammonites and belemnites are abundant. Some of the ammonites are covered by laminations of limonite-rich limestone (oncolites), which are also attributable to algae. The presence of stromatolites and oncolites indicates a clear well illuminated sea floor. Shallow, turbulent conditions are reflected in the numerous erosion surfaces and by the intraformational conglomerates (fragments of condensed limestone incorporated with the same sequence).

In some condensed sequences the benthos was very diverse, especially as regards the suspension feeders. The bivalves included thick-shelled forms (*Liostrea* and *Plicatula*), and the byssally attached *Oxytoma* and *Inoceramus*; the swimmer *Bositra [Posidonia]* is also present. Large numbers of brachiopods (mostly rhynchonellids) were unusual in Lias clays and sands, but the clear water environments with no land-derived detritus, as in these condensed carbonates, proved more favourable to them.

Gastropods were also relatively common in condensed limestones, particularly the thicker-shelled genera (*Amberleya, Pleurotamaria, Coelodiscus* and *Lewisella*). Modern gastropods are scavengers, carnivores or grazers. The grazers today rasp the substrate for algal and bacterial material and it is probable that many of those in some condensed limestones fed on the algae.

Many modern algal flats are intertidal, and show shrinkage cracks and other signs of exposure at low tide. None of these signs of desiccation are seen in the British Lower Jurassic, and we must conclude that the stromatolitic limestones here accumulated

Fig. 68 Condensed Limestone Community
a *Dactylioceras* (Mollusca: Cephalopoda: Ammonoidea)
b *Hildoceras* (Mollusca: Cephalopoda: Ammonoidea)
c algae (Prokaryota-oncolites)
d ammonite covered with algae as in c
e belemnite (Mollusca: Cephalopoda: Coleoidea)
f *Bositra* (Mollusca: Bivalvia: Pterioida)
g *Amberleya* (Mollusca: Gastropoda: Mesogastropoda)
h serpulids (Annelida — polychaete)

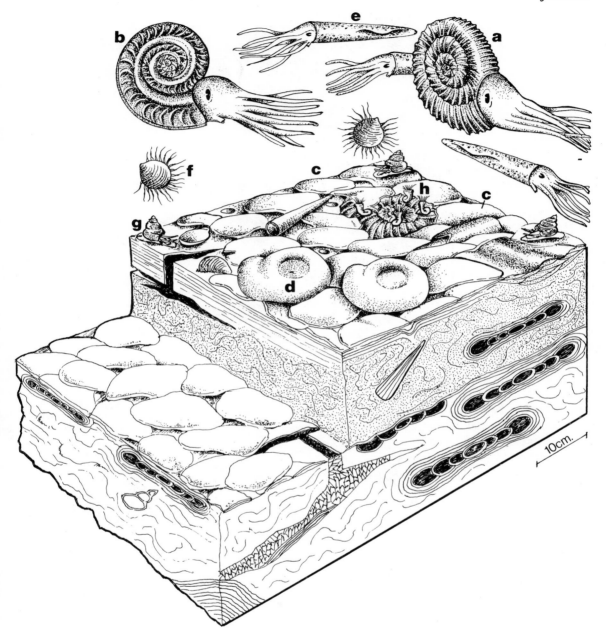

entirely under water. Not all of the condensed beds are of this
type and others may contain abundant nodules formed around
individual ammonites. The Lias Junction Bed which is exposed between Seatown and Burton Bradstock on the Dorset coast provides
an excellent example of such a condensed limestone.

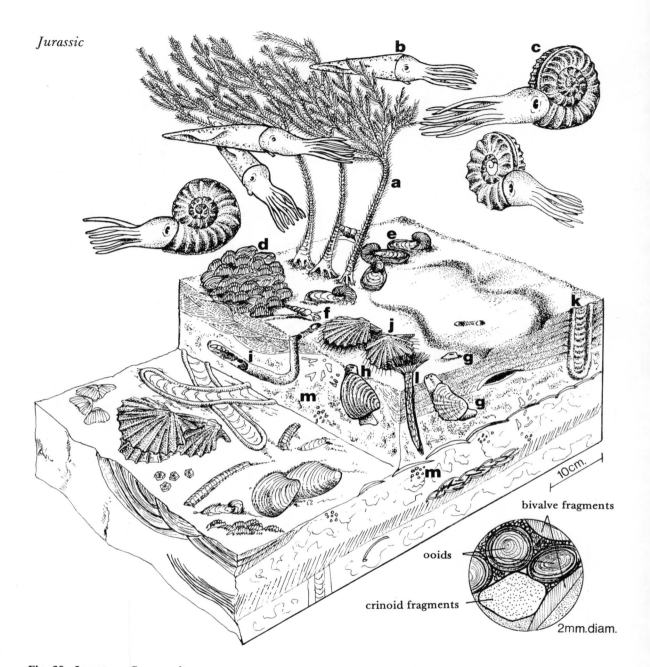

Fig. 69 Ironstone Community

a *Pentacrinus* (Echinodermata: Crinozoa)
b belemnite (Mollusca: Cephalopoda: Coleoidea)
c *Asteroceras* (Mollusca: Cephalopoda: Ammonoidea)
d rhynchonellids (Brachiopoda: Articulata: Rhynchonellida)
e *Gryphaea* (Mollusca: Bivalvia: Pterioida-oyster)

f *Procerithium* (Mollusca: Gastropoda: Mesogastropoda)
g *Pholadomya* (Mollusca: Bivalvia: Anomalodesmata)
h *Cardinia* (Mollusca: Bivalvia: Veneroida)
i *Rhizocorallium* (trace-fossil — crustacean)
j *Pseudopecten* (Mollusca: Bivalvia: Pterioida — pectinid)
k *Diplocraterion* (trace-fossil — annelid or crustacean)
l terebellid (Annelida)
m *Chondrites* (trace-fossil — annelid)

69 Ironstone Community

Many marine ironstones had a rich fauna dominated by suspension feeders. The rocks may contain abundant sphaeroidal structures (ooids) composed of a variety of iron materials (chamosite, limonite, siderite). The ooids were probably formed on a turbulent sea floor but the iron minerals probably resulted from the alteration after burial of an originally iron-rich sediment. Like condensed limestones, ironstones often accumulated on swells.

The dominant species were suspension-feeding bivalves, but many different types were present. Infaunal suspension feeders included *Pholadomya, Pleuromya* and *Hippopodium* and, in some beds, *Astarte* and *Cardinia*; the large semi-infaunal *Pinna* sometimes occurred, and the epifaunal bivalves included oysters (*Gryphaea, Liostrea*), thick-shelled pectinids (*Pseudopecten, Lima*) and many others.

Brachiopods (all of them suspension feeders) occurred commonly in 'nests'. Each 'nest' consisted of numerous individuals of one species (monospecific clusters) though within one bed many different brachiopods (rhynchonellids, terebratulids and spiriferinids) may be present. Many individuals appear to have preferred shells of their own species as an anchorage. In general, Jurassic brachiopods preferred beds without much sand or mud, so they are more common in ironstones and certain limestones rather than in clastic sediments.

The gastropods included scavengers (procerithids) and grazers (*Amberleya* and *Pleurotomaria*); the grazers probably fed on algae on the floor of the shallow and therefore well lit sea. There is little good evidence of algae, but some beds contain large ooids (1cm) which may have been produced by algal growths.

The trace fossils also reflect shallow turbulent conditions on the sea floor; they included *Diplocraterion, Rhizocorallium* and terebellid worm tubes. The terebellid worms lined their burrows with mucus to which shell fragments became attached; the fossil burrow is thus recognized by a cylindrical arrangement of shell fragments surrounding material that is finer grained than the host sediment.

Echinoderms were represented by both crinoids and echinoids; the former were fixed on the sea floor and disintegrated after death; the latter often burrowed and are sometimes well preserved.

Ammonites and belemnites are usually common fossils in ironstones. In some beds ammonites of several zones are present within a single band a few metres thick. These ironstones must have accumulated very slowly (Hallam, 1963, 1966). They developed most easily in shallow 'swell' or shoal areas away from land, where turbulence was coupled with a lack of terrestrial detritus (i.e. sand and clay). The reason why the iron minerals were precipitated is still not understood. If physical conditions were paramount, chamosite and siderite would be more stable in sea water with low oxygen concentrations. But we know from the fossils that the composition of the sea was normal, and from the cross-bedding that the sea floor was often subject to turbulent conditions. It may be that iron concentration occurred organically; both algae and bacteria are known to be able to precipitate iron minerals.

Examples of oolitic ironstones may be seen in the Upper Lias of Raasay, Scotland; the Middle Lias Marlstone of Dorset, central England and Yorkshire and in the Lower Lias at Frodingham in Lincolnshire. Some of the Middle Lias ironstones are capped by basal Upper Lias condensed limestones.

70   Calcarenite Community

Calcarenites are limestones made up of sand-sized fragments; they occurred in shallow turbulent marine areas surrounding low islands, or on 'swells'.

The dominant faunas of calcarenites were suspension-feeding bivalves. Most of the epifaunal bivalves had thick shells that had been adapted specially to the turbulent conditions *(Liostrea, Gryphaea* and some pectinids), but there were also some byssally-fixed streamlined forms *(Gervillella, Modiolus)*. The epifauna also included large brachiopods (rhynchonellids and terebratulids), preserved as they lived (in 'nests'), or drifted with other shells into shell banks. The mollusca were also represented by thick-shelled grazing gastropods. Crinoid and echinoid debris was present, though complete specimens from either group are rare as fossils; the fragments include spines from cidarids and carnivorous diademoids. Many of the shells and fragments were bored by algae, sponges and bryozoans.

Corals (which are very rare in clays and sands) were abundant locally *(Oppelismilia, Styllophyllopsis* and *Isastrea)*, but they never formed substantial reefs. Like the other epifauna, they took advantage of the clear water with a rich supply of suspended food, and the temporary hard substrates of drifted shells provided suitable attachment. Many colonial corals were overturned, either by large animals or, more probably, by storm action. The corals were often bored and encrusted and while some of the boring may have taken place while the coral was alive, the encrusting serpulids, oysters and bryozoans probably grew on dead and drifted corals.

The infauna included suspension-feeding bivalves *(Pholadomya, Pleuromya, Cardinia, Astarte* and *Pinna)*. Infaunal crustaceans and worms left a variety of trace fossils including *Diplocraterion, Rhizocorallium* and *Thalassinoides*.

Ammonites and belemnites are sometimes abundant as fossils;

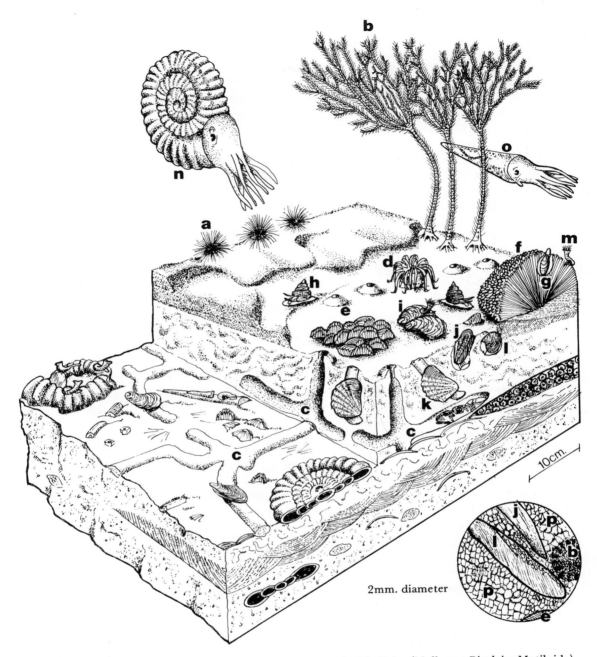

Fig. 70 Calcarenite Community

a   pseudodiadematid (Ectinodermata: Echinozoa — sea urchins)
b   *Pentacrinus* (Echinodermata: Crinozoa)
c   *Thalassinoides* (trace-fossil — crustacean)
d   *Oppelismilia* (Coelenterata: Anthozoa: Scleractinia)
e   rhynchonellids (Brachiopoda: Articulata: Rhynchonellida)
f   *Isastrea* (Coelenterata: Anthozoa: Scleractinia)
g   *Lithophaga* (Mollusca: Bivalvia: Mytiloida)
h   *Pleurotomaria* (Mollusca: Gastropoda: Archaeogastropoda)
i   *Liostrea* (Mollusca: Bivalvia: Pterioida — oysters)

j   *Modiolus* (Mollusca: Bivalvia: Mytiloida)
k   *Pholadomya* (Mollusca: Bivalvia: Anomalodesmata)
l   *Astarte* (Mollusca: Bivalvia: Veneroida)
m   serpulid (Annelida — polychaete)
n   *Uptonia* (Mollusca: Cephalopoda: Ammonoidea)
o   belemnite (Mollusca: Cephalopoda: Coleoidea)
p   cement of calcite (in inset)

225

Jurassic

like the other shells, they were often encrusted by serpulids after their death.

Some hardgrounds may exist within the calcarenites; the communites on these are described below. In the Lower Jurassic of Britain, calcarenites are found in the littoral areas of Glamorgan and the Mendips, and on local temporary 'swells', as in Dorset. In the littoral areas they often contained pebbles of older rocks upon which the Lower Jurassic rests unconformably. Splendid examples can be seen in Normandy where the Lower Jurassic overlaps the Armorican Massif.

71 Hardground Community

Ancient cemented sea floors can be recognized by the presence of borings and encrustations. They often occur in beds that show other signs of slow depositions, and the hardgrounds themselves marked non-depositional or even erosive phases. They can occur above beds of variable lithology and also at unconformities above older rocks. Although they were more common in the limestones of the British Middle and Upper Jurassic, they could also occur in mudstones, sandstones and limestones of the Lower Jurassic.

In calcareous or sandy sediments, large areas of the sea floor may have been hardened, especially in calcarenites at the top of coarsening upward sequences. If deposition was very slow, the dissolution of aragonite shells on or just below the sea floor was followed by precipitation of the calcium carbonate within the sediment; this then acted as a cement for the hardground (Purser, 1969). The Liassic hardgrounds provide no evidence of exposure to the air.

The fauna on hardgrounds was dominated by suspension feeders. Encrusting epifaunas consisted of bivalves (especially cemented oysters), serpulids and bryozoans, while byssal bivalves (pectinids and pterioids), crinoids and corals also attached themselves to the hard substrates.

If the hardground surface was irregular, small pockets could develop and these were sometimes filled with shell debris including eroded and abraded ammonites, belemnites and gastropods. Often these shell particles were overgrown and encrusted with serpulids and bryozoans. Pebbles of the hardground material are sometimes found near the base of the overlying bed; these too, may be encrusted and bored. The size of the pebbles (up to 20cm) indicates the strength of the currents that flowed over hardgrounds and it seems likely that their formation was favoured by shallow water conditions.

226

Fig. 71 Hardground Community
a *Pentacrinus* (Echinodermata: Crinozoa)
b *Liostrea* (Mollusca: Bivalvia: Pterioida — oyster)
c *Montlivaltia* (Coelenterata: Anthozoa: Scleractinia)
d echinoids (Echinodermata: Echinozoa — sea urchin)
e serpulids (Annelida — polychaete)
f boring worms (Annelida)
g *Lithophaga* (Mollusca: Bivalvia: Mytiloida)
h bryozoan (Bryozoa: Ectoprocta)
i patellid (Mollusca: Gastropod: Archaeogastropoda)
j encrusting foraminiferida (Sarcodina)
k *Amaltheus* (Mollusca: Cephalopod: Ammonoidea)
l belemnite (Mollusca: Cephalopod: Coleoidea)
m *Pseudopecten* (Mollusca: Bivalvia: Pterioida — pectinid)
n *Tetrarhynchia* (Brachiopoda: Articulata: Rhynchonellida)
x bedrock (pre-Jurassic)
y hardened Jurassic sediment

The infauna included specialized borers from many varied animal groups; all of them were suspension feeders, each animal with its own particular form of boring. The bivalve *Lithophaga* created a flask-shaped 'crypt' with a narrow entrance at the top; it spent its whole life in the same place, gradually enlarging its crypt as it grew by secreting a chemical to dissolve the surrounding rock. Another bivalve (*Pholas*) had strong anterior ribs which ground the rock when the animal rotated. Some cirripedes, sponges, algae and bryozoans also bored, each with a different pattern closely comparable to those made by their modern descendents, from which these inferences have been made.

227

Most hardgrounds formed in quite shallow clear water, but some occurred in deeper, more turbid water where the faunal diversity was much reduced.

Liassic clays often contain calcareous concretions which formed a few centimetres below the sea floor. Normally, once formed, these were never re-exposed, but on rare occasions they were exhumed and planed off by later erosion; their flat upper surfaces later provided a suitable medium for borers like *Lithophaga* and encrusting oysters whereas serpulids selected less exposed sites beneath and between the nodules (Hallam, 1969). Very occasionally these surfaces supported solitary corals.

Hardgrounds are present in Dorset, central England and Normandy.

Fig. 72 Carbonate Mud Community

a *Uptonia* (Mollusca: Cephalopoda: Ammonoidea)
b belemnite (Mollusca: Cephalopoda: Coleoidea)
c diademids (Echinodermata: Echinozoa — sea urchin)
d pectinid (Mollusca: Bivalvia: Pterioida)
e *Zeilleria* (Brachiopoda: Articulata: Terebratulida)
f pliosaur tooth (Vertebrata: Reptilia: Euryapsida)
g *Chlamys* (Mollusca: Bivalvia: Pterioida — pectinid)
h *Entolium* (Mollusca: Bivalvia: Pterioida — pectinid)

q

3m long

i *Inoceramus* (Mollusca: Bivalvia: Pterioida)
j *Diplocraterion* (trace-fossil — crustacean)
k *Chondrites* (trace-fossil — annelid)
l[1] living *Plagiostoma* (Mollusca: Bivalvia: Pterioida)
l[2] fossil *Plagiostoma* with tooth marks (Mollusca: Bivalvia: Pterioida)
m serpulid (Annelida — polychaete)
n *Gryphaea* (Mollusca: Bivalvia: Pterioida — oyster)
o *Thalassinoides* (trace-fossils — crustacean)
p *Schizospharella* (Coccolithophorida) — inset
q pliosaur (Vertebrata: Reptilia: Euryapsida) — inset

72 Carbonate Mud Community

Carbonate mudstones (or calcilutites) are fine grained limestones, and their fauna is comparable to that of restricted clays (Fig. 64), but trace fossils appear to be more diverse. Slow deposition sometimes led to the development of a hardground or near-hardground, with a resulting increase in diversity of the epifauna.

In the Lias, calcilutites often alternate with clays to form rhythms 1 to 1.5m thick. The lime mud of the calcilutites partly originated from early coccolithophorids, sub-microscopic calcareous algae which floated near to the surface of the sea. When they died their skeletons dropped to the sea floor to be mixed with mud laid down from bottom currents. This coccolith 'rain' was probably fairly continuous. Besides the coccoliths there were other floating and swimming organisms (ammonites, belemnites, ostracodes), fish and reptiles. Ichthyosaur and plesiosaur bones are rare. Pliosaurs (Fig. 72, inset q) evolved from the long-necked plesiosaurs in the Middle Jurassic.

The epifauna consisted of suspension feeders, most of which were adapted to living on a soft substrate. Bivalves were represented

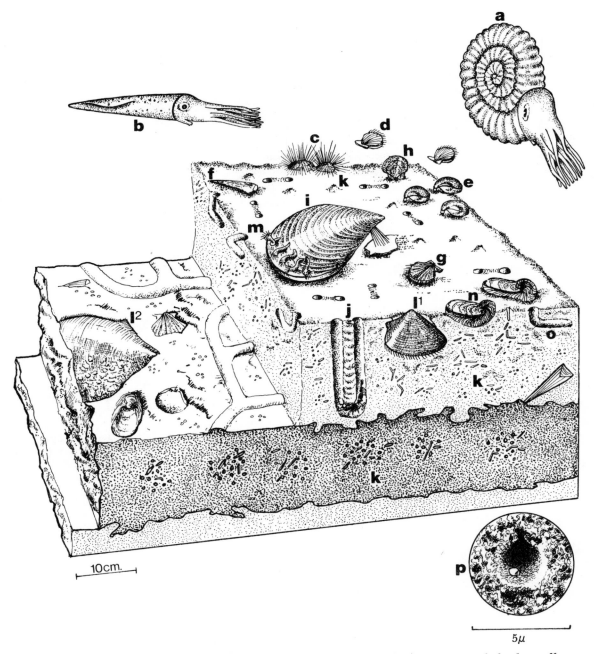

by thin-shelled pectinids, oysters *Plagiostoma*, and the byssally attached *Inoceramus* and *Oxytoma*. *Inoceramus* had a thin prismatic shell and a broad flattened shape and these features were
probably adaptations to prevent it sinking into the muddy sea
floor. Brachiopods were generally rare but became abundant at
some horizons where the substrate was more stable. The commonest brachiopods were rhynchonellids (*Tropiorhynchia*, *Calcir-
hynchia*) and terebratulides.

229

Jurassic

The sediments were intensively bioturbated, so that much of the substrate was probably 'soupy', except when slower sedimentation rates allowed consolidation or partial cementation of the sea floor. Trace fossils included burrows of suspension feeders and deposit feeders. The 'U' burrows (*Diplocraterion, Rhizocorallium*) often penetrated the tops of the limestone beds, while more complex burrow systems (like *Chondrites*) descended from the bases of many calcilutite beds. Both types of burrows were filled with sediment from the beds above (the 'U' burrows in the limestones were filled with clay, and the complex burrows in the clay are filled with lime mud) showing that the changes in sediment types were original and not just the result of subsequent movement of calcium carbonate in the rock. The normal 'background' sediment was probably coccolithic lime mud, its slow rate of accumulation being reflected in the abundance of unsorted ammonites in some limestones. Periodic influxes of terrigenous material, including comminuted plant debris, produced the interbedded mudstones. Such phases of more rapid deposition upset the oxygen budget of the basin and caused bituminous shales to form. With waning clay supply, there was a subsequent reversion to undiluted carbonate accumulation. This facies occurs as the Blue Lias in Dorset and in central England, and as the Belemnite Marls, again in Dorset.

LOWER JURASSIC COMMUNITIES IN EUROPE AND NORTH AMERICA

All the Jurassic communities from Britain which have been described lived in the shallow sea which spread over parts of north-west Europe during the early Jurassic.

In western North America, the Lower Jurassic commenced with a major phase of marine transgression and in general the western states (California, Oregon and Nevada) remained dominantly marine. The seaway extended northwards through western Canada into southern Alaska and southwards into north-west Mexico (Fig. m). Eastwards from Nevada and Oregon, the marine facies pass into dune-bedded continental sands and lacustrine sediments which contain remains of terrestrial dinosaur faunas. The shallow seas immediately to the west of these areas contain sediment and faunas similar in many respects to those of northern Europe. However, the bivalves *Gryphaea* and *Oxytoma* are latecomers in North America, arriving in the late Lower Jurassic via

230

Canada. Conversely, the bivalves *Trigonia* and *Myophorella* appear in the early Jurassic of North America, well before their arrival in Europe during the early Middle Jurassic. These latter two genera appear to have evolved in the Pacific during the Trias and spread thence through western South America, North America and on to Europe.

In North America, similar shallow marine facies often contain identical genera to their European counterparts. For example, the coral *Styllophyllopsis* occurs both in the shallow marine limestones of the Sunrise Formation in Nevada and in the similarly deposited calcarenites of South Wales (Fig.70), (Hallam, 1965). However, the great suites of volcanic rocks which occur to the west of the shallow marine areas have no counterparts in northern Europe, but represent island-arc and marginal oceanic environments on the margins of the Jurassic Pacific Ocean (Stanley et al. 1971).

In Europe, there is a transition from northern areas (like Britain) which were mostly receiving a continuous supply of clastic sediments from adjacent landmasses, to southern areas where little land-derived material was deposited. The shallow water sediments in these southern areas were mostly carbonates (limestones and dolomites), with some salt deposits occurring in the more isolated saline areas. Some of these carbonates indicate a warm sea, as well as a sea free from land-derived detritus. Palaeomagnetic evidence suggests that southern England was at a latitude of about 40^0 north at this time, so southern Europe was in the tropics during the Lower Jurassic. To the south of the shallow water carbonates lay the Tethys Ocean (though some of the carbonates may originally have been formed to the south of the Tethys); deep water deposits laid down on the floor of the Tethys, with only pelagic faunas and floras, are found today in Mediterreanean areas.

Highly continental climates are those with the largest temperature unevenness, and with the widest variety of seasonal conditions creating corresponding fluctuations in primary productivity. The most diverse marine communities should occur on shelves in low latitudes with the most stable habitats (Valentine, 1973). Communities with lower diversities would be expected in more continental or epicontinental situations.

The shallow carbonate areas of southern Europe are best seen in the Alps and some areas of southern Spain, central Italy and the Balkans. Early Jurassic environments in this region were similar to those existing today in Florida and the Bahamas. Locally the limestone beds are very thick, and throughout the late Trias and early Jurassic high rates of organic productivity caused limestone deposition to keep pace with the rapid subsidence, and the sea was seldom very deep. The fauna of these limestones is diverse compared with that of the clastic areas. Grazing gastropods are dominant; many of them fed on the rich algal material. Stromatolites are often abundant; brachiopods, bivalves, ostracodes, foraminfers and echinoderms are all present, many of the genera being confined to these environments and absent in the northern clastic areas. Ammonites, however, were only rarely present, usually preferring more open water habitats.

The diversity of the fauna in these massive limestones is very similar to that found in Florida Bay today where more than 100 genera of gastropods and bivalves are present. Water depths are seldom greater than 3m over an area of 15,000 square km. Within this large area, however, large seasonal fluctuations in temperature, rainfall (and therefore salinity) and turbulence (periodic storms) produce high environmental stresses which eliminate less tolerant species. Thus the more open water habitats contain the most diverse communities and this seems to have been the case on the margins of Tethys.

Towards the end of the Lower Jurassic time, and apparently synchronously along much of the Alpine-Mediterranean region, rates of subsidence increased so that carbonate production could no longer keep pace with the deepening sea (Bernoulli and Jenkyns, 1974). This is seen in the sudden change from shallow water to deep water sediments. The sea became too deep to be populated by abundant bottom-dwelling animals, but there is a rich pelagic fauna of ammonites and radiolaria, which sank to the sea floor after death.

Three sedimentary facies are common. On the more elevated parts of this deep sea, condensed red limestones accumulated, rich in floating and swimming organisms (ammonites, belemnites, and *Bositra*). Around these 'swells', deeper regions received thicker deposits of bioturbated marls with rare brachiopods (*Pygope*), belemnites, and ammonite aptychi. Aptychi are the calcite jaw apparatus of some ammonites. It is known that at great depths there is a solution of calcium carbonate; it is possible that, at the depths represented by the marls, the aragonite shells were dissolved but not the calcite shells, hence the preservation of the aptychi without the main part of the ammonite shells. At greater depths still, no carbonate fossils occur at all. On the ocean floor, basaltic rocks (which underlie all oceans), were interbedded with radiolarian cherts.

During the phases of subsidence, the surface of the carbonate platforms were extensively colonized by thick-shelled bivalves with strong hinge structures (*Opisoma, Pachymegalodon* and *Pachymytilus*), brachiopods (*Hesperithyris* and other rhynchonellids) and crinoids. Lenses of crinoid debris occur. These fragments sometimes drifted into crinoidal sand waves on the surface of collapsing Tethyan sea mounts (Jenkyns, 1971).

MIDDLE JURASSIC

Towards the end of the Lower Jurassic the shallow marine environments began to be replaced from north to south by increasingly variable fluviatile, deltaic and lagoonal conditions. These changing European environments resulted from local uplift round the central North Sea where rifting was occurring: a phase of earth movements which was associated with local volcanism (Sellwood and Hallam, 1975). Prior to rifting, uplift took place and clastic sediments were derived from the east as powerful rivers rapidly stripped the new land surfaces and enveloped northern and eastern areas of Britain in fluviatile and deltaic sediments. Generally, away from this major source of sediment, the influence of coarse clastic detritus decreases, but the presence of major sources of fresh water had a dramatic effect upon the faunas. A complex of lagoons were established as margins to the deltaic environments, and these gave way, as the distance from the clastic sources increased, to a broad shallow shelf upon which a variety of carbonate sediments were laid down. It was only in the far west of Britain and over northern France that unrestricted marine conditions remained (Hallam and Sellwood, 1976).

The deltaic sands are poor in fossils, but driftwood is abundant and some beds include coals showing the roots and stems in their positions of growth. Freshwater ostracodes and bivalves are present in places and when they occur they are often abundant. Some of the most exciting finds are sun-cracked bedding surfaces covered with dinosaur footprints. Westwards and southwards from the area of Yorkshire deltas, euryhaline faunas are dominant in clay-rich sediments that accumulated in the complex lagoonal environments. Further towards the south, in the Cotswold Hills, there are many limestones and clays with more diverse marine faunas, though stenohaline organisms like belemnites and ammonites are still extremely rare. These limestones often acted as barriers between the lagoonal clays to the north and the more open marine environments to the south.

The deltaic sands in the north of England and the more fully marine clays in the south are much thicker than the limestones and lagoonal clays of the English Midlands. These different thicknesses suggest that differential subsidence in the underlying Palaeozoic floor was a major control on sedimentary facies. In general the lagoons and barriers of the English Midlands are sited in the region of least subsidence, and several minor local changes in environments are also linked with local swells and basins.

The various interactions of sedimentation rate, sediment type, salinity and depth of water produced a great variety of habitats. No single factor is the primary control in the development of a particular faunal assemblage, but three major environmental associations can be recognized:

1 The fluviatile and lagoonal association, which included freshwater and euryhaline animals, notably ostracodes, oysters and certain other molluscs.

2 The carbonate areas in which clear water conditions allowed the development of a relatively diverse bottom-dwelling fauna, including mollusca, corals, brachiopods and echinoderms. Locally within these areas, the diversity was reduced drastically by decrease in salinity or by the development of highly mobile substrates (for instance oolite banks or sands).

3 The open marine environments of southern England and northern France often had diverse faunas, but the diversity of the benthos was often restricted by the depth of water above the muddy sea floor. It is only in these Middle Jurassic environments that communities similar to those in the British Lias are developed.

In the Middle Jurassic, some ammonite genera began to show well-developed sexual dimorphism, a single species being represented by two forms: a small male shell (microconch) and a larger female shell (macroconch). The shells of the males have projections on their apertural margins (lappets) which are thought to have assisted attachment to the females during mating. The shells of females often had an aperture equal in size to the total diameter of the male shell but they never carried lappets.

Both epifaunal and infaunal bivalves are often diverse in the Middle Jurassic, reflecting a greater variety of marine and lagoonal environments, but restriction in the diversity of bivalves occurred when the salinity was reduced in brackish environments. In general, oysters, lucinoids and *Astarte* occurred in mixed marine environments which had the lowest salinity and were adjacent to areas of fresh water.

Fig. 73 Freshwater Communities

a dragonfly (Arthropoda: Hexapoda — insect)

b *Unio* (Mollusca: Bivalvia: Unionida)

c *Viviparus* (Mollusca: Gastropoda: Mesogastropoda)

d *Valvata* (Mollusca: Gastropoda: Mesogastropoda)

e *Chara* (Algae — characeae)

f *Equisetites* (Pteridophyta: Calamites — horsetails)

f^1 buried roots and stems of horsetails (Pteridophyta: Calamites)

g dinosaur footprints (trace-fossil)

h small insects (Arthropoda: Hexapoda)

73 Freshwater Communities

Freshwater Middle Jurassic assemblages occur within the deltaic and lagoonal sequences as transported accumulates in marine and brackish water sediments.

Characteristic Middle Jurassic freshwater mollusca included the turreted gastropod *Bathonella [Viviparus]*, the planospiral gastropod *Valvata*, and the bivalve *Unio*; their descendants live in the same habitats today. The freshwater sediments are silty clays and sands which probably accumulated in rivers and low-lying lakes. The sediments never contain marine organisms, and, as most

d — 2mm

5cm.

invertebrates which produce **complex burrow systems** are marine, these freshwater deposits were **seldom** very bioturbated.

Plants were represented by terrestrial and marsh-dwelling forms. They occur either as **drifted** fragments, as standing stems with their roots preserved or just as roots; the commonest standing plants were horsetails (Equisetales). Rippled surfaces of some sand beds show footprints of swamp-dwelling dinosaurs and polygonal desiccation cracks, an assemblage providing evidence of intermittent exposure.

235

Jurassic

Modern *Viviparus* is able to feed on both deposited and suspended material; it is one of the few gastropods that can employ a process of ciliary feeding to supplement the normal means of obtaining food. Its Middle Jurassic ancestor, *Bathonella*, is believed to have lived in similar freshwater habitats and may have had the same ability to vary its feeding mechanism according to what was available. Modern *Unio* is a filter-feeding bivalve with an incompletely fused mantle; it lives exposed on lake and stream channel floors, or at best it can burrow a small distance into the sediment. Like many freshwater bivalves, the umbonal areas on *Unio* are often eroded; the earlier formed parts of the shell, being exposed to freshwater deficient in carbonate during the life of the individual, are often dissolved.

Some freshwater beds may be rich in charophytes; these small (0.1mm) spheres are the female reproductive bodies (oogonia) of freshwater calcareous green algae. Other beds may contain freshwater ostracodes, which can sometimes be distinguished from marine ostracodes by their smooth shells.

Freshwater sediments occur in the Middle Jurassic in Britain on the Yorkshire coast, in Scotland and in the English Midlands.

74   Restricted Marine Clay Community

Between the 'delta' complexes of north-east England and the open sea to the south, lay an area with restricted euryhaline faunas and predominantly clay deposition. These inhabited lagoons were separated from the open sea by barriers that consisted of oolite shoals, mud flats and archipelagos, but their isolation was not complete. In general, the lagoonal clays with fewer genera of euryhaline animals are thought to have been more isolated than those with more genera. There is a complete gradation between the most restricted of these euryhaline communities and the normal diverse, open sea, communities (Figs. 79, 80 and 81). By contrast, there are only a few examples of gradations (for instance in West Scotland) between the euryhaline communities and wholly freshwater communities (Fig. 73). When viviparids, *Unio,* freshwater ostracodes or charophytes are present in these lagoonal sediments, they often occur as lenses of drifted shells, probably introduced by floods.

The commonest euryhaline fossils are oysters and in some beds *Liostrea* is the only marine fossil preserved. Clays containing oysters alone may indicate euryhaline conditions, but two other factors may also give rise to oyster-rich deposits: first, the massive calcite shell is much less soluble than the thinner or aragonitic shells of other molluscs; and second, even in a diverse assemblage

236

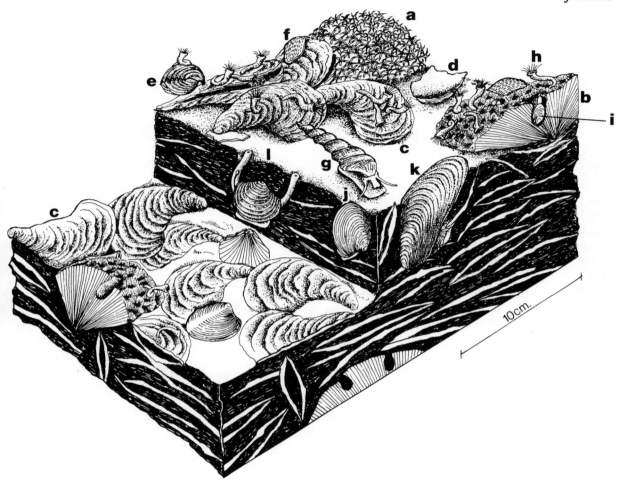

Fig. 74 Restricted Marine Clay Community
a living *Isastrea* (Coelenterata: Anthozoa: Scleractinia)
b dead *Isastrea* (Coelenterata: Anthozoa: Scleractinia)
c *Liostrea* (Mollusca: Bivalvia: Pterioida — oyster)
d *Liostrea* broken shell (Mollusca: Bivalvia)
e *Kallirhynchia* (Brachiopoda: Articulata: Rhynchonellida)
f bryozoan (Bryozoa: Ectoprocta)
g *Aphanoptyxis* (Mollusca: Gastropoda: ?Mesogastropoda)
h serpulid (Annelida — polychaete)
i *Lithophaga* (Mollusca: Bivalvia: Mytiloida)
j *Astarte* (Mollusca: Bivalvia: Veneroida)
k *Modiolus* (Mollusca: Bivalvia: Mytiloida)
l *Lucina* (Mollusca: Bivalvia: Veneroida)

237

clumps composed solely of oysters may have grown as discrete clusters. This second association is indicated by the presence of encrusting serpulids, bryozoans and other fully marine epifauna on the oyster shells.

Modern oysters require stable substrates and are often particularly abundant when burrowers are absent. Faunas dominated by oysters normally occur in very shallow water and the surrounding sediments are often fine grained. The fine material was trapped by the animals themselves; the sediment is therefore often much finer than other sediments characteristic of very shallow environments.

Rhynchonellids and ostracodes often occurred with the oysters and as the salinity approached normal, marine *Modiolus, Astarte,* and a variety of other bivalves occurred. Corals, terebratulids, gastropods and a few echinoids (for instance *Clypeus*) lived in some of the less restricted lagoonal environments, especially within the barrier areas of the English Midlands. Two fossil groups which are seldom found (except as rare, isolated, drifted individuals) are the strictly stenohaline ammonites and belemnites.

Lagoonal coral faunas were usually limited to three genera (*Isastrea, Thamnasteria* and *Microsolena*) which occurred in small patches on the lagoon floor and were often bored and encrusted by bivalves and serpulids.

Modern corals and echinoderms are sensitive to reduction in salinity below $27^{0}/00$ (parts per thousand), but some corals can tolerate raised salinities (up to $40^{0}/00$), and it may well be that some lagoons had restricted faunas because the salinity was above (rather than below) that of the open sea (about $35^{0}/00$).

Salinity is not the only factor causing the restriction of faunas in modern lagoonal environments. The variability of the environment is perhaps the main one; very shallow, partially isolated water is drastically affected by variations of temperature and by turbulence resulting from seasonal changes and from periodic storms. Corals are particularly sensitive to these changes, and are thus likely to be absent in areas subject to storms and floods.

In these restricted communities, the bivalves comprised the bulk of the infaunal and epifaunal suspension feeders. The echinoids were probably predators and scavengers, while corals were carnivorous, the individual polyps within a colony feeding on zooplankton in the water.

Restricted clays occur in the Great Estuarine Group of western Scotland, and the Great Oolite Group of Yorkshire and central England.

75 Communities in Ironstones and Marine Sands

Marine ironstones in the Middle Jurassic were similar in many respects to those in the Lower Jurassic (Fig. 69), and the same difficulties arise in the interpretation of their formation. When little quartz sand was present, the fauna was dominated by infaunal and epifaunal suspension-feeding bivalves. Suspension-feeding brachiopods are occasionally present and certain beds contain nerineid gastropods, some of which were probably grazers. Many of the ironstones are not very fossiliferous, but when fossils do occur they are often in lenses as drifted shells. In these cases the shells are seldom worn and are probably of local origin.

Although algal filaments have not been recognized in the ferruginous ooids which make up the sediment, the manner of

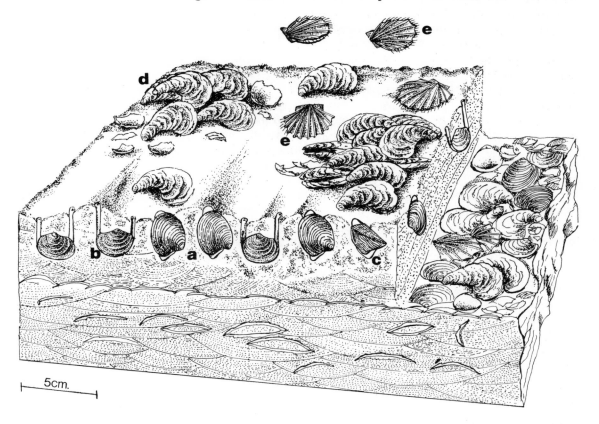

Fig. 75 Communities in Ironstones and Marine Sands
a *Astarte* (Mollusca: Bivalvia: Veneroida)
b *Lucina* (Mollusca: Bivalvia: Veneroida)
c *Trigonia* (Mollusca: Bivalvia: Trigonioida)
d *Liostrea* (Mollusca: Bivalvia: Pterioida — oyster)
e pectinid (Mollusca: Bivalvia: Pterioida)

growth of the ooids suggests that algae may have been responsible for trapping the iron-rich material. If this interpretation is correct, the algae would have been a source of food for the herbivorous nerineids.

The abundance of suspension feeders in the ironstone shell beds indicates relatively shallow water, with sufficient turbulence to carry suspended organic matter. The sporadic occurrence of terebratulid and rhynchonellid brachiopods indicates that the water was clear and that the sea floor was stable enough for pedicle attachment. Faunal diversities in the ironstones were sometimes high, with free-living epifaunal bivalves (*Varamusium*), thicker-shelled pectinids, byssally attached streamlined bivalves (like *Gervillella*) and infaunal suspension feeders (like *Astarte* and *Trigonia*). Scavengers and carnivores were represented by echinoids (*Acrosalenia*), and rarely by ammonites and belemnites.

The ironstones sometimes contain large amounts of quartz sand in addition to the iron-rich ooids and shell debris, and in the more sandy beds faunas were sparser. Like many pure quartz sands deposition rates may have been too high and the substrates too unstable for many epifaunal and infaunal animals to have become established.

Fully marine sands with diverse faunas were rare in the Middle Jurassic of Britain. More common, especially in the English Midlands, were sands that contained restricted faunas. Gradations also existed between sand and mud, sand and calcarenite, and sand and ironstone.

Diverse faunas occurred in sands in the west of Scotland where oysters, thick-shelled pectinids, and a variety of other bivalves are associated with some brachiopods. When conditions were fully marine, rich faunas of ammonites became established. As in the Lower Jurassic (Figs. 66, 67), a small amount of mud in the sands provided a more stable habitat for burrowers, which included bivalves (pholadomyoids and veneroids), crustaceans (*Thalassinoides, Rhizocorallium, Diplocraterion*), and worms (*Chondrites*).

Sands in the English Midlands contain restricted faunas. Oysters are the most widely distributed fossils; they occurred in deposits where the salinity was low. With increasing clay content infaunal bivalves (*Astarte, Lucina*) flourished. The community depicted is from a Midlands sand with a clay component.

Sandy environments were present in western Scotland and in the English Midlands during the Middle Jurassic; ironstones are present in Yorkshire, Lincolnshire and Northamptonshire.

76 Oolitic Limestone Communities

Oolitic sediments contain ellipsoidal or spherical grains up to 1mm in diameter which consist of a series of concentric carbonate laminations surrounding a nucleus. In modern carbonate areas like the Bahamas, these ooids are formed in the shallow turbulent areas between the open sea and the calm hypersaline lagoons. Modern ooids are composed of aragonite but ancient ooids are normally recrystallized into calcite.

Unbroken and indigenous fossils in the Middle Jurassic oolites are rare, generally consisting of thick-shelled oysters, pectinids and echinoderms. Most of the material is fragmented and many of the grains have been overgrown by carbonate laminae. The limestones are often cross-bedded in layers ranging up to 2m in thickness and the original sediments appear to have formed dunes with mobile surfaces. This surface instability probably accounts for the scarcity of the fauna and the rarity of unbroken shells.

Fig. 76 Oolitic Limestone Communities

a *Tigillites* (trace-fossil — annelid)

b *Purpuroidea* (Mollusca: Gastropoda: Mesogastropoda)

c shell of *Purpuroidea* replaced by calcite (Mollusca: Gastropoda)

d opening of *Diplocraterion* (trace-fossil)

e *Diplocraterion* showing response to oolite deposition

f *Pygaster* (Echinodermata: Echinozoa)

g buried *Pygaster* part-filled with sediment and sparry calcite

h oyster fragment

i pectinid (Mollusca: Bivalvia: Pterioida)

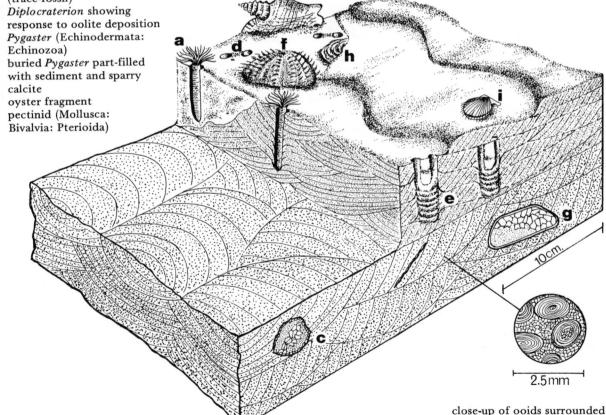

10cm.

2.5mm

close-up of ooids surrounded by void-filling calcite spar

Jurassic

The scavenging shallow-burrowing echinoid *Pygaster* and the grazing gastropod *Purpuroidea* could survive these conditions and some pectinids were also tolerant of the unstable substrates and turbulent water. This assemblage compares closely with that from a modern oolite sand from the Bahamas.

During phases of dune stabilization the sediment was sometimes penetrated by the vertical burrows of suspension feeders. Some of these burrows contain clays that accumulated during slack water episodes. In the Midlands, interbeds of clays within the oolites sometimes contained material derived from nearby rivers such as drifted logs, reptilian bones and teeth, and freshwater ostracodes (McKerrow et al. 1969).

In sublittoral bays and lagoonal areas where the water was too turbulent to allow the development of continuous algal crusts (stromatolites), blue-green algae grew around shell fragments and trapped carbonate sediment to form algal 'snowballs' that are termed oncolites or pisolites. As these 'snowballs' grew their well illuminated upper surfaces tended to grow faster and because current action periodically rolled them around their internal laminations are discontinuous. These structures are much larger than ooids and often exceed 15mm in diameter. Oncolitic limestones seem to have formed in fairly turbulent waters and they contain few complete shells. However, they sometimes contain the fossils of epifaunal suspension feeders such as encrusting and cementing bryozoans and corals. Some of the echinoids also were able to live under these conditions, particularly the rapid burrowing forms (irregulars) or thicker-shelled surface-dwelling cidarids (regulars). Grazing gastropods were present too, feeding upon the algae.

These environments developed mostly in the Midlands and south of England, but they spread periodically into Lincolnshire and Yorkshire.

77 Low Diversity Temporarily Stable Calcarenite Community

This sediment consisted of finely broken shell debris which was the result of a combination of predation and mechanical erosion. It is likely that originally the sediment surface was rippled and provided a fairly mobile substrate. Thus, the fauna which inhabited it consisted of fairly specialized animals all adapted to withstand the somewhat rigorous environmental stresses. Many of the colonizing animals took refuge within the sediment and, consequently, many of the original sedimentary structures have been totally obliterated by their burrowing activities.

242

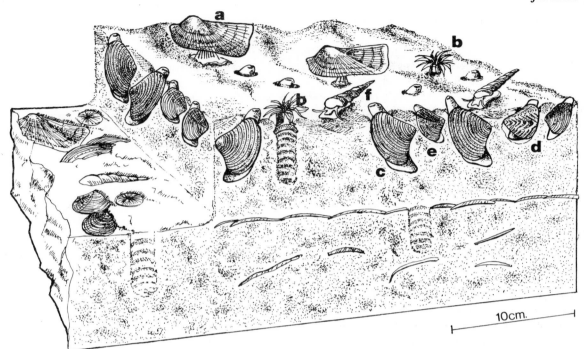

Fig. 77 Low Diversity Temporarily Stable Calcarenite Community

a *Parallelodon* (Mollusca: Bivalvia: Arcoida)

b *Chomatoseris* (Coelenterata: Anthozoa: Scleractinia)

c *Eocallista* (Mollusca: Bivalvia: Veneroida)

d *Vaughonia* (Mollusca: Bivalvia: Trigonioida)

e *Protocardia* (Mollusca: Bivalvia: Veneroida)

f nerineid gastropod (Mollusca: Gastropoda: ?Mesogastropoda)

The dominant genera were the streamlined shallow burrowers *Eocallista* and *Protocardia*; they were relatively mobile bivalves and were capable of ploughing their way through the sediment, with pauses from time to time in places where there was a good supply of suspended food material. The *Trigonia*-like bivalve *Vaughonia* probably lived in a similar fashion.

Also prominent within this sediment was the mobile burrowing coral *Chomatoseris* which, unlike many corals, did not produce an extensive skeletal frame, but merely grew up from a disc of carbonate that may have aided its burrowing activities. These carnivores could move either up or down in response to deposition or erosion on the sea floor.

The epifauna was very restricted; the only epifaunal bivalve was the rather rare suspension-feeding *Parallelodon* that was probably byssally fixed to fragments of shell debris. This bivalve was probably capable of exhuming itself if covered by deposited sediment. The only other epifaunal organisms in this community were high-spired nerineid gastropods that grazed over the surface of the sediment. Their main food source in this environment probably consisted of the bacterial linings around the sediment grains, although some modern nerineids are capable of suspension feeding by a ciliary process and are sedentary within shallow burrows.

This fauna is present in the Middle Jurassic of the English Midlands.

243

78   Diverse Calcarenite Community

As in the previous environment, the sediment was composed of sand grade shell debris which was likely to have been rippled at the time of deposition, but here too the action of burrowers has led to the total obliteration of the original cross-bedding. There is often a gradation between the last community and this type, with the development of this diverse fauna having been allowed by the reduction in both turbulence and sedimentation rates. The sediment may also show a concomitant increase in the content of mud.

As the sediment became more stable, so the diversity of the epifauna showed a marked increase, and the more sensitive infaunal suspension feeders (*Pleuromya* and *Pholadomya*) also made an appearance. These deep-burrowing bivalves with their long siphons would have been excluded previously by high rates of deposition. Also within the sediment were the shallow-burrowing and more mobile suspension feeders *Astarte* and *Trigonia* while at much greater depth (a metre or more, and thus too far down to be included correctly in our diagram) were crustaceans producing *Thalassinoides* burrows. Sometimes, some of these crustaceans can be found within their burrows (for instance *Glyphaea* as depicted in our diagram). Occasional irregular echinoids (*Nucleolites*) were deposit feeding within the sediment.

With lower sediment mobility, and probably decreased turbidity, brachiopods appeared. These consisted of both rhynchonellids and terebratulids. The rhynchonellids occurred in clusters or 'nests' composed of a single species while the terebratulids lived either in clusters or as isolated individuals. *Ornithella*, which may have had a relatively long pedicle, and *Cererithyris*, with its nearly planar commissure between the valves, are both characteristic of these environments.

Epifaunal bivalves included the byssally attached *Meleagrinella* which was anchored to shell particles on the sea-floor. Its flattened shape was ideal for it to withstand strong currents as it filtered the sea water for its food. The smooth and fairly flattened pectinid *Entolium* may have been free-living. The filter-feeding oyster *Lopha* with its thick and strongly ribbed shell was attached to shell particles and other individuals of its own kind by a carbonate cement.

The sea water above the bottom was of sufficiently normal marine character to support belemnites, nautiloids and ammonites living as scavengers and predators.

In England the Inferior Oolite and Great Oolite Groups of Dorset, the Midlands and Yorkshire contain limestones showing this community (for instance, the Cornbrash).

Jurassic

Fig. 78 Diverse Calcarenite Community

a *Clydoniceras* (Mollusca: Cephalopoda: Ammonoidea)
b *Cenoceras* (Mollusca: Cephalopoda: Nautiloidea)
c *Ornithella* (Brachiopoda: Articulata: Terebratulida)
d *Meleagrinella* (Mollusca: Bivalvia: Pterioida)

e *Lopha* (Mollusca: Bivalvia: Pterioida — oyster)
f rhynchonellids (Brachiopoda: Articulata: Rhynchonellida)
g *Cererithyris* (Brachiopoda: Articulata: Terebratulida)
h *Entolium* (Mollusca: Bivalvia: Pterioida — pectinid)
i *Nucleolites* (Echinodermata: Echinozoa — sea urchin)
j *Trigonia* (Mollusca: Bivalvia: Trigonioida)
k *Pleuromya* (Mollusca: Bivalvia: Anomalodesmata)
l *Pholadomya* (Mollusca: Bivalvia: Anomalodesmata)
m *Astarte* (Mollusca: Bivalvia: Veneroida)
n *Thalassinoides* (trace-fossil) with burrowing *Glyphaea* (Crustacea) inside

245

79 Shelly Lime Mud Community

The sediment in between the shells of this community consists of mud grade calcium carbonate (micrite), and at the time of deposition the substrate is likely to have had a pasty consistency. In view of the richness of the epifauna, the bottom was unlikely to have been very soft and the assemblage of sediment and fauna represents a shallow, but quiet water, bay association comparable to modern equivalents from Florida and the Bahamas.

Much of the shell material has been infested by microscopic boring algae and fungi; it is likely that the water was well lit and that the sediment surface supported a large quantity of lowly plant life which is not now preserved. The delicate stromatolic structures produced by some algal growths would have been totally obliterated by the burrowing activities of the infauna.

The epifauna consisted of grazers, suspension feeders and carnivores, while the infauna was dominated by suspension feeders. On the surface, the nerineid *Fibula* grazed or ingested surface detritus to digest the bacteria lining the sediment, while the mobile regular echinoid *Acrosalenia* scavenged or preyed upon soft-bodied animals that lived near the sediment surface. The teeth marks from echinoids sometimes scar shell material but the echinoids themselves are normally only represented as fossils by isolated spines and plates. These stable surfaces, covered by water with a relatively low turbidity, supported clumps of terebratulid brachiopods and in our diagram we have depicted *Epithyris*. Bromley and Surlyk (1973) have shown that brachiopod pedicles may etch a circle of dots a millimetre or more in diameter on other shells, each dot etched by a separate strand of the pedicle. These attachment scars may be commonly found on shell fragments and other epithyrids in this facies.

Epifaunal bivalve suspension feeders were represented by the byssally attached or free-living *Camptonectes* and *Pseudolimea*, while the streamlined *Costigervillia* probably swung freely, possibly attached to anchored gorgonians that have left no fossil record. The oysters (*Liostrea*) were attached by a carbonate cement to shell and other hard debris upon the substrate.

Within the sediment, helping to destroy the internal structure, were a variety of invertebrates. *Modiolus* were sometimes common, living semi-infaunally with a byssal attachment to buried shell debris, while fat sluggish burrowers (*Anisocardia* and *Sphaeriola*) filter fed through short siphons. Clusters of terebellid worms extended their filamentous tentacles into the water, and lined their tubes below this crown on tentacles with shell and fish-scale debris. Sometimes their tubes (trace fossils) were removed and redeposited parallel to the current.

Crustaceans also produced complex *Thalassinoides* burrow

Fig. 79 Shelly Lime Mud Community
a   *Epithyris* (Brachiopoda: Articulata: Terebratulida)
b   *Camptonectes* (Mollusca: Bivalvia: Pterioida — pectinid)
c   *Liostrea* (Mollusca: Bivalvia: Pterioida — oyster)
d   *Pseudolimea* (Mollusca: Bivalvia: Pterioida)
e   *Costigervillia* (Mollusca: Bivalvia: Pterioida)
f   *Acrosalenia* (Echinodermata: Echinozoa)
g   *Modiolus* (Mollusca: Bivalvia: Mytiloida)
h   *Fibula* (Mollusca: Gastropoda: Archaeogastropoda)
i   *Thalassinoides* (trace-fossil) with burrowing *Glyphaea* (Crustacea)
j   *Anisocardia* (Mollusca: Bivalvia: Veneroida)
k   terebellid worms (Annelida)
l   gorgonian (Coelenterata: Octocorallia) hypothetical in this reconstruction

247

systems within the sediment, sometimes descending to considerable depths (more than 1m) and may well have caused the original sediment surface to have been irregularly marked by mounds of excavated material (not shown); the burrows were often partially filled with faecal pellets called *Favreina*.

This fauna exhibits the same characteristics as modern assemblages of bay or lagoonal environments in most respects, except for the greatly decreased diversity of modern brachiopod faunas. Restricted circulation probably prevented the more stenohaline ammonites and belemnites from making an appearance while these sediments were being deposited at various times over the Midlands area in England.

80 Muddy Lime Sand Community

Many of the Middle Jurassic limestones contain carbonate mud (micrite) and there are many gradations between those that contain pure micrite and those that contain detrital clay. The cleaner and less muddy limestones also show many gradations both laterally and vertically into less well sorted muddy sediments, often within one quarry exposure.

Much of the micrite of these Middle Jurassic limestones is assumed to have been formed in the same way that the majority of modern micrite is produced, namely by the degradation of algal material, probably red and green algae with a very low preservation potential.

The environment probably consisted of an area covered mainly by mud deposits into which lime sand was periodically introduced. The sandier layers promoted a more stable substrate condition which probably accounts for the increased infaunal diversity by comparison with the Shelly Lime Mud Community described earlier.

The epifauna was similar to that of the latter community (Fig. 79) containing cemented oysters and byssally attached *Camptonectes*. It also included *Gervillella* and *Isognomon,* again byssally attached to shell fragments, and living as fairly closely attached suspension feeders. No clusters of terebratulids are found, but instead there are isolated brachiopods which were the only forms able to live there. It is probable that the slightly increased rates of sand accumulation were enough to inhibit the development of brachiopod clusters.

Gastropods such as *Alaria* grazed over the surface, again feeding upon presumed algal films or upon bacterial slimes around deposited grains. Within the sediment, the diversity of infaunal suspension feeders was quite high. Deep-burrowing *Homomya*

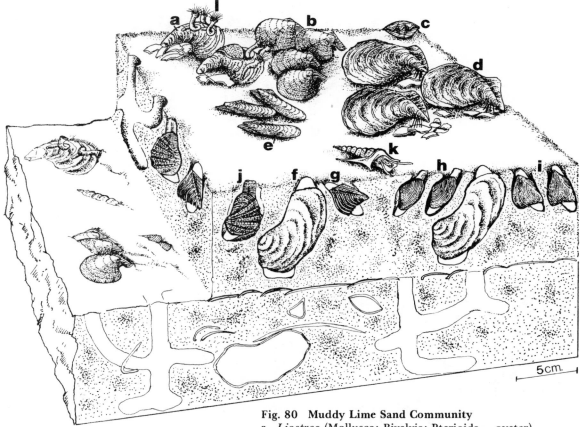

Fig. 80 Muddy Lime Sand Community
a *Liostrea* (Mollusca: Bivalvia: Pterioida — oyster)
b *Camptonectes* (Mollusca: Bivalvia: Pterioida — pectinid)
c *Epithyris* (Brachiopoda: Articulata: Terebratulida)
d *Isognomon* (Mollusca: Bivalvia: Pterioida)
e *Gervillella* (Mollusca: Bivalvia: Pterioida)
f *Homomya* (Mollusca: Bivalvia: Anomalodesmata)
g *Trigonia* (Mollusca: Bivalvia: Trigonioida)
h *Pleuromya* (Mollusca: Bivalvia: Anomalodesmata)
i *Pseudotrapezium* (Mollusca: Bivalvia: Veneroida)
j *Pholadomya* (Mollusca: Bivalvia: Anomalodesmata)
k *Alaria* (Mollusca: Gastropoda: Mesogastropoda)
l serpulids (Annelida)

appeared (we have had to show it at shallow depths) associated with *Pleuromya* while the shallow-burrowing and very mobile trigonids and *Pseudotrapezium* took advantage of the slightly coarser and more stable substrate.

This assemblage occurs in many horizons of the Middle Jurassic in the English Midlands; however towards the west and south, where salinity was more normal, some ammonites and nautiloids are also present.

249

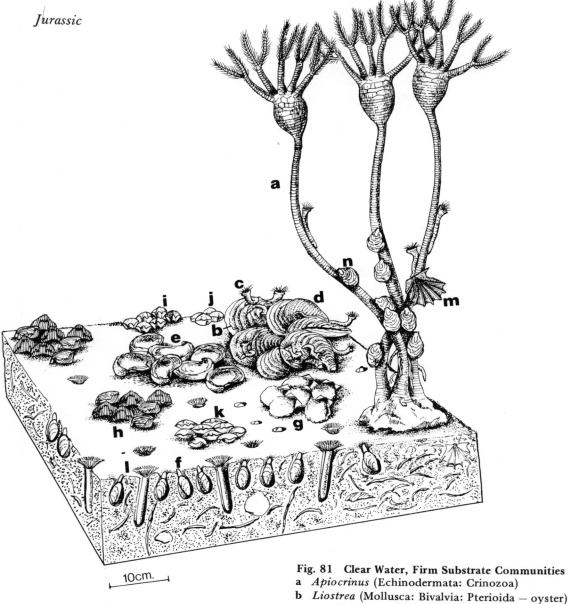

⊢ 10cm. ⊣

Fig. 81 Clear Water, Firm Substrate Communities

a *Apiocrinus* (Echinodermata: Crinozoa)
b *Liostrea* (Mollusca: Bivalvia: Pterioida — oyster)
c serpulids (Annelida)
d bryozoan (Bryozoa: Ectoprocta)
e *Exogyra* (Mollusca: Bivalvia: Pterioida — oyster)
f *Lithophaga* (Mollusca: Bivalvia: Mytiloida)
g *Radulopecten* (Mollusca: Bivalgia: Pterioida — pectinid)
h *Rhynchonella* (Brachiopoda: Articulata: Rhynchonellida)
i *Dictyothyris* (Brachiopoda: Articulata: Terebratulida)
j ornithellids (Brachiopoda: Articulata: Terebratulida)
k *Avonothyris* (Brachiopoda: Articulata: Terebratulida)
l tubular worm (Annelida)
m *Oxytoma* (Mollusca: Bivalvia: Pterioida)
n *Liostrea* (Mollusca: Bivalvia: Pterioida — oyster)

81   Clear Water, Firm Substrate Communities

Stable and firm substrates developed in a variety of ways and at
many times in the shallow marine and marginal marine areas of
central and southern England. These substrates offered attachment
sites for a broad spectrum of cementing, encrusting and boring
organisms that were otherwise restricted in their distribution to

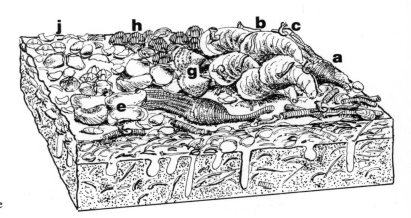

Fig. 81a   Death Assemblage

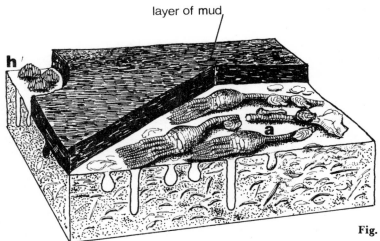

Fig. 81b   Assemblage after Burial by Clay

5cm.

the surfaces of scattered skeletal debris in other environments. Indeed, the recognition of fossil borings and a diverse array of associated encrusting organisms is usually a strong indication that the sea bottom was firm and the waters above clear of large amounts of detrital materials at the time of deposition.

Slow rates of deposition allowed consolidation of the substrate, and sometimes the interstitial precipitation of a carbonate cement led to the formation of hardgrounds which, after differential erosion of uncemented sediment, provided irregular and even undercut surfaces for colonization. In these cases, the faunas inhabiting the lower surfaces were frequently different from those inhabiting the more exposed upper surfaces (Palmer and Fursich, 1974). Unfortunately, many of the characteristic genera were extremely small organisms that are too tiny to be represented on our diagram.

The exposed upper surface epifauna included foraminifera like *Nubeculinella* (not shown) that inhabited both hardened substrate and shell debris, and certain types of serpulid worms. These, like many of the organisms, lived as suspension feeders. Cemented oysters (*Liostrea, Exogrya* and *Lopha*) were also present, while other suspension-feeding bivalves such as *Radulopecten* and *Pseudolimea* were attached by byssal threads. Surfaces and shell debris were often encrusted by various types of bryozoans and these, like the serpulids and foraminiferida frequently grew upon living shell material.

Terebratulides and rhynchonellid brachiopods flourished in the clear waters and often formed 'nests' comprising hundreds of individuals. This community includes thecideidine brachiopods, which lived on the undersurfaces of overhangs and within cavities, but are too small to be shown.

The largest and most spectacular organisms that attached themselves to these hardened surfaces were crinoids, and even where the crinoid stalks have been removed by later erosion, the hold fasts of these creatures may still be preserved, themselves encrusted by bryozoans and foraminiferida.

Beneath overhangs and within fissures in the sediment, a nestling fauna became established, often containing different serpulids, sponges and bryozoa and these organisms either preferred the dark or tolerated less turbulent conditions than the organisms on upper surfaces.

The infauna too was markedly different from that found upon softer bottoms. If the substrate was not actually hard, but was merely stiff, normal burrowing infaunas were developed, but when it became cemented only specialized boring organisms thrived. Of these, the most characteristic was the bivalve *Lithophaga* which bored into both upper and lower surfaces. They produced 'crypt' shaped cavities, particularly in carbonate substrates, and fed as suspension feeders. Other borers included worm-like organisms which produced straight and inclined *'Trypanites'* borings and which in our figure are drawn as worms with filamentous tentacles.

When the Middle Jurassic rests directly on older carbonates (for instance the Carboniferous Limestone in Somerset) the fauna may be identical to that of a hardground. However, if the older rocks are not calcareous, as in Normandy (where the Middle Jurassic rests on Lower Palaeozoic sandstones and shales), only the encrusting components of this hardground community may be present.

There are all gradations between a hardground community and Diverse Calcarenite Community (Fig. 78). A typical intermediate assemblage occurs in the Boueti Bed of Dorset, where the substrate of shell debris became firm enough for the attachment of abundant brachiopods (notably *Goniorhynchia boueti, Avonothyris, Ornithella* and *Dictyothyris*). Most of these brachiopods were encrusted by serpulids and bryozoans and byssally-fixed bivalves also flourished.

Figs. 81a and 81b show a flourishing hardground assemblage which was inundated by a clay influx. Clay influxes were the result either of increased rates of terrigenous deposition or of the sudden arrival of airborne tuffs from distant volcanoes.

82 Lime Mud Coral Community

In environments where salinities were normal or only slightly hypersaline, patch coral communities developed. The local colonization and proliferation of corals probably required relatively clear water with not too much suspended detritus. Some modern corals are able to cope with modest and temporary incursions of sediment into their habitats but continuous and copious amounts of turbid water normally cause their eventual demise.

Middle Jurassic coral faunas in central England are of low diversity, rarely containing more than three genera of colonial corals. *Isastrea* was the commonest genus in either its rounded *'Isastrea'* form or in the branching form that has often been termed *'Thamnasteria'*. The colonies of corals, like those of their

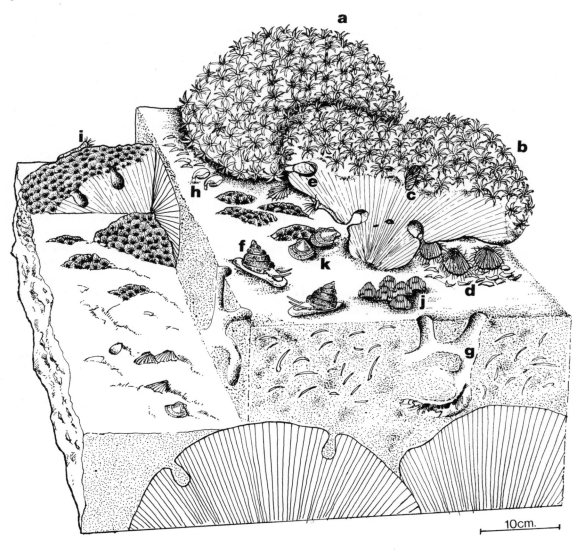

Fig. 82 Lime Mud Coral Community

a rounded *Isastrea* (Coelenterata: Anthozoa: Scleractinia)
b branching *Isastrea* (Coelenterata: Anthozoa: Scleractinia)
c *Eonavicula* (Mollusca: Bivalvia: Arcoida)
d *Plagiostoma* (Mollusca: Bivalvia: Pterioida)
e *Lithophaga* (Mollusca: Bivalvia: Mytiloida)
f *Pleurotomaria* (Mollusca: Gastropoda: Archaeogastropoda)
g *Thalassinoides* (trace-fossil) made by *Glyphaea* (Arthopoda: Crustacea)
h *Ornithella* (Brachiopoda: Articulata: Terebratulida)
i serpulid (Annelida)
j rhynchonellids (Brachiopoda: Articulata: Rhynchonellida)
k *Entolium* (Mollusca: Bivalvia: Pterioida — pectinid)

modern relatives, were made up of a number of carnivorous polyps. In between the coral fronds, 'nestling' in shell debris, were byssate arcid bivalves such as *Eonavicula*, which, like other arcids, probably had fairly short siphons and lived as a suspension feeder in the protected niches between the corals. There were also free-living or byssally attached limids (*Plagiostoma*) nesting beneath coral heads.

The coral skeletons themselves provided firm substrates in which *Lithophaga* bored, and in our sketch we have shown the nearer specimen cut away to reveal the living lithophages boring into the dead under-surfaces of the living coral heads.

The sediment consisted partly of shell debris derived from the degradation of local coral and other shell material. The rest was composed of fine grained micrite taken out of suspension by the molluscs and corals and deposited initially as pellets that are no longer preserved. The sediment thus had a muddy surface, and in places a Shelly Lime Mud Community (Fig. 79) became established (not shown here). Grazing gastropods, like *Pleurotomaria*, lived upon the surface, while within the soft sediment crustaceans such as the shrimp *Glyphaea* produced *Thalassinoides* burrows and *Favreina* faecal pellets.

In the coral rubble, too small to be shown in the picture, were encrusting organisms typical of hardgrounds. These included small calcareous sponges, serpulids, thecideidine brachiopods and bryozoans.

Coral communities with lime mud or clay occur in the Middle Jurassic of the Cotswold Hills and Oxfordshire.

MIDDLE JURASSIC ENVIRONMENTS IN EUROPE AND NORTH AMERICA

Southwards from Britain into western Europe, the faunal diversity gradually increased, particularly among the stenohaline groups of organisms. In France, ammonites and belemnites are found, accompanied by many other invertebrate genera seldom abundant in Britain. In general they occur in shallow water limestones with some sands and shales. In southern Europe (Fig. n, p. 206), however, the picture is totally different, fine grained pelagic facies being dominant and containing *Radiolaria,* planktonic foraminiferida, *Bositra* and ammonite aptychi.

These fine grained facies represent sediments that, for the most part, accumulated in relatively deep water upon collapsed contintal margins bordering the edges of the opening Tethys Ocean. Not all the sediments are basinal deposits, and at various times parts of the sea floor in this offshore area were uplifted above the level of the surrounding basins; under these conditions condensed and nodular ammonite-rich limestones formed. These condensed ammonite-limestones accumulated slowly upon starved submarine 'swells' and the rock produced is sometimes termed 'Kondensationskalk'.

The collapse of the Lower Jurassic carbonate platforms was not complete everywhere and isolated platforms sometimes released oolitic and other shallow water carbonate components. Periodically these were carried down by turbidity currents into the adjacent basins. In other areas, such as Spain and Sicily, the Middle Jurassic was marked by the eruption of submarine volcanoes.

The fauna of southern Europe differed from that of the north in being far more diverse. The ammonites show the greatest increase, in diversity, while many other invertebrate groups, including foraminifera, sponges, corals and coccolithophorids, also show a greater variety of species.

The northern part of the northern hemisphere, which included Britain and northern Europe with their relatively restricted faunas, was included in the Boreal Realm while the Tethyan faunas occupied the rest of the world (Hallam, 1972; 1975). The recognition of these major realms was based initially upon the differences between ammonites; subsequent work has shown that within other groups relevant differences also existed. Within these realms provincial differences can be seen but are sometimes subtle. Throughout the Jurassic, the dissimilarities between Tethyan and Boreal communities increased, so much so that Waagen wrote in 1864 almost in despair: 'The higher we climb in the Jurassic series, the greater become the difficulties, either of recognizing or separating individual beds, or of correlating' (Arkell, 1956).

For nearly a century opinions have differed, first over the recognition of 'Realms', and second over their cause. Arkell, in his masterly review (1956) did not actually put forward an explanation for faunal realms, but one is left with the impression that he favoured the theory of a broadly 'climatic' control. At that time the idea that continents were mobile was quite unfashionable and all his arguments and reviews of evidence indicating Jurassic climatic belts were based upon present-day continental distributions.

Today, knowing more about continental motions and the complex variables that control the distribution of plants and animals, we are too cautious to assume that a single environmental parameter, such as temperature, could be the primary controlling factor in faunal realm development and perpetuation. Climate was very relevant but other factors, such as the variability in the entire environment, are likely to have been as important, or even more so (Hallam, 1975).

Arkell realized that the gradual decrease in the amount of limestones in sediments from south to north across Europe was probably related to decreasing temperatures. He also noted that coniferous plants, at least in the north, showed well developed growth rings which could only be brought about by seasonal changes. However, as we stated earlier, no evidence has yet come to light of Jurassic glacial sediments on the earth even in the areas that are assumed, from modern geophysical studies, to have lain over the Jurassic poles (Fig. m, p. 205).

Modern ecological ideas suggest that environmental stability is a key factor in establishing a diverse population (Valentine, 1973). Where there are strong seasonal differences and high environmental stresses, as in shallow shelf seas, diversities are relatively low in comparison with those in adjacent oceanic areas where seasonal variations tend to have less effect. This may help to explain the diversity differences that occurred during the Jurassic between the Boreal faunas, which were essentially shelf faunas, and the Tethyan faunas that lived in the Tethys and the Super Pacific Oceans and on their margins.

In North America (Fig. m), the sea renewed its advance during the Middle Jurassic and spread over large areas of the Western Interior, probably from the north (Imlay, 1957). This seaway was partially enclosed, and at certain times dolomitic and gypsiferous beds were laid down as a result of evaporation. At the margin of the basin, the lagoonal and restricted sediments passed laterally into red beds that represent subaerial deposits, probably in arid and semi-arid continental areas. All the Middle Jurassic sediments of the Western Interior seem to have been deposited in very shallow water, and many of the limestone beds contain abundant molluscan faunas including *Liostrea, Gryphaea* and nerineid gastropods. Corals sometimes occur and in many cases the faunal assemblages seem to be comparable to those of the Middle Jurassic of Britain. The beds with evaporites tend to have extremely low diversities or, more often, no faunas at all.

During the Middle Jurassic, North American ammonite faunas became markedly differentiated into Boreal and Tethyan groups. As in Europe, the dominantly Boreal faunas occurred in the north while those with more Tethyan affinities occurred in the south (Imlay, 1965).

UPPER JURASSIC

The late Middle Jurassic witnessed a phase of deepening seas accompanied by marine invasion of the earlier Middle Jurassic lagoons and deltas. The resulting environmental changes led to the deposition of more uniform marine sediments over much of northwestern Europe and provided sharp environmental contrasts with those that had preceded them. Although major marine invasion began in the late Middle Jurassic, as defined by ammonite zones, for convenience we have included these latest Middle Jurassic facies within our Upper Jurassic section because of environmental comparability.

The initial phase of deepening produced a fairly widespread clay facies, the Oxford Clay, but this was succeeded by a temporary shallowing that produced sands and coral-rich limestones during the Upper Oxfordian (the Corallian Facies). Later, renewed deepening and further transgression spread relatively uniform clay facies, the Kimmeridge Clay, once more over much of Britain and northern France. During the latest period of the Jurassic a further and substantial regressive phase occurred and this led to uplift in northern Britain and to the development of extremely varied environments in the south that were comparable in many respects to those that had developed during the Middle Jurassic in central Britain.

Animals in stagnant basins (bituminous mud associations) (not illustrated)

Several times during the Upper Jurassic, bituminous mud facies spread extensively over north-west Europe. The facies is very similar to that developed in the Lower Jurassic (Fig. 63, p. 210), and are similar enough for a separate diagram to be unnecessary. However, it is worth while mentioning here the main differences between the Upper and Lower Jurassic bituminous mud associations.

The faunas were dominated by plankton, epiplankton and nekton with an absence of benthic forms. Whole bedding surfaces are often crowded with *Bositra* (a bivalve which may have been an active swimmer) and ammonites that have inevitably been crushed flat during compaction.

At times, and particularly toward the late Upper Jurassic, the minute planktonic algae called coccolithophorids were sufficiently abundant to produce coccolithic limestones. In the laminated and bituminous shales, where anaerobic bottom conditions excluded burrowing organisms, coccolithic laminae could sometimes develop. Coccolithic sediments were sometimes concentrated on submarine highs and this material was occasionally carried by turbidity currents into the stagnant basins.

Examples of Upper Jurassic bituminous shales are found in Britain on the Dorset coast, in Yorkshire and in west Scotland. In east Scotland, a series of Kimmeridgian sandy bituminous shales occur interbedded with spectacular conglomerates and breccias. These conglomerates contain a shallow water shelly benthos mixed with large fragments of Devonian and older rocks brought from further north; the conglomerates are slumped masses of shallow and near-shore sediments which slipped from the north down the scarp of a fault which was active during Kimmeridgian times.

83 Muddy Sea Floor Communities

Restricted clay communities were similar to their counterparts in the Lower Jurassic. Besides thin-shelled pectinids, protobranchs and lucinoids, they also contained certain specialized deep-burrowing infaunal bivalves (like *Thracia*) which, although suspension feeders, became adapted to life in environments with a low food supply from which other infaunal suspension feeders were excluded. Cerithid gastropods were still common as surface or shallow-infaunal detritus feeders and were often accompanied by *Dicroloma*, an aporrhaid gastropod with an inflated flange around its aperture. In modern aporrhaids this flange separates the inhalent and exhalent currents while the snails forage as scavengers or detritus feeders on the sea floor.

Some species of *Pinna* also spread into this facies while the large and inflated *Gryphaea dilatata,* and other oysters, were sometimes common. The *Gryphaea,* as well as being able to clap their valves free of mud, probably had a large gill area capable of driving powerful currents through their mantles. This would enable them to derive adequate food supplies from these tranquil waters while the inflated mantle margin efficiently separated their inhalent and exhalent currents.

During phases of slow deposition in relatively deep shelf waters, the sea floor often became colonized by thickets of sponges (*Rhaxella* and *Pachastrella*). These organisms were epifaunal suspension feeders consisting of a bag of cells supported by a siliceous frame and were able to live in the deeper waters from which other filter feeders were excluded by the low concentrations of food material. It is also possible that dense clumps of these sponges consumed so much of the food material suspended in the sea that other filter feeders were often excluded. After death, the

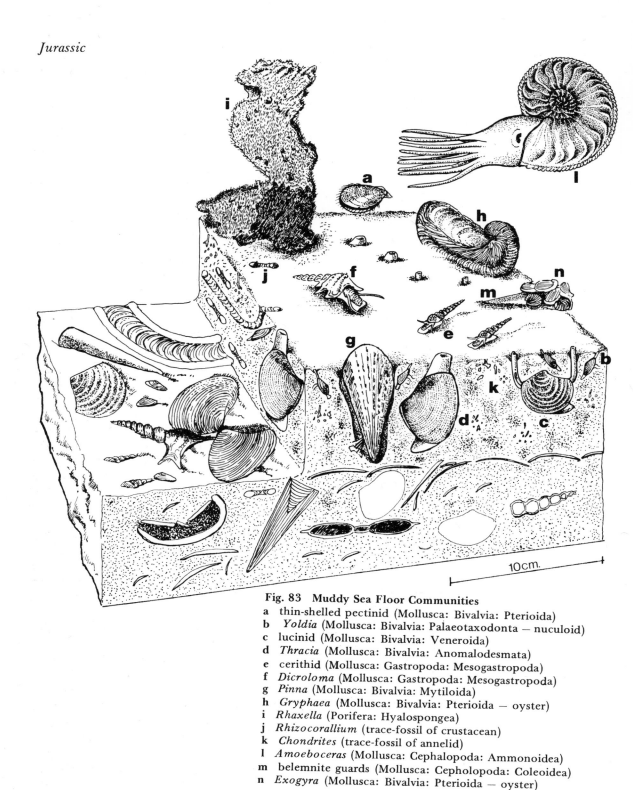

Fig. 83 Muddy Sea Floor Communities

a thin-shelled pectinid (Mollusca: Bivalvia: Pterioida)
b *Yoldia* (Mollusca: Bivalvia: Palaeotaxodonta — nuculoid)
c lucinid (Mollusca: Bivalvia: Veneroida)
d *Thracia* (Mollusca: Bivalvia: Anomalodesmata)
e cerithid (Mollusca: Gastropoda: Mesogastropoda)
f *Dicroloma* (Mollusca: Gastropoda: Mesogastropoda)
g *Pinna* (Mollusca: Bivalvia: Mytiloida)
h *Gryphaea* (Mollusca: Bivalvia: Pterioida — oyster)
i *Rhaxella* (Porifera: Hyalospongea)
j *Rhizocorallium* (trace-fossil of crustacean)
k *Chondrites* (trace-fossil of annelid)
l *Amoeboceras* (Mollusca: Cephalopoda: Ammonoidea)
m belemnite guards (Mollusca: Cepholopoda: Coleoidea)
n *Exogyra* (Mollusca: Bivalvia: Pterioida — oyster)

sponges collapsed and their siliceous spicules dropped down to the sediment which then constituted a spiculite.

Often, the only other organisms associated with these spiculites were deposit-feeding burrowers, particularly crustaceans which produced meandering horizontal 'U' burrows (*Rhizocorallium*) and mined the particularly nutritious layers of sediment for food (Seilacher, 1967).

Where hard substrates were afforded by the shells of dead ammonites, either clusters of small *Exogyra* (oysters) became established or knotted 'skeins' of the serpulid worm *Glomerula* produced small mounds on the sea floor before being overwhelmed with mud.

Examples of these clays are exposed in Britain on the Dorset coast, and in central England, Yorkshire and Scotland.

84 Marine Sand and Muddy Sand Communities

These communities bore some similarities to their earlier Jurassic counterparts and they often had a rather limited benthos. However, the sediments were often intensively bioturbated, particularly by crustaceans that produced *Rhizocorallium* and *Thalassinoides* burrows. These burrow systems made by infaunal suspension feeders were constructed within cross-bedded and rippled sands and this particular combination of burrows and sedimentary structures indicates deposition under relatively shallow, agitated water conditions. Occasionally another, very distinctive burrow-type was present — *Ophiomorpha* — a complex crustacean burrow similar in many respects to *Thalassinoides*, but with its various galleries lined with pellets that provide a nodular grape-like texture to the surfaces of the tubes after fossilization. In modern sediments, the shrimp *Callianassa* produces comparable burrow systems, particularly in inshore and shallow subtidal environments. In our diagram, we have not been able to indicate the fact that these burrows often extend over a metre down into the sediments. The vertical 'U' burrow *Diplocraterion*, which housed an infaunal suspension feeder may also be present.

Associations of rippled sands with clay laminations indicate alternate turbulent and slack water episodes. Such associations are found today in some tidal areas and Wilson (1968) has interpreted some of these Upper Jurassic associations in this light. It is probable that high rates of sedimentation in strongly fluctuating environmental conditions inhibited colonization of these substrates by a diverse shelly benthos. The sand depicted in our diagram represents one of the low-diversity assemblages. With slower rates of sedimentation, however, the sediments became more diversely

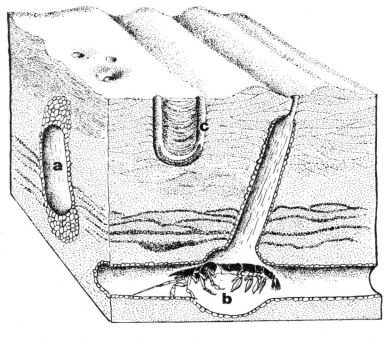

Fig. 84 Marine Sand and Muddy Sand Communities
a *Ophiomorpha* (trace-fossil) produced by
b a 'ghost shrimp' (Arthropoda: Crustacea); seldom preserved
c *Diplocraterion* (trace-fossil of annelid or crustacean; annelid shown)

⊢————— 3cm. ————⊣

inhabited. Deep-burrowing bivalves like *Pholadomya* and *Goniomya* appeared accompanied by the byssally-fixed *Gervillella*, which, in the uncemented beds, may all occur as moulds where the shell material has been dissolved.

Thalassinoides burrows may contain abundant small oysters (*Liostrea* and *Exogyra*) and serpulids (*Glomerula*) that may have lived within the burrows alongside the shrimps, or were perhaps collected as debris by the crustacean from the sediment surface. Other bivalves that lived in these slightly more diverse sand environments included the semi-infaunal *Pinna*, large *Liostrea*, the thick-shelled *Ctenostreon* cemented to other shells and the pectinids *Chalmys* and *Camptonectes*. The more diverse communities also included scavenging and detritus-feeding gastropods like *Pseudomelania* together with *Natica* which today is a carnivore.

Heart-urchins such as *Nucleolites* lived in shallow depths within the sediment and the fully marine characteristics of the higher diversity communities are confirmed by the presence of ammonites and belemnites.

Examples of these sands occur on the Dorset coast, in central England and Yorkshire and compose the reservoir of the Piper oil field in the North Sea.

85 Calcarenite Community

In the Upper Jurassic of England, clays, sands and limestones commonly occur in coarsening upward cycles which have often been interpreted as shallowing sequences. The limestones capping these cycles are often oolitic (Fig. 85a), and like those in the Middle Jurassic, many oolites contain few indigenous organisms. Burrows are sometimes present, particularly the straight vertical *Skolithos* which is perpendicular either to dune surfaces or to individual cross-bedded laminae. These burrows, along with occasional *Diplocraterion,* were produced by suspension-feeding organisms inhabiting very shallow turbulent waters. Another type of vertical 'U' burrow, *Arenicolites,* which lacks reworked spreite between the tubes, represents the burrow of a detritus feeder. Examples of these true oolite communities occur in Dorset and in central England.

Transported ooliths, however, are a common component in other Upper Jurassic limestones which developed adjacent to the sites of oolite formation, and which accumulated under more stable conditions and thus contain a richer fauna. Sometimes the ooid sand dunes became stabilized and were then colonized by an abundant and diverse fauna. Epifaunal bivalves took advantage of such surfaces, and these bivalves included large cemented oysters, the pectinid *Camptonectes* and the byssally attached clam *Isognomon,* this last being well adapted to life in turbulent conditions, first because of its thick prismatic shell and second because of its streamlined shape. In modern seas, this bivalve lives swinging freely from plants (guessed at in our diagram) and other exposed features on the sea floor (Kauffman, 1969). These epifaunal shells are seldom preserved in their original articulated form, but have usually been dispersed and broken by currents.

Patches of coral and clusters of the serpulid worm *Glomerula* also sometimes became established, while grazers were represented by the gastropod *Aptyxiella* and other snails.

Soft-bodied animals burrowed in the sediment, as did the bivalves *Myophorella* and *Trigonia*. These bivalves were sluggish shallow burrowers; although they lacked long siphons they were infaunal suspension feeders with a heavily ornamented shell which aided their burrowing and anchorage within the sediment (Stanley, 1970). Clusters of coral sometimes became established temporarily (not shown here) and these patches of coral along with shell debris were usually bored both by algae (on a fine scale) and by the mytiloid *Lithophaga.* Algal borings penetrated the surface of much of the shell material, as in modern clear tropical seas where algal infestation rapidly creates patinated surfaces on both living and dead shell material. The quick-burrowing *Thracia* and other siphonate suspension-feeding bivalves also indicate shallow water conditions with stabilized substrates.

Fig. 85 Calcarenite Community
a *Rhaxella* (Porifera: Demospongea)
b *Isognomon* (Mollusca: Bivalvia: Pterioida)
c *Trigonia* (Mollusca: Bivalvia: Trigonioida)
d *Camptonectes* (Mollusca: Bivalvia: Pterioida — pectinid)
e serpulids (Annelida)
f conjectural algal fronds
g *Arenicolites* (trace-fossil of annelid)

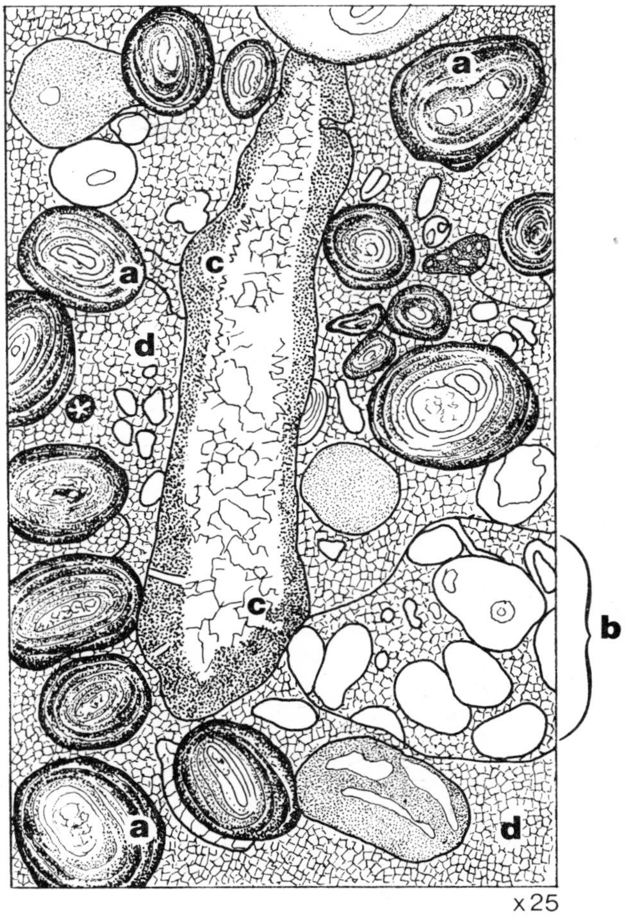

Fig. 85 a
Sketch of thin section of calcarenite
a ooids
b intraclast
c replaced shell
d calcite cement

×25

Modern oysters are tolerant of variable salinity, and Barthel (1969) has suggested that even Jurassic trigonids and *Pholadomya* were able to withstand salinity fluctuations (and particularly hypersaline conditions). In the highest Jurassic limestones of southern England, although the enormous titanitid ammonites occur, belemnites are never found, and it may well be that many of these limestones were deposited under hypersaline conditions to which some ammonites had adapted themselves.

A contribution to the sediment is made by the spicules of two kinds of siliceous sponge: *Rhaxella* and *Pachastrella*. These animals are seldom found intact, but when alive they may have acted as sediment baffles and trapped material. Sometimes whole areas became so infested with sponges that other suspension feeders were unable to compete.

Examples of this community occur in Britain on the Dorset coast, in central England and in Yorkshire.

265

Jurassic

BIOHERM COMMUNITIES (86–87)

Reefs are organically produced mounds which are resistant to wave and current action. To recognize an old reef we not only have to show that a patch of fossils produced a mound, but also that the mound was resistant to erosion. In the Upper Jurassic we can recognize a number of bioherms (organic mounds), some of which were predominantly corals and others that were produced by groups of organisms not usually thought of as reef builders (for instance sponges, bivalves and algae); some of these mounds were wave-resistant, others were not.

86 Oyster-algal Bioherms Community

Stabilized surfaces of ooid sand were sometimes colonized by communities of oysters (*Liostrea*) which, once established, often grew upon each other, either spreading out over the sea floor or upwards, thus producing shelly mounds. These thick-shelled epifaunal bivalves with their extremely high tolerance of salinity fluctuations were capable of producing bioherms of up to 2m in height. The oyster shells were often encrusted with other epifaunal suspension feeders that took advantage of the hard substrates. These other organisms included the oyster-like scallop *Plicatula*, serpulid worms and bryozoans. Red algae such as *Solenopora* were sometimes associated with these oyster bioherms, and the growth of algae and other encrusting organisms helped to bind the oyster mounds together. The crevices between the shells were filled with the debris derived from the shell and algal masses either by erosion or by predators feeding upon the shells.

The crevices with their fillings provided a new habitat for organisms including thecideidine brachiopods (too small to be shown). Lithophage and pholadid bivalves are commonly found in crypts penetrating both shells and laminated algal masses, and presumably the detritus provided by these borers was incorporated in the crevice fillings. The shells of grazing gastropods (*Aptyxiella* and *Pleurotomaria*) which fed upon the *Solenopora* and other algae occur mixed with the shell debris.

Surprisingly, although the foundation sediment for the bioherms was often sand grade oolite, the non-shelly interstitial sediment within each bioherm was commonly composed of lime mud (micrite). Modern oysters, like some other epifaunal bivalves, collect mud grade material as they feed and the discarded material is aggregated into sand grade pseudofaeces which rapidly decompose after deposition to produce mud. The micritic matrix in these

266

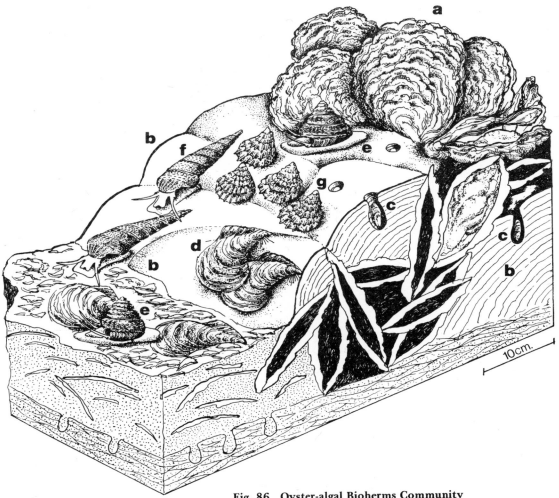

Fig. 86 Oyster-algal Bioherms Community
a   *Liostrea* (Mollusca: Bivalvia: Pterioida — oyster)
b   *Solenopora* (Algae)
c   *Lithophaga* (Mollusca: Bivalvia: Mytiloida)
d   *Isognomon* (Mollusca: Bivalvia: Pterioida)
e   *Pleurotomaria* (Mollusca: Gastropoda: Archaeogastropoda)
f   *Aptyxiella* (Mollusca: Gastropoda: ?Mesogastropoda)
g   *Plicatula* (Mollusca: Bivalvia: Pterioida)

Jurassic

Jurassic oyster-rich mounds may have been formed in a similar way.

Sometimes the perimeters of the bioherms were marked by shell debris from free-swinging epifaunal suspension feeders like *Isognomon* which were attached to the surface of the shell mounds. However, it is by no means clear whether these mounds were wave-resistant or not.

Examples in Britain of these oyster-algal mounds occur in Dorset.

87 Coral-algal Patch-reef Community

This community was dominated by massive and branching corals that produced an unbedded rock. The main genera of corals are *Thecosmilia* and *Isastrea* (with its *Thamnasteria* growth form) producing individual masses up to 1.5m in diameter. Like modern corals, these genera were adapted to life in shallow, well lit wave agitated waters. Comparisons between these Jurassic scleractinian corals and their modern descendants are fairly reliable, and it is quite probable that they were hermatypic, with zooxanthellae algae living within their tissues in a symbiotic relationship. Modern polyps are wholly carnivorous and feed upon microscopic zoo-plankton which they paralyse with the aid of stinging cells upon their tentacles. The coral polyps were never preserved as fossils, although the massive calcareous structures which they secreted are commonly found. However, their skeletons were often made of aragonite, and this mineral form of calcium carbonate is easily dissolved during diagenesis. Thus, rocks that were originally coral-rich are now cavernous or the cavities have subsequently been filled with sparry calcite, with the corals as mounds.

Modern corals are particularly intolerant of reduced salinities but can withstand high salinities (over $40^\circ /00$) and although some larger polyps can remove irritating sand grade sediment with their tentacles many modern genera are unable to withstand very turbid habitats. The tropical reef-builders also require warm ($18-36^\circ$ C) waters and firm substrates on which to grow. Diverse coral communities are found in shallow situations fringing the open sea and communities become more restricted towards lagoonal environments where food supply is less and where the seasons are most strongly marked by environmental fluctuations.

The organisms associated with the Jurassic patch reefs included many forms with modern analogues. The coral heads were often

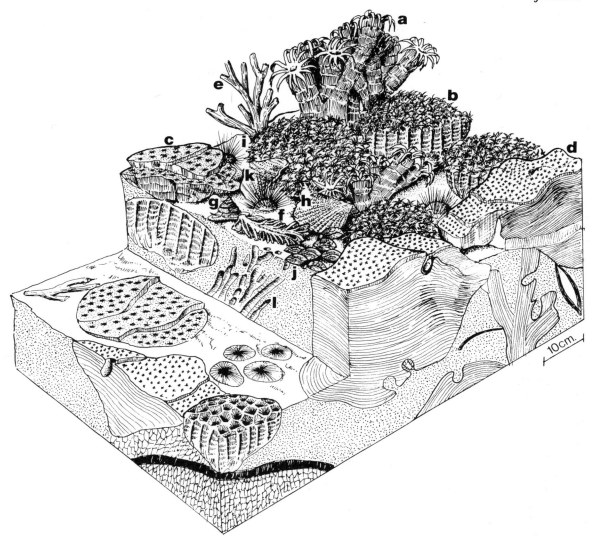

Fig. 87 Coral-algal Patch-reef Community

a *Thecosmilia* (Coelenterata: Anthozoa: Scleractinia)
b *Isastrea* (Coelenterata: Anthozoa: Scleractinia)
c *Thamnasteria arachnoides* (Colenterata: Anthozoa: Scleractinia)
d *Thamnasteria concinna* (Colenterata: Anthozoa: Scleractinia)
e *Rhabdophyllia* (Coelenterata: Anthozoa: Scleractinia)
f *Lopha* (Mollusca: Bivalvia: Pterioida)
g trochid (Mollusca: Gastropoda: Archaeogastropoda)
h *Chlamys* (Mollusca: Bivalvia: Pterioida — pectinid)
i *Cidaris* (Echinodermata: Echinozoa — sea urchin)
j terebratulids (Brachiopoda: Articulata: Terebratulida)
k bryozoan (Bryozoa: Ectoprocta)
l *Cladophyllia conybeari* (Coelenterata: Anthozoa: Scleractinia)

drilled by boring bivalves (*Lithophaga*), algae, bryozoans and sponges (clionids). Other organisms, particularly pectinid bivalves and thecideidine brachiopods inhabited fissures between the coral heads. Nests of terebratulid brachiopods occasionally became established, probably growing within crevices amongst the older, dead portions of the coral patches. This phenomenon has been observed in modern coral associations by Jackson et al. (1971).

In modern reefs and patch reefs, waves and predators (particularly fish) attack the corals and produce sand grade sediment which helps to fill the fissures between the coral heads. The Jurassic coral patches were similarly attacked and both fissures and patch-reef margins were marked by coral detritus. These fragments were frequently encrusted by suspension-feeding serpulids and *Exogyra* oysters which could not establish themselves on the living coral heads because the carnivorous coral polyps preyed upon their spat.

The dead portions of modern patch reefs are frequently overgrown with red and blue-green algae and bryozoans which help to bind the frame of the bioherm. In the Jurassic patch reefs much of the shell material is riddled and sometimes overgrown with algal infestations.

Other common predators on modern reefs are asteroid starfish, but although known in the Jurassic these organisms have a low preservation potential and are found only exceptionally as fossils. However, algal grazers are frequently represented by the thick-shelled gastropod *Bourgetia* and by trochids.

The detrital sands derived from reef erosion provided habitats similar to those of the stable ooid sands (Fig. 86) and in them a variety of infaunal burrowers are found, including the carnivorous sea-urchins *Cidaris, Nucleolites* and *Pygaster*. These beds, unlike the massive patch-reef limestones, sometimes contain ammonites.

The diversity of coral communities generally increased southwards into Europe and decreased northwards into Scotland where only *Isastrea* occurred. This distribution probably reflected a combination of climatic and ecological factors including increasing turbidity, salinity fluctuations and climatic instability towards the north. Influxes of volcanic ash sometimes killed off fauna on parts of the sea floor and the resulting clay blankets help to show the low relief of the sea floor even when it was covered by flourishing growth.

Examples of the patch-reef communities occur in Britain in Dorset, central England and Yorkshire and in east Scotland blocks of slumped coral occur in basinal Kimmeridgian shales.

Oyster Lumachelle Community (not illustrated)

Oyster communities which did not form part of patch reefs occurred high in the Jurassic and sometimes mark the base of the Cretaceous (Casey, 1971) in southern Britain. These communities were dominated by *Liostrea* which occur either fragmented into single disarticulated valves or as complete individuals in position of growth. Assemblages dominated by oysters often occur in reduced salinity and also indicate deposition at very shallow depths (Kauffman, 1969). These epifaunal suspension feeders occur in beds where there is a matrix of limey clay between the shells, and this sediment is probably the remains of pseudofaeces deposited by the animals (see Fig. 86, p. 267).

Oysters often comprised more than 90 per cent of the fauna; other organisms included encrusting suspension feeders such as serpulids, and infaunal mobile bivalves like *Trigonia* and *Protocardia*. Thick spines of the regular echinoid *Hemicidaris* are sometimes found and this scavenger probably lived upon the material from dead bivalves. The presence of *Trigonia* and *Hemicidaris* indicates that salinity was probably no lower than in normal marine environments, but the scarcity of other marine organisms suggests that they were affected by salinity, perhaps hypersalinity. The oyster beds often show internal low-angle cross-bedding which may be the original growth surfaces. The topmost surfaces of the beds are richer in articulated valves. Epifaunal bivalves rapidly disintegrate after death as the ligament rots, and it is not entirely unexpected that many individuals in these beds are often lacking their upper valves. Many specimens have also suffered disturbance of their lower valves after death, either from current or biological action, but although some movement of the valves has occurred, often the ratio of left to right valves in a single unit is approximately the same, indicating only limited disturbance.

As we have seen earlier (Fig. 86), *Exogyra* beds may occur in deeper water clay-rich sediments where substrate conditions restricted the other faunas. The beds in which these deeper water *Exogyra* communities occurred were frequently dolomitic (Townson, 1975), and heavy brines flowing down from shallow water may also have played a part both in restricting the fauna and in causing dolomitization of the sediment.

Examples of these oyster communities occur in Britain in Dorset and also, rather poorly exposed, in central England.

LAGOONAL COMMUNITIES (88–90)

In southern Britain, the Jurassic ended with a phase of lagoonal conditions in which a variety of communities lived in habitats ranging from dominantly marine or hypersaline to wholly fresh-water. Some of the oyster and oolite communities already considered are also relatively restricted.

88 & 89 Intertidal and Subtidal Algal Mat Communities

Some Late Jurassic formations contain laminated lime mud result-ing from the growth of blue-green algae (e.g. *Girvanella*). These algae trapped and bound mud grade carbonate in sheet-like mats.

Laminated lime muds (micrites) contain scattered shells, particularly those of minute bivalved crustaceans (ostracodes) which must have teemed in the shallow water above the mats. Often, the mats were full of burrows, particularly *Thalassinoides* produced by crustaceans, and the burrowing was sometimes so in-tense that only remnants of algal laminations remain. Fragments of the burrowing crustacean *Callianassa* are found associated with *Thalassinoides*. Modern intertidal algal mats (Fig. 88) are high stress environments, and burrowing animals that destroy algal lam-inations are progressively eliminated towards the highest intertidal regions, particularly in arid climates.

Where the lamination is best preserved, a series of polygonal cracks may be seen on the bedding surfaces (Fig. 89e), and these desiccation cracks indicate exposure to the sun at the time of de-position. The algal laminations often contain gypsum pseudo-morphs now replaced by calcite (Fig. 89f). The original gypsum crystals were probably also produced during phases of desiccation.

Gastropod, bivalve and, rarely, ammonite shells are embedded within the porcellaneous sediment. The bivalves include the in-faunal deposit feeder *Nucula* and the infaunal siphonate suspension feeder *Pleuromya*. Weakly siphonate or non-siphonate mobile sus-pension feeders included *Trigonia*, *Isocyprina* and *Protocardia* while the epifaunal bivalves consisted of pectinids (*Camptonectes*), rare oysters and byssally fixed *Mytilus*. Grazers and scavengers were represented by gastropods (*Procerithium* and *Aptyxiella*).

This association of organisms contains too many marine forms for the habitat to have been isolated from the sea; however, the association of marine organisms, including ammonites, with sedi-mentary features indicating exposure (sun cracks and gypsum pseudomorphs) requires some explanation. There are few normal marine environments in which desiccation occurs; in general the only marine environments which habitually suffer exposure are intertidal ones. Where carbonate deposition occurs on modern coasts algal mats are a feature of the intertidal zone (Ginsburg, 1975). Here, the mats below the low tide level commonly suffer disruption by *Callianassa* whereas in the intertidal and supratidal zones, little burrowing occurs, particularly if evaporation is causing concentration of marine salts.

In the well laminated Jurassic micrites, the shells are usually disarticulated and often occur in layers, which implies that they were drifted. However, in the bioturbated micrites the bivalves are frequently preserved in their growth positions. Thus it may be that the Upper Jurassic laminated micrites are inter- and supratidal algal flat deposits which were periodically inundated by storm tides that carried and dumped reworked shells. The bioturbated micrites probably indicate the tidal channel and shallow subtidal zones from which the shells were derived. The gastropod grazers, like their modern counterparts, may have been able to withstand limited phases of desiccation.

The aragonite-shelled fauna has mostly been dissolved, leaving only a rock with cavities showing both the internal and external moulds of the gastropods, bivalves and ammonites.

Examples of this community are found in Britain on the island of Portland and in other parts of Dorset (e.g. Purbeck).

Hypersaline Lagoon Communities

The fauna and flora of the hypersaline lagoons were extremely restricted. The sediments consist of laminated, stromatolitic lime mud which may have been associated with evaporite minerals. The salts have often been removed by solution, sometimes resulting in collapse of the beds (West, 1960, 1964, 1975). As well as the laminations produced by blue-green algae (Fig. 89c, e and f), there were large numbers of small pellets probably produced by grazing gastropods and crustaceans (Fig. 89a and d). In the diagrams we show a general view of the reconstructed environment with broad low-lying hinterlands basking under a blazing sun. The supratidal area would have been very similar to that of today in the Persian Gulf (Fig. 88). Even if tidal movements were small, the low relief would have provided wide intertidal areas which would have been colonized by blue-green algae. Cutting through this intertidal area would have been tidal channels (Fig. 89). The insets show various surface features found in modern sabkha environments in the intertidal zone and their general location in the pattern: the entrance to a crab burrow (a) with a mound of pellets (a, d); an algal surface near a channel margin with the trails produced by grazing cerithid gastropods (b_1); the entrances to *Callianassa* burrows from

Jurassic

Fig. 88 View from the sea of a salt flat (sabkha) coast near Abu Dhabi, on the south shore of the Persian Gulf. The background is several km from the viewer

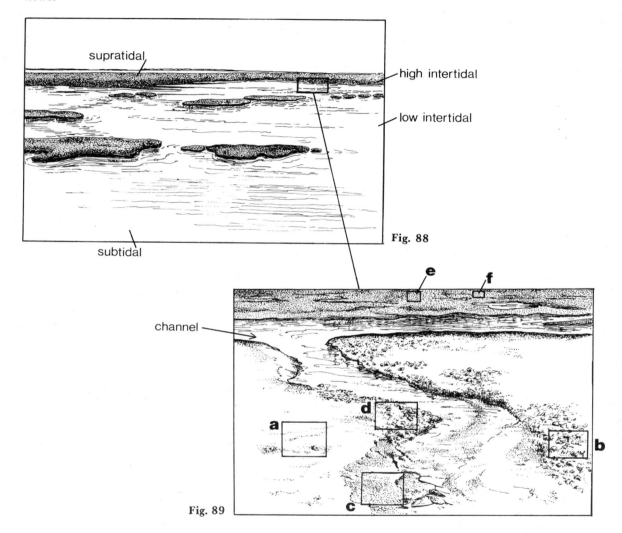

supratidal

high intertidal

low intertidal

subtidal

Fig. 88

e

f

channel

d

a

b

c

Fig. 89

the same zone (b_2); the irregular surface of the algal flat with the leathery algal crusts partially broken by contraction during desiccation (c and e); while f shows the type of internal structure obtained on digging a section through the mat. All the surface pictures are based upon photographs from the present-day sabkha

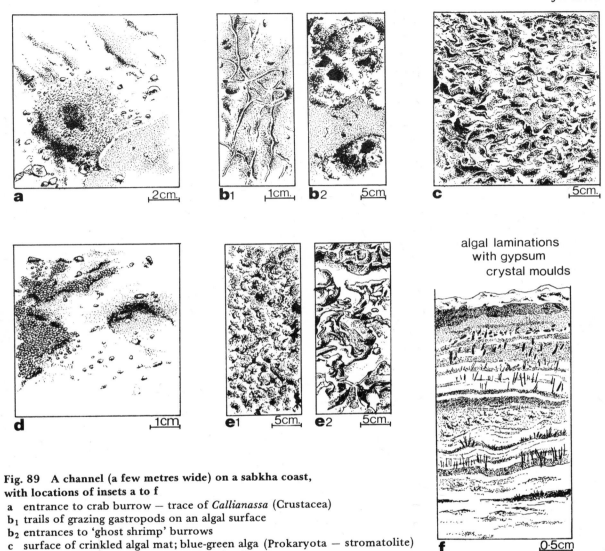

algal laminations
with gypsum
crystal moulds

**Fig. 89   A channel (a few metres wide) on a sabkha coast,
with locations of insets a to f**

a   entrance to crab burrow — trace of *Callianassa* (Crustacea)
b₁ trails of grazing gastropods on an algal surface
b₂ entrances to 'ghost shrimp' burrows
c   surface of crinkled algal mat; blue-green alga (Prokaryota — stromatolite)
d   surface covered with crustacean faecal pellets
e₁ pustular algal mat
e₂ algal mat cut by desiccation cracks
f   stromatolitic laminations with gypsum pseudomorphs and vugs

at Abu Dhabi on the Trucial Coast while the section (f) is based
upon a specimen from the Purbeck. Structures comparable to the
modern ones (a—e) can be seen in parts of the Purbeck Formation
in Dorset.

275

90 Freshwater Lagoon Communities

As the marine influence decreased, so the lagoons of the Upper Jurassic began to contain a variety of bivalves, gastropods, ostracodes and vertebrates which were adapted to life under freshwater conditions. The faunas often have modern counterparts and descendants and their modes of life can be interpreted fairly confidently. The dominant benthic animals were the molluscs, particularly the suspension-feeding bivalve *Unio* which, like its ancestors in the Middle Jurassic (Fig. 73, p. 235) and its modern descendants, probably lived free on stream and lake beds. *Unio* was a mobile bivalve which lacked siphons and its modern descendant is usually found slightly covered by deposited sediment. It has a thick proteinaceous outer covering to the shell (the periostracum) which protects the shell from solution by running fresh water.

Other molluscs associated with *Unio* were mostly gastropods. These included the mucosal filter feeder *Viviparus* (see Fig. 73) and a variety of other genera that were predominantly grazers. Many of these gastropods were pulmonate, lacking gills but breathing air through a mantle modified into a simple lung. This is a common feature of modern pond snails. Pulmonates included the smooth planispiral *Planorbis*. The smooth-shelled ostracodes complete the list of freshwater invertebrates which are preserved as fossils.

Fossils of the vegetation upon which the gastropods and other soft-bodied grazers fed include fragments of the calcareous algal *Chara*, and presumably other forms of aquatic vegetation were present but apart from spores these have not been preserved. Diverse spores and pollen assemblages may be found in these sediments after the matrix has been dissolved with strong acids.

Many of the freshwater beds contain vertebrate remains. Typically aquatic forms include a variety of crocodiles (*Goniopholis* and many others) some of which were small genera a mere 450mm in length (e.g. *Nannosuchus*). Larger vegetarians were represented by turtles (*Tertosternum*) and dinosaurs (*Iguanodon*). Some beds were traversed by the footprints of these large reptiles.

A large fauna of ganoid fish has been recorded and these presumably formed the stable diet of the crocodile population.

In addition, insects are often represented by fragments in a remarkable state of preservation. These include the wings and wing-cases of beetles, cockroaches, grasshoppers, dragonflies and various other types. The insects were the staple diet of the shrew-like mammals *Amblotherium* and *Triconodon*.

The environments suggest a low-lying swampland probably similar in some respects to the modern Everglades in Florida, with a wealth of freshwater habitats at or very close to sea level; hurricanes and storms sometimes temporarily spread sea-water many

Fig. 90 Freshwater Lagoon Communities

a   *Iguanodon* (Vertebrata: Reptilia: Archosaur — dinosaur)
b   *Amblotherium* (Vertebrata: Mammalia — panthothere)
c   *Goniopholis* (Vertebrata: Reptilia: Archosauria — crocodile)
d   *Equisetites* (Pteridophyta: Calamites — horsetails)
e   cycads (Gymnospermae)
f   surface covered with blue-green algae partially sun-dried (stromatolitic)

kilometres over the marsh. In the freshwater regions of the Everglades the dominant floras are grasses, algae, cypresses and
occasional 'hammocks' of dense mahogany forest. Of this flora,
only the algae and conifers were represented in the Jurassic.

UPPER JURASSIC COMMUNITIES IN EUROPE AND NORTH AMERICA

Through the Jurassic, facies distributions were controlled by the
interplay between two dominant factors: first a eustatic rise in sea
levels, and second the regional uplift or down-warp of certain areas
of the earth's crust. In the Upper Jurassic, continuing eustatic rise
and a diminution in the supply of coarse clastic sediments from an
ever-reducing land area led to the spread of calcareous deposits
northwards over Europe. Terriginous clastic material deposited in
Britain and in the North Sea was derived only from local and
ephemeral island areas (Fig. n, p. 206).

In Northern Europe the later Upper Jurassic (Kimmeridgian)
is represented by an extremely widespread and uniform facies of
shales that are often bituminous and appear, in the North Sea at
any rate, to be the main source rocks for the 'North Sea oil' being
piped ashore at the moment. These fine laminated shales accumulated in anaerobic basins that were probably not very deep but
were bounded by active faults. Great slides of debris occurred
periodically, bringing terriginous clastic material, and sometimes
shallow marine shelly communities, into the slightly deeper areas.

The late Jurassic was a time of sponge reef development in
Bavaria, barrier and back reef lagoonal environments becoming
established there. Coral and hydrozoan reefs also developed at
this time, and coccoliths became more diverse. The development
of coccoliths caused some bizarre facies like that of the Solnhofen
Limestone, which formed in some inter-reef areas as very fine-
grained and laminated micrites. The Solnhofen fauna is sparse, but
diverse, and it includes some of the most spectacularly preserved
fossils ever found. These include ammonites, insects, dinosaurs,
shrimps, and of course the earliest fossil bird (*Archaeopteryx*). It
appears that the Solnhofen Limestone was another type of anaerobic sea bottom; this would account for the splendid preservation of
fossils and laminations alike. It seems that the deaths of at least
some of the swimming organisms were caused by coccolithophorid
and dinoflagellate blooms poisoning the water. Rapid deposition
helps to account for the good preservation of very active animals
like the dinosaurs and arthropods.

Further to the south in Europe, the Upper Jurassic was characterized by a series of pelagic deposits that would have been similar

to our modern-day 'oozes' when they were first laid down. Now they have become massive and bioturbated muddy or chalky limestones that sometimes contain irregular beds of chert. In many parts of the 'Alpine Belt', late Jurassic pelagic deposits of a different type occur; these are nodular, sometimes red, limestones full of corroded ammonites. In Italy this rock type is known as the Ammonitico Rosso and the term has come to be applied to the facies wherever (and, often erroneously, whenever) it appears throughout the belt. These ammonite-rich beds are often extremely condensed, and rich in the bivalve *Bositra*. They seem to represent isolated sea mount deposits where current activity was too strong to allow the accretion of coccolithic material and only cephalopod shells and specialized bivalves could accumulate in any number.

In the Western Interior basin of the United States, the sea retreated at the very end of the Jurassic from a huge area of Colorado, Utah, Wyoming, Montana, New Mexico and Arizona. Here the sediments consist of sandstones, with interbedded green clays (bentonites) containing an entirely terrestrial or freshwater fauna famous for its dinosaurs, and containing also turtles and mammals. These beds occur almost at the top of the Jurassic sequence. They were deposited during a time of world-wide facies changes taking place towards the end of Jurassic time. Until the Oxfordian-Kimmeridgian period, world-wide sea levels had been rising, but the Jurassic was brought to a close by an important phase of regression that was associated with a major series of earth movements. The terrestrial Morrison Formation with its impressive dinosaur faunas succeeds a shallow marine phase of Upper Jurassic deposits in which beds contain many ammonites similar at the generic level to those in northern Europe.

The late Jurassic was also a time when the marine faunas of the world showed striking provincialism in that the division between the two major marine 'Realms' of the Tethyan and Boreal persisted. The Tethyan Realm included the ancestral Pacific margins as well as the traditionally Tethyan Alpine-Himalayan Belt, while the Boreal Realm mostly included the areas of the inland or epeiric seas covering the great northern continent (Fig. m, p. 205). A most likely explanation of this provinciality is that the shelf seas provided less stable habitats than the oceanic areas, and thus supported less diverse assemblages of more tolerant organisms. This may explain the great dominance of bivalves in the Boreal benthos, and the reduced diversity of groups like the brachiopods (Hallam, 1975).

Cretaceous

In broad terms, the faunas of the Cretaceous were very similar to those of the Jurassic. In marine environments, there were new families of ammonites, including heteromorphs (where the coiling of the whorls depart from the regular planospiral), and two important groups of bivalves: the rudists, which were important rock builders in tropical regions, and the inoceramids, which were among the commonest bivalves in many environments. On land, although many Jurassic plants continued to flourish, the development and radiation of the angiosperms (the flowering plants) was important during the later part of the period.

During the Cretaceous, the area of Britain and western Europe lay between 30° and 45° N (Smith and Briden, 1977). As in the Jurassic, the now shrinking Tethys Ocean lay to the south (Fig. o) but to the west, lay the Atlantic. In the southern Atlantic initial rifting led first to the development of deep salt basins which were destroyed by marine transgression during the Albian (see table below). Some time between 150 to 80 million years ago (Triassic

Table VIII *The stages of the Cretaceous Period*

Upper Cretaceous:	Maastrichtian Campanian Santonian ⎤ Coniacian ⎦ Senonian Turonian Cenomanian
Lower Cretaceous:	Albian Aptian Barremian Hauterivian Valanginian Berriasian/Ryazanian

Fig. o. *The world during the Cretaceous. Positions of the continents after Briden et al. 1974.*

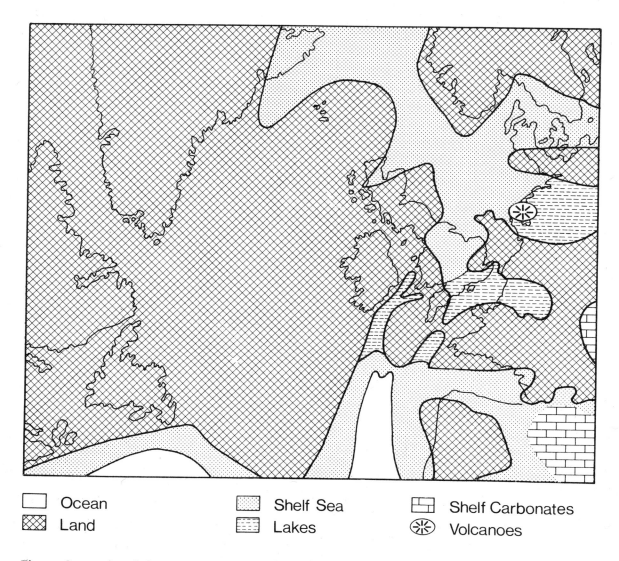

	Ocean		Shelf Sea		Shelf Carbonates
	Land		Lakes		Volcanoes

Fig. p. *Geography of the North Atlantic region during the Early Cretaceous (Berriasian). Later in the Cretaceous the shelf seas became more extensive. (After Hallam and Sellwood, 1976).*

or early Jurassic) there was both a break-up of the North Atlantic continents and a rotation of the Iberian Peninsula, which caused the opening of the Bay of Biscay (see Wilson, 1975, with bibliographic references). Reconstructions show, however, that even during late Cretaceous times, the west of Ireland was situated only a short distance east of Newfoundland (about 900–1200km). Rapid spreading, leading to the development of the present North Atlantic seaway was thus a late Cretaceous and Tertiary event, as is shown both by direct sea floor spreading data (e.g. magnetic stripes on the ocean floor) and by the widespread latest Cretaceous and Tertiary volcanicity in Ireland, Scotland, Greenland, Iceland and the North Sea.

What is now western Europe was thus at the dawn of the Cretaceous a more or less landlocked epeiric sea (Fig. p), whilst at the close of this period, there was an open, if narrow, ocean to the west. These changes in fundamental geography were accompanied by major sea level fluctuations, perhaps themselves a result of fluctuating sea floor spreading, and minor tectonic activity.

At the close of the Jurassic, the areas of shallow current-swept marine sands in Norfolk and Lincolnshire which are now the lower parts of the Sandringham Sands and the Spilsby Sandstones (Casey, 1973) were separate from the lakes, lagoons and supratidal saline flats of the Purbeck basin to the south. At the beginning of the Cretaceous, a brief marine incursion spread across the English Midlands and linked the two areas, leading in southern England to the development of oyster reefs in what became, albeit briefly, saline bays and estuaries. This incursion was followed by a return to brackish and fresh water conditions in the south, although marine conditions continued in the north, and the facies and broad faunal assemblages of the earliest Cretaceous are comparable to those of the latest Jurassic. This period was terminated by a phase of regional uplift which led to the rejuvenation of rivers and the building up of thick sequences of mud plain, delta and lacustrine deposits.

In the early Aptian, fully marine faunas in the Lower Greensand Group show that the sea covered parts of southern England, This transgression marked the earliest phase of one of the greatest known periods of submergence of continental areas, a fact recognized first by Edouard Suess (1885—1901). This transgression started in the Barremian, proceeded most rapidly in the Albian and Cenomanian, and extended into the late Campanian with only minor interruptions and regressions. As a result, the geography of Europe changed from the landlocked basins dominated by terrigenous clastic influences, which characterized the beginning of the period, to a wide-spread epicontinental sea with only scattered islands emerging from it (Fig. o). Lower Cretaceous sediments are largely terrigenous; those of the Upper Cretaceous are predominantly pelagic carbonates: the chalks (Latin — *creta*) which gave their name to the system.

The disposition of basins and massifs in western Europe during the Cretaceous was broadly similar to that of the Jurassic, and was a consequence of the geological history of the region extending back to the Precambrian (Sutton, 1968). During the mid to late Cretaceous, these massifs dominated sedimentation, facies and hence faunal distributions in several ways. Many were islands,

283

standing above the Cretaceous sea for the whole period while others were encroached upon and submerged as the sea transgressed. Some, such as the Anglo-Brabant Massif, influenced sedimentation long after their submergence and relatively deep burial.

Between these island-massif areas were basinal regions which were permanently submerged from mid Cretaceous times onwards, and subsided evenly so that the Cretaceous sedimentary successions are complete. In some areas, subsidence exceeded sediment supply, and there was actual deepening. There are two major basinal areas in north-west Europe: the North Sea Basin, bounding an area of continuous subsidence since the Jurassic and regarded by some as a failed rift system, and the Anglo-Paris (or, more correctly, Wessex-Paris) Basin to the south. Within basins and on the submerged flanks of massifs are areas of minor variation in thickness which seem to be the result of basement 'highs' movements of salt plugs, and later displacement along pre-Mesozoic faults and fold axes. These areas are generally marked by thinning, condensation and re-working, as they were in the Jurassic (Sellwood and Jenkyns, 1975).

During the early Cretaceous, there was a limited tectonic activity in Europe (see Ziegler, 1975, 1975a); in the Upper Cretaceous it was minimal.

BIOGEOGRAPHY

Throughout much of the Cretaceous, as in the Jurassic, the whole of Britain and much of western Europe fell within the Boreal Realm which was characterized by the rarity or absence of a number of major invertebrate groups including many important ammonite families, larger Foraminiferida and rudistid bivalves (Casey and Rawson, 1973; Hallam, 1973). The composition of the faunas of the Boreal Realm was probably not controlled simply by temperature and salinity, but more likely by prevailing patterns of environmental stability. The Boreal Realm was probably an area of high stress, with fluctuations of temperature, salinity and environmental energy while the Tethyan Realm had relatively stable environments. This is supported by the known palaeogeography of the Jurassic and early Cretaceous and the differences in faunal diversities and densities between the two realms. These realms persisted throughout much of the early Cretaceous, but from the Cenomanian onwards the differences declined.

STRATIGRAPHY

The standard stages of the Cretaceous are summarized above. As has been mentioned, the ammonite faunas of western Europe were restricted, and, for much of the interval, correlation with areas outside this region is difficult, particularly during the early Cretaceous. An added difficulty in the late Cretaceous is the absence of rarity of ammonites in chalk facies, due either to their ecological exclusion, or to pre-burial dissolution of their skeletons. In consequence, the standard Upper Cretaceous zones are based on various groups, including echinoids, crinoids, belemnites, bivalves and others. In non-marine facies, subdivision and correlation is based primarily on ostracodes and plant microfossils.

LOWER CRETACEOUS

91 Lower Cretaceous Terrestrial Communities

In parts of northern Europe the end of the Jurassic witnessed a phase of marked earth movements which caused some of the old subsiding marine basins to become swamps and marshlands. These early Cretaceous flood-plain regions were periodically inundated by seasonal rivers. In shallow pools along the watercourses and in the waterlogged parts of the flood plains horsetails (Equisetales) grew. One genus of this group of pteridophytes (*Equisetum*) still survives, and like its early Cretaceous ancestors, possesses a ramifying underground stem and tuber system which binds the sediment. It is not particularly tolerant of salt water and generally grows in fresh water on acid soils, in water depths of less than about 1m. In places, perhaps where the early Cretaceous soils were better drained, tree-ferns like *Tempskya* grew. The horsetails with their coarse foliage seem to have been the staple diet of the larger herbivores like *Iguanodon*. In Belgium, fissure fillings have been discovered containing a number of *Iguanodon* that were trapped as a group, suggesting that perhaps the animals were gregarious. These slow, lumbering dinosaurs contrasted with the smaller and more agile bipedal dinosaurs like *Hypsilophodon*, which were also herbivores and may have been able to climb trees. Flesh-eating dinosaurs were represented by *Megalosaurus* which were probably slow moving carrion feeders.

The passage of *Iguanodon* herds and of solitary megalosaurs was recorded on soft surfaces by footprints and occasionally by the impressions of their hides. In the skies above there were

Fig. 91 Lower Cretaceous Terrestrial Communities

a *Iguanodon* (Vertebrata: Reptilia: Archosaur — dinosaur)
b *Megalosaurus* (Vertebrata: Reptilia: Archosaur — dinosaur)
c *Hypsilophodon* (Vertebrata: Reptilia: Archosaur — dinosaur)
d *Acanthopholis* (Vertebrata: Reptilia: Archosaur — dinosaur)
e *Equisetites* (Pteridophyta: Calamites — horsetails)

pterosaurs and birds. The terrestrial invertebrates included insects, other arthropods and worms; they were a source of food for mammals similar to those of the latest Jurassic. The pools sometimes supported unionid bivalves, ostracodes and rarer pulmonate gastropods.

Early Cretaceous sediments in this 'Wealden' Facies are found in south-east England (the Weald), the Isle of Wight, the Isle of Purbeck and under the Celtic Sea. On the continent of Europe a 'Wealden' facies flanks the southern margin of the Ardennes massif and runs south from Belgium. Current ideas about the 'Wealden' depositional environment have been published by Allen (1976).

92 Lower Cretaceous Lake Communities

The waters of the lakes of the Wealden basins of southern England lapped against shorelines of low-lying coastal plains like those represented in the previous figure. These lakes fluctuated markedly in salinity, with indications of fresh, brackish-marine and possibly coastal salt marsh conditions at times. Kilenyi and Allen (1968) noted two broad molluscan assemblages: a freshwater one dominated by the gastropod *Viviparus* but also including unionids, and a much less common assemblage with *Filosina, Cassiope, Melanopsis, Nemocardium, Ostrea* and other forms. Locally, as at the top of the Wealden shales (Barremian) echinoid debris is recorded (Casey, 1961) whilst some of the subordinate sandstones in the Weald Clay have yielded the trace fossil *Ophiomorpha*, the burrow of arthropods (normally considered to be marine). The former freshwater assemblage is usually associated with plants known to be intolerant of saline waters, including stoneworts (*Cirtonitella*), liverworts (*Hepaticites*) and horsetails (*Equisetites*), which are sometimes found with in situ rootlets and rhizomes, (Batten, 1975), whilst work on oxygen and carbon isotopes (Allen et al. 1973) and on microfaunas, notably ostracodes, has confirmed that this was a freshwater environment. Freshwater Wealden microfaunas are dominated by the weakly ornamented genus *Cypridea*, and the brackish water faunas by a 'non-*Cypridea*' assemblage of *Theriosynoecum, Mantelliana, Rhinocypris,* darwinulids, and *Dicrorygma* (Anderson, 1963, 1967, Kilenyi and Allen, 1968). Salinity fluctuations are recorded in the lake clays of the Wealden; but in the sandier Hastings Beds these fluctuations appear to have been lower. Definite salinity cycles are indicated by the following:

Cassiope (a gastropod; most marine) — *Filosina* (bivalve) — *Paludina* (gastropod) — *Cypridea* (ostracode) — *Equisetites* (horsetails; freshwater).

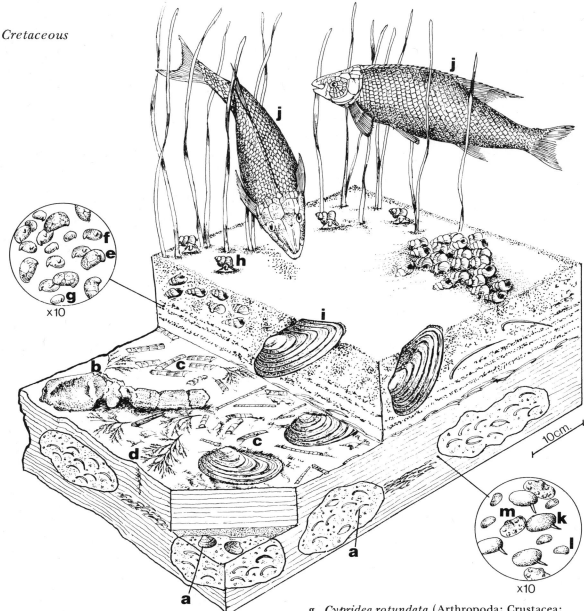

Fig. 92 Lower Cretaceous Lake Communities

a *Filosina gregaria* (Mollusca: Bivalvia: Veneroida)

b *Iguanodon* fragments (Vertebrata: Reptilia: Archosaur — dinosaur)

c *Equisetites lyelli* (Pteridophyta: Calamites — horsetail)

d *Onychiopsis mantelli* (Pteridophyta: Fern)

e *Cypridea valdensis* (Arthropoda: Crustacea: Ostracoda)

f *Cypridea spinigera* (Arthropoda: Crustacea: Ostracoda)

g *Cypridea rotundata* (Arthropoda: Crustacea: Ostracoda)

h *Viviparus sussexiensis* (Mollusca: Gastropoda: Mesogastropoda)

i *Pseudunio valdensis* (Mollusca: Bivalvia: Unionoida)

j *Lepidotes mantelli* (Vertebrata: Osteichthyes)

k *Sternbergella cornigera* (Arthropoda: Crustacea: Ostracoda)

l *Mantelliana mantelli* (Arthropoda: Crustacea: Ostracoda)

m *Theriosynoecum fittoni* (Arthropoda: Crustacea: Ostracoda)

The illustration summarizes two such faunal associations. At the base of the sequence are ironstone nodules developed at the level of a shell bed rich in *Filosina*, a burrowing, suspension-feeding heterodont bivalve. The lower inset illustration shows the ostracodes occurring in shales at this level as a 'non-*Cypridea*' association with *Sternbergella, Mantelliana* and *Theriosynoecium*.

In contrast, the bedding plane shown in the middle of the sketch shows a drifted association of entirely freshwater and terrestrial elements. The semi-infaunal suspension-feeding bivalve *Pseudunio*, drifted remains of the semi-aquatic horsetail *Equisetites,* and some dinosaur vertebra and bones (*Iguanodon*) perhaps from a decayed drifted corpse. The ostracodes in the shale at this level (shown in the upper inset illustration) are again freshwater, being entirely species of *Cypridea.* A similar surface assemblage is shown, with unionids such as *Pseudunio* here accompanied by the freshwater gastropod *Viviparus*, either live, or drifted after death into shell heaps, strands of *Equisetites* and the freshwater holostean fish *Lepidotes*, whose rhomboidal scales occur frequently in many Wealden shales and bone beds. This broad association thus differs little from the fresh-brackish association of the Upper and Middle Jurassic already described. Wealden lake deposits are best exposed in Britain in south-eastern England, the Isle of Wight and eastern Dorset.

APTIAN SAND BOTTOM COMMUNITIES (93—95)

These three examples are taken from different horizons in the Lower Greensand Group of the Isle of Wight, Hampshire in England where marine faunas of Aptian age are most easily studied (Casey, 1961). The Lower Greensand is a complex sequence of clays, silts, sands and sandstones, sometimes glauconitic, with occasional conglomerates, limestone and chert in some areas of southern England, whilst in the Midlands there are calcareous sponge gravels. Much of the sequence consists of cross-bedded sands (for instance the Folkestone Beds, Woburn Sands and much of the Sandrock of the Isle of Wight) which represents the build-up of submarine dune complexes. This facies, now widely exploited for glass and building sand, is generally barren of fossils, owing to the mobile substrates and also to post-depositional leaching out of all originally calcareous shells. It is thus only at certain levels in the sequence, notably in clays, limestones, silty and argillaceous burrowed sands, or in calcareous concretions which formed at an early stage, that fossils are common.

Cretaceous

Although the examples shown are taken from the coastal outcrops in the Isle of Wight, the faunas described below can also be found in Britain in the Weald and to a lesser degree in eastern Dorset. Broadly similar, although not identical, types of fauna can be found in similar facies of the Ryazanian-Aptian of Lincolnshire and Norfolk.

93   Aptian Sand Community (1)

The marine sands and clays of the Lower Greensand (Aptian) of southern England begin with a unit known as the *Perna* Bed, which is notable for the occurrence of hermatypic corals; indeed this is the only level in the British Lower Cretaceous where hermatypic corals are numerous. Only one genus, *Holocystis*, is common, occurring in the form of hemispherical masses, sometimes attached to shell debris. Associated epifaunal elements are byssate bivalves such as the large digitate genus *Mulletia*, and *Isognomon,* the latter living attached to other shells. Shells also acted as a base for the development of large 'nests' of brachiopods, both rhynchonellids (*Sulcirhynchia*) and terebratulids (*Sellithyris*), which may contain hundreds of individuals, some distorted by crowding. Other elements of the epifauna are oysters, notably the large exogyrid *Aetostreon,* which lived cemented to shell debris or to others of its species, or lying loose. Byssate semi-infaunal bivalves are represented by elongate gervillellids, two species of which are shown here, partly buried in sediment. The preserved infaunal elements are mainly thick-shelled siphonate suspension-feeding bivalves of which the genera *Sphaera, Protocardia* and *Venilicardia* are illustrated. Indications of other infaunal elements are given by various other burrow systems; indeed the sands at this level have generally been homogenized by burrowing infauna. Decapod burrows (*Thalassinoides*) are particularly conspicuous.

Nektonic elements of the *Perna* Bed fauna are rather limited; fish are common, and are usually represented by bone debris and teeth. Ammonites are scarce, but the coarsely-ribbed nautiloid *Cymtoceras* is fairly common. This fauna is found in southern England in the Isle of Wight and the Weald.

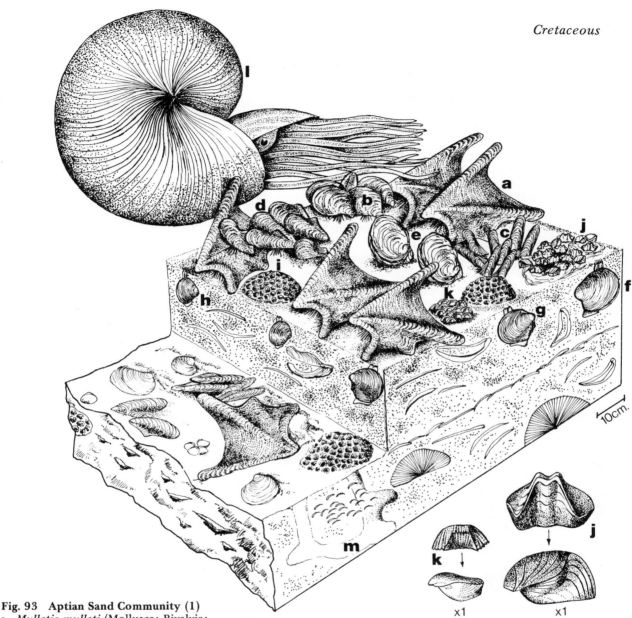

Fig. 93 Aptian Sand Community (1)

a *Mulletia mulleti* (Mollusca: Bivalvia: Pterioida)

b *Isognomon ricordeana* (Mollusca: Bivalvia: Pterioida)

c *Gervillella sublanceolata* (Mollusca: Bivalvia: Pterioida)

d *Gervillella alaeformis* (Mollusca: Bivalvia: Pterioida)

e *Aetostreon latissima* (Mollusca: Bivalvia: Pterioida — oyster)

f *Sphaera corrugata* (Mollusca: Bivalvia: Veneroida)

g *Protocardia sphaeroidea* (Mollusca: Bivalvia: Veneroida)

h *Venilicardia protensa* (Mollusca: Bivalvia: Veneroida)

i *Holocystis elegans* (Coelenterata: Anthozoa: Scleractinia)

j *Sellithyris sella* (Brachiopoda: Articulata: Terebratulida)

k *Sulcirhynchia hythensis* (Brachiopoda: Articulata: Rhynchonellida)

l *Cymatoceras radiatum* (Mollusca: Cephalopoda: Nautiloidea)

m *Thalassinoides* (Trace-fossil — Crustacea)

291

Cretaceous

94 Aptian Sand Community (2)

The Crackers and Lobster Beds are one of the most fossiliferous units in the Lower Greensand. The former consists of several lines of calcareous concretions in fine grained, bioturbated clay sands (Casey, 1961), the latter of silts and silty clays with small calcareous nodules. The exceptional fossil preservation in these beds allows us to reconstruct a diverse assemblage of organisms, although only a fraction of it is illustrated.

Ammonites are diverse and often abundant; two genera of ribbed forms are shown here, the typically compressed *Deshayesites* and the inflated genus *Roloboceras*. Benthic elements include the crab *Mithracites* and the lobster-like *Hoploparia longimana,* which may also have been a burrower; the prawn *Meyeria* (the crustacean which gives its name to the Lobster Beds) is abundant particularly within small nodules at this level.

Other elements of vagrant benthos include diverse gastropods (the deposit-feeding *Tessarolax* and *Anchura,* which have long marginal digitations, are shown here); infaunal gastropods include *Turritella.* Infaunal bivalves are very common, including several species of trigoniids, heterodonts (*Thetironia*) and the deep-burrowing *Panopea,* commonly found in its life position. Semi-infaunal elements are again represented by the elongate *Gervillella,* whilst bioturbation suggests the presence of a range of soft-bodied infauna, and *Thalassinoides* (perhaps produced by some of the decapods present) are common.

Fig. 94 Aptian Sand Community (2)
a *Meyeria magna* (Arthropoda: Crustacea: Malacostraca — decapod)
b *Mithracites vectensis* (Arthropoda: Crustacea: Malacostraca — decapod)
c *Hoploparia longimana* (Arthropoda: Crustacea: Malacostraca — decapod)
d *Deshayesites forbesi* (Mollusca: Cephalopoda: Ammonoidea)
e *Roloboceras hambrovi* (Mollusca: Cephalopoda: Ammonoidea)
f *Gervillella sublanceolata* (Mollusca: Bivalvia: Pterioida)
g *Yaadia nodosa* (Mollusca: Bivalvia: Trigonioida)
h *Thetrionia minor* (Mollusca: Bivalvia: Veneroida)
i *Panopea gurgitis* (Mollusca: Bivalvia: Myoida)
j *Anchura* (Mollusca: Gastropoda: Mesogastropoda)
k *Tessarolax fittoni* (Mollusca: Gastropoda: Mesogastropoda)
l *Turritella (Haustator) dupiniana* (Mollusca: Gastropoda: Mesogastropoda)
m *Thalassinoides* (Trace-fossil — Crustacea)

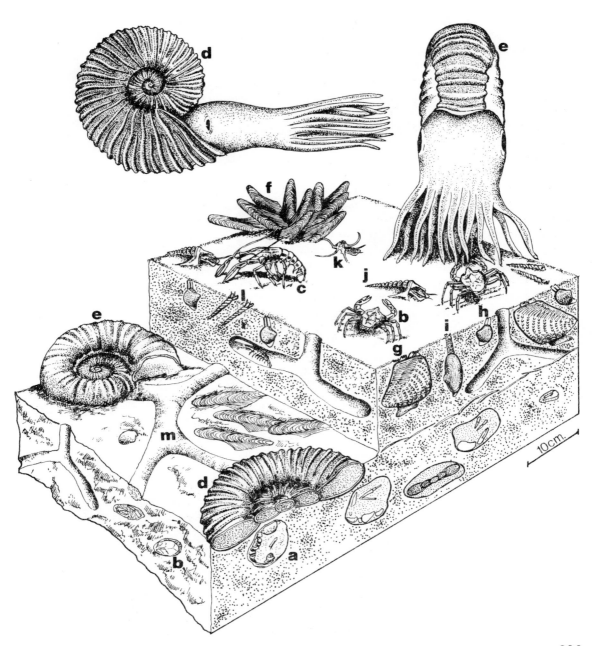

95   Aptian Sand Community (3)

This is a further example of a Lower Cretaceous sand bottom fauna, and shows many similar elements to those already discussed, including clusters of brachiopods (*Sellithris*), epifaunal decapods (*Hoploparia*) and large thick-shelled coiled oysters (*Aetostreon*). Infaunal elements are represented by the ubiquitous decapod burrow *Thalassinoides*, and two bivalves, the deep burrower *Panopea* and the costate *Pterotrigonia*. The ammonites shown include two normally coiled genera: the rather compressed ribbed genus *Deshayesites*, and *Cheloniceras*. Juveniles of the *Cheloniceras* are spinose (they are here shown bedded in sediment) whilst the adults have strong, coarse ribs (they are shown swimming). Three loosely coiled forms are illustrated: the giant *Australiceras* and *Epancyloceras*, and the smaller *Toxoceratoides*. Loosely coiled hetermorph ammonites of this type are known from a few genera in late Triassic and mid-Jurassic rocks, but in the latest Jurassic and throughout the Cretaceous they are an important component of most ammonite faunas (e.g. Figs. 102, 103, 105, 107). Many genera and families have long time-ranges and show complex evolutionary patterns: some even re-coil and assume a 'normal', planispiral shell form. The *Deshayesites* and *Cheloniceras* (Fig. 95, a, b, f) are both descended from early Cretaceous heteromorph ancestors. The large Aptian heteromorphs (Fig. 95, c, d) reach sizes of nearly a metre across; they commonly form the core of concretions; other concretions may contain masses of juvenile *Cheloniceras* (Fig. 95, a).

Fig. 95   Aptian Sand Community (3)
a   *Cheloniceras* (juvenile) (Mollusca: Cephalopoda: Ammonoidea)
b   *Cheloniceras* (Mollusca: Cephalopoda: Ammonoidea)
c   *Australiceras gigas* (Mollusca: Cephalopoda: Ammonoidea)
d   *Epancyloceras* (Mollusca: Cephalopoda: Ammonoidea)
e   *Aetostreon latissima* (Mollusca: Bivalvia: Pterioida — oyster)
f   *Deshayesites grandis* (Mollusca: Cephalopoda: Ammonoidea)
g   *Toxoceratoides* (Mollusca: Cephalopoda: Ammonoidea)
h   *Panopea gurgitis* (Mollusca: Bivalvia: Myoida)
i   *Pterotrigonia mantelli* (Mollusca: Bivalvia: Trigonioida)
j   *Hoploparia longimana* (Arthropoda: Crustacea: Malacostraca — decapod)
k   *Sellithyris sella* (Brachiopoda: Articulata: Rhynchonellida)
l   *Thalassinoides* (Trace-fossil — Crustacea)

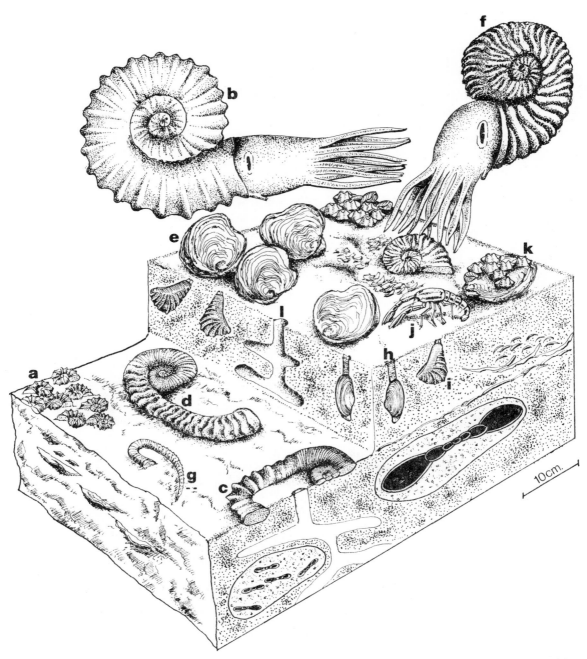

ALBIAN CLAY BOTTOM COMMUNITIES (96–97)

The Albian Gault Clay of south-east Britain is a sequence of black and grey clays with lines of phosphatic nodule beds and a rich and diverse marine fauna, often preserved with original aragonitic shell material. As they are traced westwards the clays pass laterally into the Upper Greensand, a complex of terrigenous silts, sands, limestones and similar facies, (see Figs. 97, 98). In Norfolk, Lincolnshire and Yorkshire they pass into a condensed red, sometimes nodular limestone, the Red Chalk. Gault faunas (Jukes-Browne and Hill, 1900, Smart et al. 1966) are rich and diverse; a series of quite distinctive associations can be recognized, reflecting variations in general depth-related factors and substrate conditions. The illustrations show two common situations: one with a soft clay bottom, the other with harder substrates, formed either by shell debris or by phosphatic nodule beds. These phosphates are a particular feature of the Gault, and indeed other, Cretaceous clays. Many nodules are moulds of whole and fragmentary fossils, or burrow fillings; others are irregular and of unknown origin. A combination of subsurface and sea floor alteration of cemented carbonate-rich mudstone seems to be the most likely origin for these nodules, and they are commonly associated with non-sequences, condensed beds and periods of slow or non-deposition. Rather similar nodule beds and faunas occur in other Lower Cretaceous clays in Britain, in Lincolnshire and Yorkshire.

96 Phosphatic Nodule Bed Community

The illustration is based on a phosphatic nodule bed and associated clays found near the base of the Gault in southern England, known as the *dentatus-spathi* nodule bed, after the ammonites *Hoplites dentatus* and *Hoplites spathi* which are found there. As in Jurassic clays, deposit-feeding protobranch bivalves are particularly common and, as examples, species of *Nucula* and the radially ribbed *Acila* are shown. These are accompanied by the byssate, semi-infaunal suspension-feeding arcoid *Nannoavis carinata* and the burrowing, suspension-feeding *Linotrigonia fittoni*, a rather rare form. Other elements of infauna include scaphopods, carnivorous gastropods and turritellids. Trace fossils are prominent, and the sediment is bioturbated with *Planolites*, *Chondrites* and *Thalassinoides*.

Epifaunal organisms are of two types: those living on the hard surfaces of nodules and shells, and those living on soft or stiff mud. The former include several types of cemented bivalves, such as *Atreta, Spondylus* and *Pycnodonte,* and serpulid polychaetes.

Fig. 96 Phosphatic Nodule Bed Community

a *Hoplites dentatus* (Mollusca: Cephalopoda: Ammonoidea)
b *Anahoplites planus* (Mollusca: Cephalopoda: Ammonoidea)
c *Heteroclinus nodosus* (Mollusca: Cephalopoda: Ammonoidea)
d *Neohibolites minimus* (Mollusca: Cephalopoda: Coleoidea — belemnite)
e *Inoceramus concentricus* (Mollusca: Bivalvia: Pterioida)
f *Nucula (Pectinucula) pectinata* (Mollusca: Bivalvia: Palaeotaxodonta — nuculoid)
g *Nucula (Leionucula) ovata* (Mollusca: Bivalvia: Palaeotaxodonta — nuculoid)
h *Acila bivirgata* (Mollusca: Bivalvia: Palaeotaxodonta — nuculoid)
i *Linotrigonia fittoni* (Mollusca: Bivalvia: Trigonoida)
j *Pycnodonte* (Mollusca: Bivalvia: Pterioida)
k *Plicatula gurgitis* (Mollusca: Bivalvia: Pterioida)
l *Atreta nilssoni* (Mollusca: Bivalvia: Pterioida)
m *Spondylus* (Mollusca: Bivalvia: Pterioida)
n *Entolium orbiculare* (Mollusca: Bivalvia: Pterioida — pectinid)
o *Nanonavis carinata* (Mollusca: Bivalvia: Arcoida)
p *Ichtyosaurus* (Vertebrata: Reptilia: Euryapsida — ichthyosaur tooth)
q serpulids (Annelida)

x1.3

10cm.

Cretaceous

Soft bottom epifauna include *Plicatula*, a reclining form, and *Entolium*, a free-living, and perhaps occasionally swimming pectinid. *Inoceramus* are particularly common living in clusters attached by byssal threads to each other and to shell debris. The species shown here, ornamented by fine concentric growth striae and ribs is known as *Inoceramus concentricus;* it ranges throughout much of the Gault.

The nekton consists chiefly of cephalopods. Of the forms with normal coiling, hoplitids are dominant: *Hoplites* itself, a strongly ribbed genus with a sulcate venter, and *Anahoplites*, a compressed, feebly ornamented genus. Heteromorphs are represented by the ribbed and spinose *Heteroclinus*, and coleioids by the small belemnite *Neohibolites*. Other nekton such as fish and saurians are usually represented by fragments only, like the ichthyosaur tooth shown here. The fauna can be found in southeast England and the south Midlands.

97   Upper Gault Clay Community

This illustration shows an entirely soft bottom faunal association from the Upper Gault Clay. Many elements are identical to, or have evolved from, those of the previous fauna. Hoplitid ammonites are still a major element of the nekton, with the long-ranging species *Anahoplites planus* common. It is accompanied by *Epihoplites* and *Dimorphoplites*, coarsely ribbed and tuberculate forms with a flat venter, and *Euhoplites*, a ribbed tuberculate form with a distinctive deep and narrow ventral groove. Other ammonites are the diminutive *Hysteroceras*, a loosely coiled genus with rounded whorls and strong simple ribs and the larger, ribbed, tuberculate and keeled *Mortoniceras*. The specimen (f) in the shell bed in the lower part of the figure is an adult *Mortoniceras* and shows a horn-like extension or rostrum to the ventral part of the aperture, a structure associated with sexual maturity, possibly housing some accessory reproductive organ.

Other elements of nekton are the long ranging belemnite *Neohibolites*, and various vertebrates such as sharks, here represented by teeth. A diverse epifauna is shown: the small crab *Notopocorystes stokesi*; the crinoid *Nielsenicrinus*, which presumably lived rooted in mud; and various gastropods, including the aphorrhaid *Anchura*, which perhaps fed on plant debris, *Pleurotomaria*, a sponge feeder, and the superficially similar but spinose *Nummocalcar*. *Gyrodes*, here shown on the surface, was a carnivore which probably spent much of its time burrowing, and fed on

Fig. 97   Upper Gault Clay Community

a   *Hysteroceras orbignyi* (Mollusca: Cephalopoda: Ammonoidea)
b   *Anahoplites planus* (Mollusca: Cephalopoda: Ammonoidea)
c   *Epihoplites* (Mollusca: Cephalopoda: Ammonoidea)

298

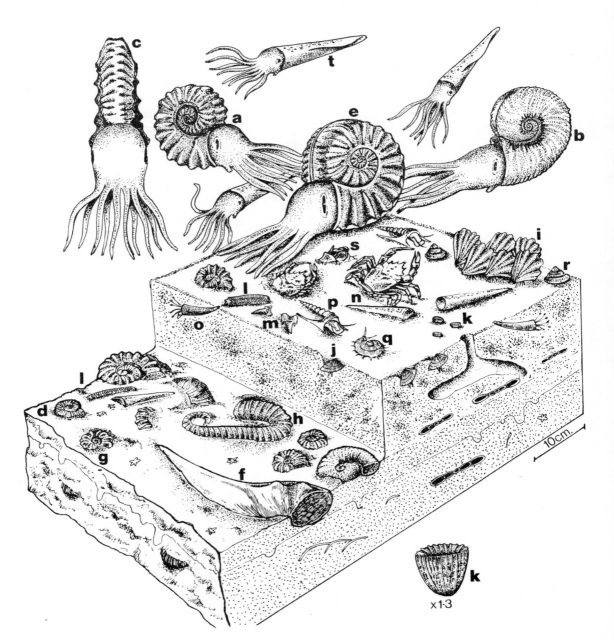

x1.3

d *Euhoplites* (Mollusca: Cephalopoda: Ammonoidea)

e, f *Mortoniceras inflatum* (Mollusca: Cephalopoda: Ammonoidea)

g *Dimorphoplites* (Mollusca: Cephalopoda: Ammonoidea)

h *Hamites* (Mollusca: Cephalopoda: Ammonoidea)

i *Actinoceramus sulcatus* (Mollusca: Bivalvia: Pterioida)

j *Nuculana (Pectinucula) pectinata* (Mollusca: Bivalvia: Palaeotaxodonta — nuculoid)

k *Trochocyathus harveyanus* (Coelenterata: Anthozoa: Scleractinia)

l *Nielsenicrinus cretaceus* (Echinodermata: Crinozoa)

m fish teeth (Vertebrata: Chondrichthyes)

n *Notopocorystes stokesi* (Arthropoda: Crustacea: Malacostraca — decapod)

o *Dentalium (Fissidentalium) decussatum* (Mollusca: Scaphopoda)

p *Anchura carinata* (Mollusca: Gastropoda: Mesogastropoda)

q *Nummocalcar fittoni* (Mollusca: Gastropoda: Archaeogastropoda)

r *Pleurotomaria* (Mollusca: Gastropoda: Archaeogastropoda)

s *Gyrodes genti* (Mollusca: Gastropoda: Mesogastropoda)

t *Neohibolites minimus* (Mollusca: Cephalopoda: Coleoidea — belemnite)

299

bivalves by boring small circular holes in the valves with its specialized radula. Byssally attached inoceramids are common in clusters, as in the Lower Gault, but the ribbed *Inoceramus concentricus* is replaced by *Actinoceramus sulcatus*, a descendant with coarse radial ribs. Coelenterates are represented by the tiny, solitary, ahermatypic cup coral *Trochocyathus*.

Deposit feeders dominate the infauna, and the sediment is bioturbated, with traces of arthropods (*Thalassinoides*) and the ubiquitous *Chrondrites* and *Planolites*. The deposit-feeding protobranchs are represented by *Nuculana*, and the scaphopods by *Dentalium*, which probably hunted through the sediment for Foraminiferida.

Much of the Gault fauna retains its original aragonitic shell material; although often crushed, the ammonites show nacreous lustre. In other cases, early cementation of clay infillings resulted in the preservation of the original shape; ammonite phragmocones are commonly filled with iron sulphides as they are in many Jurassic clays. Exposures with this fauna may be found in south-east England and the south Midlands.

ALBIAN COMMUNITIES OF SILTY-SANDY AND GLAUCONITIC SEDIMENTS (98–100)

These three faunas are all drawn from the Upper Greensand formation of south-east and south-west England, a terrigenous and glauconitic equivalent of the Gault Clay, which presumably accumulated in shallower water higher energy environments. Some or all of these faunas can be found in every part of the outcrop of the Upper Greensand, and comparable faunas occur in some of the lithologically similar facies of the Cenomanian which are to be found in south-west England, in Scotland and in Northern Ireland.

98 Exogyra-pectinid Community

The *Exogyra* Rock is a quartrose glauconitic and calcarenitic sand crowded with interlocking calcareous nodules which occurs over much of Dorset and south-east Devon. It takes its name from the abundance of the coiled oyster *Exogyra*, and was broadly contemporaneous with the Upper Albian Gault Clay shown in the previous illustration. Nektonic elements are represented chiefly by rare ammonites; a specimen of *Mortoniceras* is shown here.

Benthos was of rather low diversity, the commonest animals being clusters of *Exogyra* cemented to each other or on to dead

Fig. 98 Exogyra-pectinid Community

a *Mortoniceras inflatum* (Mollusca: Cephalopoda: Ammonoidea)
b *Cardiaster fossarius* (Mollusca: Cephalopoda: Echinoidea)
c *Chlamys aspera* (Mollusca: Bivalvia: Pterioida — pectinid)
d *Rotularia concava* (Annelida — polychaete)
e *Entolium orbiculare* (Mollusca: Bivalvia: Pterioida — pectinid)
f *Exogyra obliquata* (Mollusca: Bivalvia: Pterioida — oyster)
g *Torquesia granulata* (Mollusca: Gastropoda: Mesogastropoda)
h *Neithea gibbosa* (Mollusca: Bivalvia: Pterioida — pectinid)
i *'Spongeliomorpha' annulatum* (Burrow)

301

shells, often turritellid gastropods. Epifaunal pectinid bivalves are also present with the free-living, perhaps occasionally swimming, smooth, orbicular *Entolium*, the radially ribbed and spinose *Chlamys aspera*, and a large species of coarsely ribbed *Neithea*. Small coiled serpulids *Rotularia concava* are also part of the benthos, commonly occurring in clusters. The infauna includes turritellid gastropods and the heart-shaped shallow-burrowing *Cardiaster*. Trace fossils indicate other animal groups, including arthropods and polychaetes; a trace common in sediments of this type is the burrow *'Spongeliomorpha' annulatum*, a branching, clay-lined cylinder, with ridges on the outer surface, produced either by arthropods or polychaetes.

The reconstruction of these faunas is based on specific occurrences on the Dorset coast; the *Exogyra* Rock association can be found in Britain throughout Dorset and Devon, and broadly similar associations occur in sandy Albian and Cenomanian facies elsewhere in south-west England, Scotland and Northern Ireland.

99   Diverse Molluscan-sponge Community

The Blackdown Hills between Honiton and Wellington in Devon are capped by white, light yellow and red sands, with white siliceous concretions. In the last century, the siliceous concretions were worked for whetstones, and very large fossil faunas, usually silicified, were brought to light. Scores of species are known, and they make up some of the more diverse Upper Albian sandy bottom faunas.

The nekton includes ammonites, such as the *Mortoniceras* already illustrated (Figs. 97-98), various hoplitids, and the small ribbed *Hysteroceras*, shown in this diagram. Epifauna includes bivalves, serpulids and other animals similar to the examples shown in Figs. 98, 100 and 101, and also diverse gastropods, including *Rostellaria* and species of *'Fusus'*, *'Phasianella'* and their relatives. Sponges, including *Siphonia*, are also common. The infauna includes some gastropods, notably the carnivorous naticid *Gyrodes* (shown here as empty shells) and the particulate feeder *Turritella*; the latter often occurs in vast numbers and forms shell beds. Bivalves are diverse and include shallow burrowers such as *Glycimeris*, *Epicyprina*, *Eriphyla*, *Cyprina* and *Protocardia*, the rostrate *Pterotrigonia*, and deeper burrowers such as *Panopea*. Intensive bioturbation, with trace-fossils such as *Thalassinoides*, *Chondrites* and *Planolites*, indicates the presence of diverse arthropods and polychaetes.

Fig. 99   Diverse Molluscan-sponge Community

a   *Siphonia tulipa* (Porifera: Demospongea)
b   *Torquesia granulata* (Mollusca: Gastropoda: Mesogastropoda)
c   *Epicyprina angulata* (Mollusca: Bivalvia: Veneroida)
d   *Pterotrigonia aliformis* (Mollusca: Bivalvia: Trigonioida)
e   *Rostellaria* (Mollusca: Gastropoda: Mesogastropoda)
f   *'Fusus'* (Mollusca: Gastropoda: Neogastropoda)
g   *'Phasianella'* (Mollusca: Gastropoda: Archaeogastropoda)
h   *Gyrodes genti* (Mollusca: Gastropoda: Mesogastropoda)

i   *Cucullaea glabra* (Mollusca: Bivalvia: Arcoida)
j   *Eriphyla* (Mollusca: Bivalvia: Veneroida)
k   *Cyprina cuneata* (Mollusca: Bivalvia: Veneroida)
l   *Glycymeris* (Mollusca: Bivalvia: Arcoida)
m   *Panopea* (Mollusca: Bivalvia: Myoida)
n   *Protocardia hillana* (Mollusca: Bivalvia: Veneroida)
o   *Hysteroceras varicosum* (Mollusca: Cephalopoda: Ammonoidea)

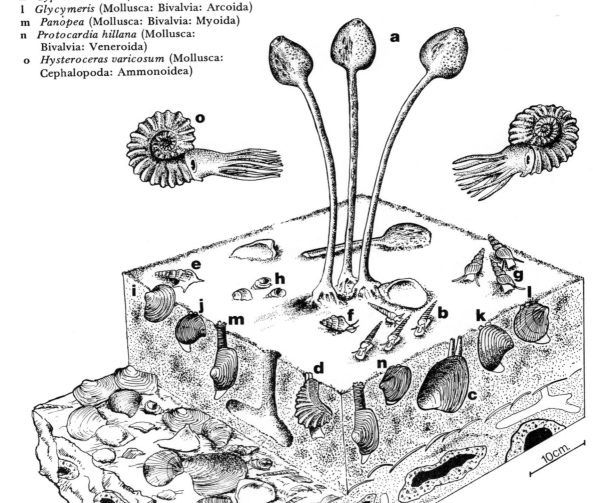

10cm.

303

Cretaceous

The classic Blackdown sections have long vanished, but representative elements of this fauna are widespread in the lower parts of the Upper Greensand of Dorset and Devon.

100 Sponge Community

At various localities in southern and south-west England, the glauconitic sands of the Upper Greensand of Albian and Cenomanian age have yielded diverse sponge faunas. These are commonly dominated by lithistid sponges, a group of Demosponges characterized by lumpy, knobbly spicules which are fused into a framework which results in good preservation of the overall body form.

Siphonia is a genus with a basal root-like anchorage, a long, slender stem, and a pear- or tulip-shaped body. There are small, slightly curved incurrent canals which extend from the surface of the sponge to its centre, and larger excurrent canals, running parallel to the surface from the base to the summit, where they open into a deep central cavity. *Hallirhoa* is a similar genus, but the sides of the body are lobed. *Doryderma* is another stemmed form, rooted at the base; some species, such as *Doryderma bennetiae,* have a body shaped like a tall wine glass while others, such as *Doryderma dichotomum,* have a distinctive dichotomously branched body. *Doryderma* has parallel vertical canals running from the base to the summit of the sponge, and smaller radial canals extending from the surface towards the centre.

Other elements of the fauna in this community are similar to those occurring in the *Exogyra* Rock and Blackdown Greensand (Figs. 97—98), such as the coiled oyster *Exogyra* and the pectinids *Chlamys* and *Neithea.* The associated sediment is bioturbated, and large *Thalassinoides* are shown here.

The best examples of rich sponge faunas of this type are found in Devon and Dorset, but elements of it occur widely in the Albian and Cenomanian sandy facies of southern England.

UPPER CRETACEOUS FAUNAS

While the British Lower Cretaceous is characterized by diverse and variable facies, the Upper Cretaceous is characterized by the

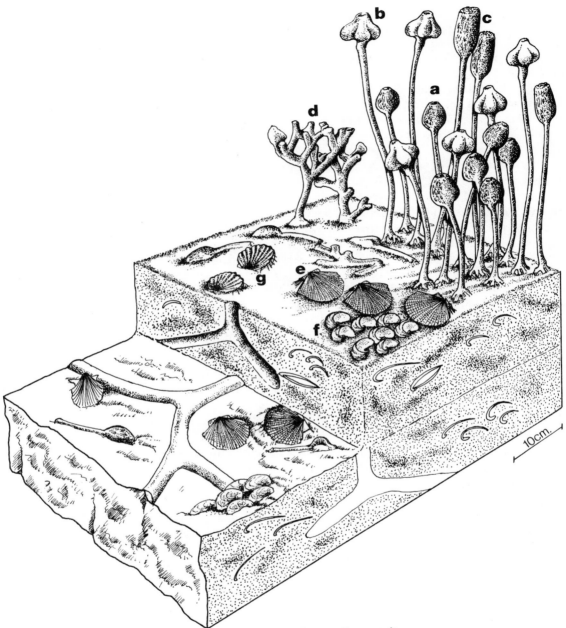

Fig. 100 Sponge Community

a *Siphonia tulipa* (Porifera: Demospongea)
b *Hallirhoa costata* (Porifera: Demospongea)
c *Doryderma bennetiae* (Porifera: Demospongea)
d *Doryderma dichotomum* (Porifera: Demospongea)
e *Chlamys aspera* (Mollusca: Bivalvia: Pterioida — pectinid)
f *Exogyra obliquata* (Mollusca: Bivalvia: Pterioida — oyster)
g *Neithea gibbosa* (Mollusca: Bivalvia: Pterioida — pectinid)

presence of fine grained pelagic limestones (chalks) over the whole of Britain, and much of western Europe. Marginal, non-chalk facies are thus limited to south-west England, Northern Ireland and Scotland, and are predominantly of Cenomanian age. Elsewhere in northern Europe, these non-chalk facies occur in areas such as Normandy, Sarthe and Touraine in France, Belgium, Germany, Czechoslovakia and Poland, each associated with the flanks of emergent massif areas. The faunas of these marginal facies commonly resemble those of the sandy Albian facies already described.

By contrast, chalks are a distinctive lithology and show many faunal peculiarities. They may be characterized as friable micrites or fine grained limestones, composed predominantly of the remains of whole and fragmentary haplophycean algae, coccolithorphorids. Coccoliths were more diverse and abundant in the Upper Cretaceous than at any other time, and because of geographical and environmental factors as yet not well understood, became sediment producers in the shelf seas of Europe and North America, although they had previously contributed chiefly to deep sea oozes, which they continued to do after this period.

In addition to coccolithophorid debris, six constituents were commonly present in the chalk at the time of deposition:

1   A fine clay fraction, which may form an appreciable part of the sediment at levels such as that of the Chalk Marl or Plenus Marl, but generally amounts to only a few per cent or less of the sediment.

2   A coarse terrigenous fraction, predominantly detrital quartz and heavy minerals. This generally forms less than one per cent of the sediment.

3   Minerals such as pelletal glauconite and phosphate, which formed on the sea floor. These occur as minor constituents throughout the whole of the chalk and are relatively common in winnowed and condensed horizons. They are essential constituents of some facies.

4   A coarse calcite skeletal fraction consisting of *Oligostegina* (calcispheres) planktonic and benthic Foraminiferida, debris of echinoderms, articulate brachiopods, bryozoans, serpulid polychaetes, ahermatypic corals and pteriomorph bivalves, notably inoceramid prisms.

5   Skeletal material, which was originally siliceous, such as sponge spicules and probably also radiolarians and diatoms.

The silica has usually dissolved, its place being taken by calcite, but it may have been the source of some or all of the silica in the flints which are so common in much of the sequence.

6 Skeletal aragonite. This has now disappeared as a result of dissolution, but evidence of its former presence is given by the casts and moulds of originally aragonitic organisms.

These chalks accumulated in an outer shelf environment in a tectonically stable region (see reviews by Hakansson et al. 1974; Kennedy and Garrison, 1975). Depths were chiefly around 300m, but could at times have been as little as 50m. Water temperatures based on deductions from oxygen isotope ratios ranged from 13.5 — 28.5° C, and salinities appear to have been normal.

Bottom conditions were very variable. The structural adaptations of some groups and sedimentological criteria suggest soft, thixotropic, pelletal sea bottoms in some places (Kennedy and Garrison, 1975), but firmer although still fairly soft bottoms in others.

There is much evidence of early lithification leading to the growth of calcareous nodules below the sediment-water interface. These in turn were reworked into conglomerates, which produced hard substrates and also semi-continuous lithified layers; when exposed, the latter formed hardgrounds which frequently became glauconitized and phosphatized. Depositional and diagenetic processes show cyclical variations, and much of the European Chalk is rhythmically bedded with soft and nodular chalks, soft chalks and hard-grounds or sharply defined burrowed firmer bottoms following one another regularly up the sequence.

The seven examples illustrated in the succeeding pages show many of the typical elements of chalk faunas. They are, however, drawn from particularly fossiliferous horizons and localities. Much of the chalk, particularly in the higher parts of the sequence, is very barren of macrofossils, presumably because of unfavourable fluid substrate conditions. Sample groups of those faunas which do occur, are commonly biased in favour of the more durable fossils since there is strong evidence that originally aragonitic shells of organisms such as ammonites, nautiloids, many gastropods and bivalves were dissolved prior to burial. The curious composition of many chalk faunas which are dominated by originally calcitic organisms, such as pteriomorph bivalves, brachiopods and echinoderms is thus the result of diagenesis rather than being an accurate reflection of the original community. A further problem involves the nature of chalk floras: were there seaweeds, both thallophytes and marine angiosperms, or not? (See Hakansson et al. 1974, Nestler, 1965, Steinich, 1967, Surlyk, 1972 and Surlyk and Birkelund, 1976 for discussions of these, and other ecological problems of chalk faunas).

101   Early Cenomanian Sand Community

This illustration is based on the sandy fauna of the early Cenomanian at Wilmington in south Devon; there are many obvious similarities to the fauna of the Albian sands described in Figs. 98—100. The whole sequence is bioturbated with spectacular decaped burrows (*Thalassinoides*); whilst other organisms responsible for the disturbance of sedimentary lamination are infaunal and semi-infaunal irregular echinoids. Those shown here are *Holaster laevis, Holaster altus,* and *Conulus castanea.* Many other echinoids are common in these sands. The benthos includes coiled oysters such as *Exogyra,* the bowl-shaped pectinid *Neithea,* the byssate *Chlamys,* and two other organisms showing a similar development of ribbing and plicate margins to counter the clogging of their feeding mechanisms by coarse debris — namely the oyster *Lopha* and the rhynchonellid brachiopod *Cyclothyris.* The other examples of benthic fauna shown are the ramose bryozoan *Ceriopora,* encrusting oysters (*Pycnodonte*) and inoceramid bivalves. The nekton shown are all ammonites, and these include heteromorphs: the costate *Turrilites scheuchzerianus* and the tuberculate and spinose *Hypoturrilites gravesianus.* Of the normally coiled forms, *Hyphoplites,* with sickle-shaped ribs and sulcate venter, is a descendant of the Albian sulcate hoplitids, whilst the ribbed, tuberculate and keeled *Schloenbachia* is descended from hoplitids with a flat or slightly elevated venter. *Mantelliceras,* with strong, alternately long and short ribs, and tubercles which may be lateral, umbilical, ventrolateral, but never siphonal, is an early member of the superfamily Acanthocerataceae which dominate the normally coiled Upper Cretaceous taxa. Early Cenomanian sands with this fauna are exposed in Dorset and Devon.

Fig. 101   Early Cenomanian Sand Community
a   *Schloenbachia varians ventriosa* (Mollusca: Cephalopoda: Ammonoide
b   *Schloenbachia varians subvarians* (Mollusca: Cephalopoda: Ammonoi
c   *Hyphoplites* (Mollusca: Cephalopoda: Ammonoidea)
d   *Schloenbachia varians varians* (Mollusca: Cephalopoda: Ammonoidea)
e   *Mantelliceras* (Mollusca: Cephalopoda: Ammonoidea)
f   *Hypoturrilites* (Mollusca: Cephalopoda: Ammonoidea)
g   *Euturillites scheuchzerianus* (Mollusca: Cephalopoda: Ammonoidea)
h   *Inoceramus conicus* (Mollusca: Bivalvia: Pterioida)
i   *Chlamys aspera* (Mollusca: Bivalvia: Pterioida — pectinid)
j   *Neithea gibbosa* (Mollusca: Bivalvia: Pterioida — pectinid)
k   *Exogyra obliquata* (Mollusca: Bivalvia: Pterioida — oyster)
l   *Holaster altus* (Echinodermata: Echinoidea)
m   *Holaster laevis* (Echinodermata: Echinoidea)
n   *Conulus castanea* (Echinodermata: Echinoidea)
o   *Cyclothyris difformis* (Brachiopoda: Articulata: Rhynchonellida)
p   *Lopha colubrina* (Mollusca: Bivalvia: Pterioida — oyster)
q   *Ceripora ramulosa* (Bryozoa: Ectoprocta)
r   *Pycnodonte vesicularis* (Mollusca: Bivalvia: Pterioida — oyster)

102 Mid-Cenomanian Condensed Fauna

This illustration represents another non-chalk fauna of the Upper Cretaceous which is widespread in south-west England. Here, the base of the chalk commonly rests on a hardground at the top of calcarenites or sandy limestones of Albian-Cenomanian age. The basal few centimetres are a phosphatic conglomerate of pebbles and cobbles, commonly derived from the sandstones and limestones below and accompanied by phosphatized fossils and fossil moulds. These levels are very condensed, and have undergone a complex history of diagenesis and reworking, comparable in some respects to the processes shown by the Gault nodule beds (Fig. 96). Because of this complex history, the faunas are varied and include forms which lived on and in the hardground, epifauna living on the phosphatic nodules, a soft bottom fauna living associated with patches of shell sand which intermittently buried shells and nodules, and diverse nekton. The example shown here is of mid-Cenomanian age.

The hardground at the top of the Upper Greensand has been mineralized by glauconite and phosphate, and bear an epifauna of serpulids, ectoprocts, cemented bivalves and other similar organisms; it has been bored by lithodomous bivalves and the sponge *Cliona*. In places these hardgrounds may also be covered by laminated crusts, now phosphatized, but perhaps originally stromatolitic. The fauna (not shown here) associated with the soft bottoms produced by sand and carbonate mud which temporarily buried the hardground includes diverse bivalves, such as trigoniids, arcids and many heterodonts, and arthropods, diverse echinoids and polychaetes. The nekton includes both ammonites and nautiloids; examples of heteromorph ammonites are the helically coiled *Turrilites* and the loosely coiled *Scaphites* and *Stomohamites*. Normally coiled forms are the keeled, ribbed and tuberculate *Schloenbachia,* which commonly forms over 90 per cent of Boreal Cenomanian ammonite faunas, and two acanthoceratids, *Acanthoceras* and *Calycoceras,* which differ from the Lower Cenomanian *Mantelliceras* in the presence of siphonal tubercles. Nautiloids are represented by the ribbed genus *Cymatoceras;* smooth genera, closer to present-day forms of *Nautilus,* also occur.

The remains of all these groups were filled with sediment, phosphatized and broken. Diverse epizoans lived on this hard conglomerate substrate, including terebratulid brachiopods, serpulids and oysters, bryozoans, sponges and other similar fauna; *Pleurotomaria* and other gastropods almost certainly grazed on algae (not shown in the illustration).

The community is present in Britain in south Devon, Dorset and southern Somerset.

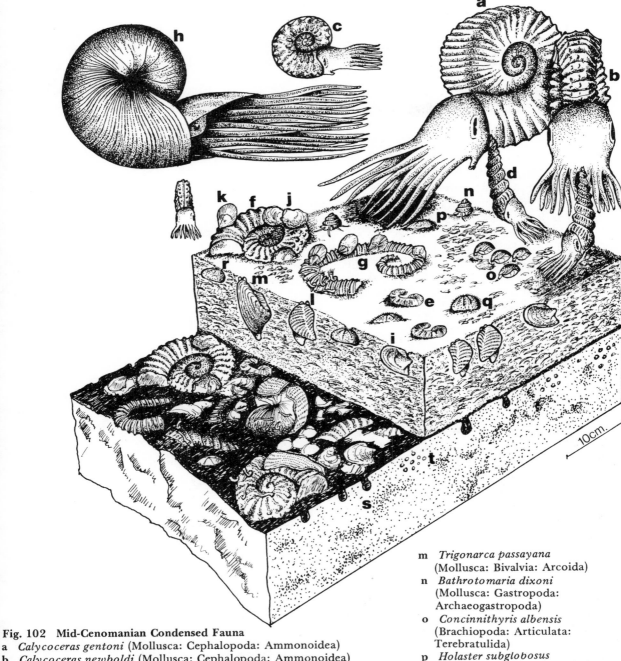

Fig. 102 Mid-Cenomanian Condensed Fauna

a *Calycoceras gentoni* (Mollusca: Cephalopoda: Ammonoidea)
b *Calycoceras newboldi* (Mollusca: Cephalopoda: Ammonoidea)
c *Schloenbachia coupei* (Mollusca: Cephalopoda: Ammonoidea)
d *Turrilites costatus* (Mollusca: Cephalopoda: Ammonoidea)
e *Scaphites equalis* (Mollusca: Cephalopoda: Ammonoidea)
f *Acanthoceras rhotomagense* (Mollusca: Cephalopoda: Ammonoidea)
g *Stomohamites simplex* (Mollusca: Cephalopoda: Ammonoidea)
h *Cymatoceras deslongchampsianum* (Mollusca: Cephalopoda: Nautiloidea)
i *Cucullaea mailleana* (Mollusca: Bivalvia: Arcoida)
j *Pycnodonte vesicularis* (Mollusca: Bivalvia: Pterioida — oyster)
k *Spondylus latus* (Mollusca: Bivalvia: Pterioida — pectinid)
l *Trigonia vicaryana* (Mollusca: Bivalvia: Trigonioida)

m *Trigonarca passayana* (Mollusca: Bivalvia: Arcoida)
n *Bathrotomaria dixoni* (Mollusca: Gastropoda: Archaeogastropoda)
o *Concinnithyris albensis* (Brachiopoda: Articulata: Terebratulida)
p *Holaster subglobosus* (Echinodermata: Echinoidea)
q *Discoides subuculus* (Echinodermata: Echinoidea)
r *Conulus castanea* (Echinodermata: Echinoidea)
s *Lithophaga* borings (Mollusca: Bivalvia: Mytiloida)
t *Cliona* borings (Porifera: Demospongea)

311

103   Mid-Cenomanian Argillaceous Chalk Community

This fauna is based on that found in the middle part of the Lower Chalk of the Kent coast, at a horizon in the Middle Cenomanian a little older than the basement bed shown in the previous figure. The nekton includes similar ammonites and the nautiloid *Cymatoceras*. The straight ammonite *Sciponoceras*, is locally abundant at this level.

By contrast, the benthos was very different from the condensed fauna; it suggests that the Chalk bottom was a soft ooze. There are large masses of siliceous hexactinellid sponges (*Plocoscyphia*), which served as a substrate for brachiopods and as food for pleurotomariid gastropods, whilst ahermatypic solitary corals (*Micrabacia*) lie loose in the sediment, as do large, thin-shelled pectinids such as *Chlamys (Aequipecten) beaveri*. Terebratulid brachiopods have branched, root-like pedicles, and can live on soft bottoms, attached to shell debris, foraminifera and other material, whilst the large, concentrically ribbed *Inoceramus crippsi* may have lived in a similar fashion, in dense clumps attached by byssal threads. All hard substrates were potential habitats for encrusting organisms, and many shells are covered by bryozoans, serpulids, bivalves (such as the *Pycnodonte* illustrated), and may be bored by clionid sponges or other organisms. Only some infauna is preserved fossil, but bioturbation is very marked, with *Thalassinoides*, *Chondrites* and *Planolites* prominent. Water depths were probably too great, and therefore light intensities too low, for the development of algal thickets. This community occurs in Britain in the Weald, the Hampshire Basin, the Chilterns and Cambridgeshire.

Fig. 103   Mid-Cenomanian Argillaceous Chalk Community
a *Inoceramus crippsi* (Mollusca: Bivalvia: Pterioida)
b *Bathrotomaria perspectiva* (Mollusca: Gastropoda: Archaeogastropoda)
c *Acanthoceras rhotomagense* (Mollusca: Cephalopoda: Ammonoidea)
d *Cymatoceras elegans* (Mollusca: Cephalopoda: Nautiloidea)
e *Schloenbachia coupei* (Mollusca: Cephalopoda: Ammonoidea)
f *Exanthesis labrosus* (Porifera: Hyalospongea)
g *Terebratulina striatula* (Brachiopoda: Articulata: Terebratulida)
h *Cocinnithyris albensis* (Brachiopoda: Articulata: Terebratulida)
i *Orbirhynchia mantelliana* (Brachiopoda: Articulata: Rhynchonellida)
j *Micrabacia coronula* (Coelenterata: Anthozoa: Scleractinia)
k *Chlamys (Aequipecten) beaveri* (Mollusca: Bivalvia: Pterioida — pectinid)
l *Pycnodonte vesicularis* (Mollusca: Bivalvia: Pterioida — oyster)
m *Sciponoceras baculoide* (Mollusca: Cephalopoda: Ammonoidea)

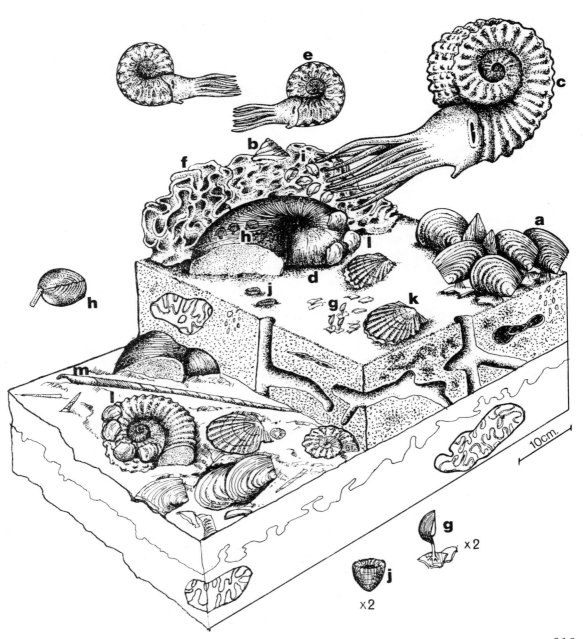

104   Early Turonian Inoceramus Shell Bed Community

This illustration is based on material found at a level low in the Middle Chalk on the Kent coast. At this level, the clay content of the chalk is very low; it consists of layers of soft, very bioturbated chalks grading up into chalks crowded with early diagenetic nodules and frequently capped by an erosion surface containing reworked nodules (sometimes bored and encrusted) and shell debris. Aragonitic fossils are rare, and all but the largest shells have probably been dissolved. Two large ammonites are shown, the oval-whorled *Lewesiceras*, and the coarsely ribbed and tuberculate *Mammites*. The benthos is dominated by the semi-infaunal, byssate inoceramid *Mytiloides* which lived in crowded shell beds, and after death contributed substantial amounts of prismatic calcite shell layer to the sediment thus frequently building inoceramid "pavements". These and other shells formed substrates for various cemented organisms such as oysters, bryozoans and serpulids, and some brachiopods may also have attached themselves, as the rhynchonellid *Orbirhynchia* shown here. The diminutive *Terebratulina* probably lived rooted to shell debris by its tufted pedicle. The benthos also includes the ubiquitous siliceous hexactinellid sponges, echinoids, notably regular, epifaunal grazing forms such as *Glyphocyphus*, and the epifaunal or semi-infaunal ploughing deposit feeder *Discoides*. Starfish are usually represented by their scattered ossicles; a specimen of the genus *Calliderma* is shown here. Faunas of this type occur in the Middle Chalk of Britain from south Devon to Kent and north through the Salisbury Plain, Berkshire Downs and Cotswolds to the Lincolnshire and Yorkshire Wolds.

Fig. 104   Early Turonian Inoceramus Shell Bed Community
a *Mammites nodosoides* (Mollusca: Cephalopoda: Ammonoidea)
b *Lewesiceras lewesiense* (Mollusca: Cephalopoda: Ammonoidea)
c *Mytiloides labiatus* (Mollusca: Bivalvia: Pterioida)
d hexactinellid sponge (Porifera: Hyalospongea)
e *Discoides dixoni* (Echinodermata: Echinoidea)
f *Terebratulina lata* (Brachiopoda: Articulata: Terebratulida)
g *Orbirhynchia cuvieri* (Brachiopoda: Articulata: Rhynchonellida)
h *Calliderma smithiae* (Echinodermata: Asterozoa — asteroid)
i *Glyphocyphus radiatus* (Echinodermata: Echinoidea)
j *Thalassinoides* (Trace fossil: Crustacea)
k *Cocinnithyris obesa* (Brachiopoda: Articulata: Terebratulida)
l *Pycnodonte vesicularis* (Mollusca: Bivalvia: Pterioida — oyster)

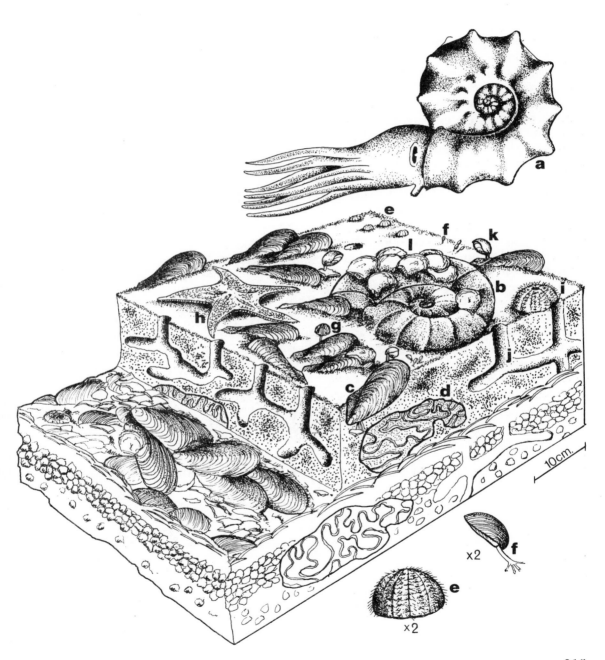

105 Late Turonian Hardground Community

The Chalk Rock is a spectacular complex of glauconitized and phosphatized hardgrounds which form a prominent rock band at the base of the English Upper Chalk over much of central England. Unlike most other faunas of the English Upper Chalk, lithification took place before the aragonite in the shells dissolved, and the fossilized assemblage closely resembles that of the living animals as it would have appeared at that time. Interpretation of the fauna is complicated, however, by the fact that both hard and soft bottoms (the latter are deposits of chalk ooze in the irregular surface of the hardground) coexisted, and the faunas are mixed. Furthermore, sedimentological and petrographical evidence (Kennedy and Garrison, 1975) suggest that the Chalk Rock formed at relatively shallow depths, and that algal thickets were probably present. The Chalk Rock thus yields diverse and sometimes rich faunas.

Organisms associated with the rocky bottom of the hardground surface include boring animals, such as the bivalve *Martesia,* the sponge *Cliona* and polychaetes; there is also extensive evidence of the presence of boring thallophytes (algae and fungi). Cemented to surfaces are bivalves, serpulids, bryozoans, ahermatypic corals, and sometimes crinoids and sponges, whilst a byssate, attached fauna includes bivalves such as *Modiolus* and *Inoceramus costellatus.* The pockets of soft sediment were inhabited by tiny bivalves including *Cardium, Nuculana* and arcids, and semi-infaunal ploughing echinoids like the holasteroid *Sternotaxis.* Gastropods are sometimes abundant at this level, and those present include sponge-grazing pleurotomariids and various trochids (*Trochus* and *Turbo*) which were probably algal feeders.

The nekton consists of diverse fish and many cephalopods, including nautiloids, and several genera of ammonites; the latter include normally coiled forms such as *Suprionocyclus* and *Lewesiceras,* and loosely coiled heteromorphs such as *Scaphites, Hyphantoceras, Bostrychoceras* and *Allocrioceras.* This community is found in Britain from Berkshire to south Norfolk, and in the northern Weald.

Fig. 105 Late Turonian Hardground Community
a *Cliona* crypts (Porifera: Demospongea)
b *Martesia ? rotunda* (Mollusca: Bivalvia: Myoida)
c *'Cardium' turoniense* (Mollusca: Bivalvia: Veneroida)
d *Inoceramus costellatus* (Mollusca: Bivalvia: Pterioida)
e *Nucula renauxiana* (Mollusca: Bivalvia: Palaetaxodonta — nuculoid)
f *Barbatia geinitzi* (Mollusca: Bivalvia: Arcoida)
g *Septifer lineatus* (Mollusca: Bivalvia: Mytiloida)
h *Calliostoma schlueteri* (Mollusca: Gastropoda: Archaeogastropoda)
i *'Turbo' gemmatus* (Mollusca: Gastropoda: Archaeogastropoda)
j *'Turbo'* (Mollusca: Gastropoda: Archaeogastropoda)

k *Bathrotomaria perspectiva* (Mollusca: Gastropoda: Archaeogastropoda)
l *Subprionocyclus neptuni* (Mollusca: Cephalopoda: Ammonoidea)
m *Subpriongcyclus branneri* (Mollusca: Cephalopoda: Ammonoidea)
n *Scaphites geinitzi* (Mollusca: Cephalopoda: Ammonoidea)
o *Allocrioceras woodsi* (Mollusca: Cephalopoda: Ammonoidea)
p *Hyphantoceras reussianum* (Mollusca: Cephalopoda: Ammonoidea)
q *Bostrychoceras woodsi* (Mollusca: Cephalopoda: Ammonoidea)
r *Holaster (Sternotaxis) planus* (Echinodermata: Echinoidea)
s serpulids (Annelida)
t *Parasmilia* (Coelenterata: Anthozoa: Scleractinia)
u *Lewesiceras mantelli* (Mollusca: Cephalopoda: Ammonoidea)

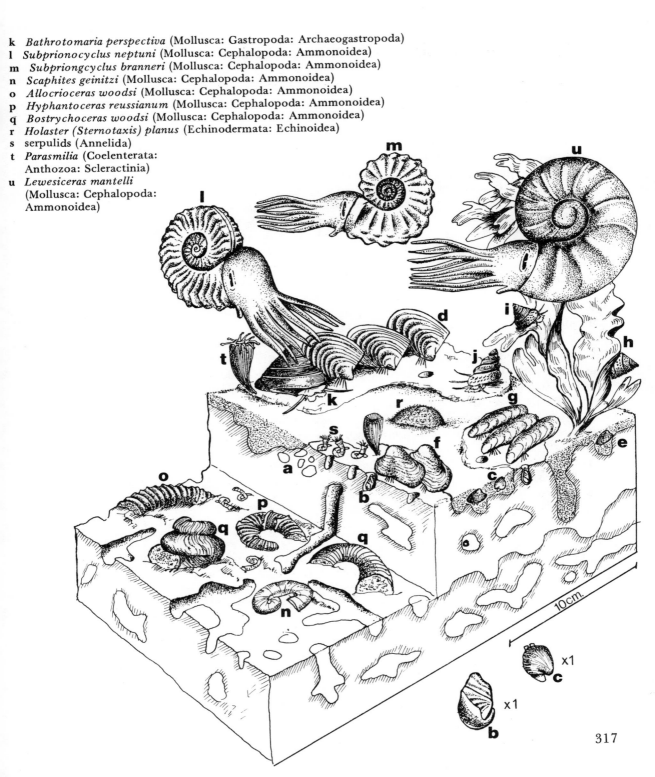

106 Santonian Micraster Chalk Community

The reconstruction of this fauna assemblage is based on a highly fossiliferous soft Santonian chalk on the Kent coast. Faunas of this general composition occur throughout much of the English Upper Chalk, but elsewhere faunal densities are very much lower.

The chalk has been intensively bioturbated, but is so pure that burrow outlines are difficult to detect. Bottoms were probably thixotropic or semi-fluid, and the faunal elements present show adaptations to this habitat. All but the largest aragonitic organisms are absent.

Benthic elements frequently adapted themselves for living on soft bottoms: they sometimes developed long spines to spread the weight of the shell, as in the case of the pectinid *Spondylus,* or club-shaped spines, as with the cidarid echinoid *Tylocidaris.* Inoceramids are scarce; when present they have often developed a bowl-like lower valve, or are vastly expanded and flattened. Shell debris provided hard substrates which were heavily encrusted by bryozoans, serpulids and oysters; they may also have provided a substrate for some brachiopods. Other brachiopods such as *Cretirhynchia,* lay loose on the sediment surface when adult. Echinoderms are a prominent element of these faunas and shown here are *Micraster,* an infaunal or semi-infaunal deposit feeder; *Conulus* and *Echinocorys,* which probably lived partly buried, ploughing through sediment and feeding on detritus; the cushion star *Mitraster* and the crinoid *Bourgueticrinus.* Nekton are represented by the giant ammonite *Parapuzosia* which reaches a diameter close to 2m, and by teeth of the mollusc-eating fish *Ptychodus,* probably responsible for much of the fragmentation of inoceramids and other shells.

This fauna occurs in the Upper Chalk of southern England, in Norfolk, Lincolnshire and Yorkshire, and in parts of the White Limestone of Northern Ireland.

Fig. 106 Santonian Micraster Chalk Community
a *Parapuzosia* (Mollusca: Cephalopoda: Ammonoidea)
b *Ventriculites* (Porifera: Hyalospongea)
c *Echinocorys scutata* (Echinodermata: Echinoidea)
d *Micraster* (Echinodermata: Echinoidea)
e *Conulus albogalerus* (Echinodermata: Echinoidea)
f *Bourgueticrinus ellipticus* (Echinodermata: Crinozoa — crinoid)
g *Mitraster hunteri* (Echinodermata: Asterozoa — asteroid)
h *Spondylus spinosus* (Mollusca: Bivalvia: Pterioida — pectinid)
i *Inoceramus* fragments (Mollusca: Bivalvia: Pterioida
j *Pycnodonte vesicularis* (Mollusca: Bivalvia: Pterioida — oyster)
k *Parasmilia centralis* (Coeleterata: Anthozoa: Scleractinia)
l serpulids (Annelida)

m bryozoans (Bryozoa: Ectoprocta)
n *Gibbithyris* (Brachiopoda: Articulata: Terebratulida)
o *Cretirhynchia* (Brachiopoda: Articulata: Rhynchonellida)
p *Ptychodus* tooth (Vertebrata: Chondrichthyes)
q *Tylocidaris clavigera* (Echinodermata: Echinoidea)

107 Maastrichtian Ostrea Lunata Chalk Community

The highest stage of the Cretaceous, the Maastrichtian, is represented in Britain in limited areas of Northern Ireland, and on the
north Norfolk coast. The huge, glacially transported masses found
in the latter area are the basis for the illustration.

The nekton is dominated by belemnites, a group common
only in the late Campanian stage of Britain, although well known
throughout the whole of Upper Cretaceous of more northerly
areas of Europe. The heteromorph ammonite *Bostrychoceras* and
the nautiloid *Hercoglossa* also occur. The benthos is similar to that
of other soft bottom chalk faunas already described, with masses
of hexactinellid and other sponges, bivalves adapted to soft substrates, for instance the plicate *'Ostrea' lunata,* the free-living
rhynchonellid *Cretirhynchia,* vagrant arthropods, and echinoids
such as *Echinocorys* and *Galerites.* Hard substrates such as shells
and belemnite guards are heavily bored, and encrusted by oysters
(*Pycnodonte*), serpulids and bryozoa.

The sediment is intensively burrowed, with a typical *Chondrites,
Planolites* and *Thalassinoides* association. This fauna is seen in
Britain on the coasts of north Norfolk and Northern Ireland.

Only the lower parts of the Maastrichtian occurs on mainland
Britain. Chalks and limestones with faunas rather similar to those
described here with regard to Britain extend into the late Maastrichtian in Denmark, Belgium and Holland. These areas show facies
changes progressing up the section of a kind that suggest a shallowing of water, associated with an overall marine regression.
Bryozoan bioherms, calcarenites and carbonate shoal facies are
also widespread in northern Europe, whilst the Maastrichtian/
Palaeocene boundary is everywhere a non-sequence. In the North
Sea Basin, chalks and chalky limestones continued to be deposited
during the early Tertiary, and this is also true in parts of Scandinavia, but these Palaeocene chalks and chalky limestones contain
different faunas from those of the Cretaceous. Ammonites, belemnites, marine saurians and many groups of bivalves, echinoderms
and foraminiferida have disappeared, and the faunas take on a
distinctive Tertiary aspect.

Fig. 107 Maastrichtian Ostrea Lunata Chalk Community
a *Belemnella lanceolata* (Mollusca: Cephalopoda: Coleoidea — belemnite)
b *Belemnella lanceolata* (Mollusca: Cephalopoda: Coleoidea — belemnite guard
c *Hercoglossa* (Mollusca: Cephalopoda: Nautiloidea)
d *Bostrychoceras* (Mollusca: Cephalopoda: Ammonoidea)
e *Enoploclytia* (Arthropoda: Crustacea: Malacostraca — decapod)
f *Seliscothon* (Porifera: Demospongea)
g hexactinellid sponges (Porifera: Hyalospongea)
h *'Ostrea' lunata* (Mollusca: Bivalvia: Pterioida — oyster)

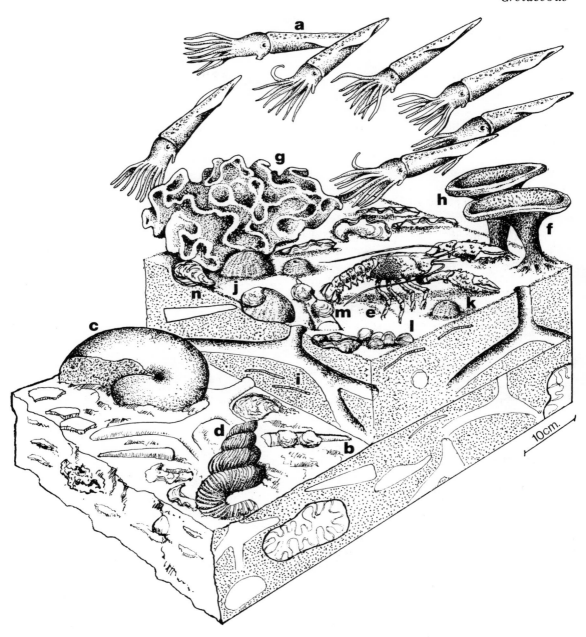

i *Inoceramus* fragments (Mollusca: Bivalvia: Pterioida)
j *Echinocorys ciplyensis* (Echinodermata: Echinoidea)
k *Galerites* (Echinodermata: Echinoidea)
l *Cretirhynchia* (Brachiopoda: Articulata: Rhynchonellida)
m *Pycnodonte vesicularis* (Mollusca: Bivalvia: Pterioida — oyster)
n *Ostrea* (Mollusca: Bivalvia: Pterioida — oyster)

CRETACEOUS FAUNAS ELSEWHERE IN EUROPE AND NORTH AMERICA

The faunas discussed in the preceding pages are typical of the Boreal Realm and can be traced as far as Transcaspia and northern Iran, facies of different ages tending to yield broadly similar faunas throughout the region (Fig. o, p. 281).

As they did in the Jurassic, these Boreal faunas merge into Tethyan (or Mesogean) associations as they progress southward. These Tethyan faunas are as diverse and variable as those of the Jurassic of the same areas, and carbonate sequences are prominent. Particularly distinctive are the spectacular occurrences of the aberrant heterodont bivalves known as the rudists, which were abundant and varied in the late Cretaceous in the tropical shallow water areas of southern Europe. Taking the place, in many respects, of hermatypic corals, they form thickets and framework structures. Other shallow water Tethyan facies are dominated by algae, or by giant Foraminiferida. Deeper water facies include pelagic marls and clays with ammonite-belemnite faunas; these are closely similar to the Jurassic clays of northern Europe. The more calcareous facies resemble the clay-rich parts of the northern chalks. Deep water deposits are also found, similar to those of the Jurassic and including radiolarian sediments; in Cyprus and some other areas where Cretaceous ocean floor sediments are exposed, deep sea chalks occur. In other areas of central and southern Europe there are widespread turbidite successions.

Rudist-algal-foraminiferidal shallow water facies occur in the Caribbean Sea, Mexico and parts of Texas. During the late Cretaceous (chiefly the Coniacian to Campanian) chalks were deposited in many areas of North America from Texas in the south to the Western Interior towards the north. However, they yield very different faunas from those of the European chalks, lacking brachiopods, bryozoa and various others, and were perhaps deposited in an epeiric, landlocked sea like that of Jurassic Europe, with reduced or perhaps fluctuating salinity. In many other respects besides this the Cretaceous successions of the North American Western Interior more closely resemble those of the terrigenous clastic facies of the European Jurassic.

In contrast, the sequences of the Atlantic seaboard, the Gulf Coast and the Mississippi embayment, which include greensands, calcarenites and chalks, are reminiscent of many of the more marginal facies of the European Cretaceous.

Cenozoic

Several widespread and fundamental changes in the composition of marine faunas occurred in the early Cenozoic. Many major components of Mesozoic communities disappeared, including ammonites, belemnites, inoceramid and rudist bivalves and many previously abundant gastropods including *Nerinea* and *Actaeonella*. Other forms such as trigoniid bivalves and the gastropod *Pleurotomaria*, common in Mesozoic shelf areas, became very restricted in distribution during the Tertiary; *Trigonia* is now confined to southern Australia, and *Pleurotomaria* to deeper waters. Brachiopods diminished considerably in importance, and many echinoid groups common during the Cretaceous also disappeared. The marine reptiles, icthyosaurs, plesiosaurs and pliosaurs became extinct, while on the land the extinction of the dinosaurs was followed by a great expansion of the mammals.

The Cenozoic saw the rise in importance of many marine benthic groups; most of these originated in the Cretaceous, but were often present then in rather subordinate numbers. A good example of this are the predatory gastropods, mainly from the order Neogastropoda, which comprise about half the species of gastropods in Eocene assemblages. Groups which probably preyed on echinoids, polychaetes, sipunculids, bivalves and gastropods were present, and there is an abundance of polychaete feeding specialists which indicates a diverse and abundant polychaete fauna, although little of this is preserved. Forms with sophisticated predatory devices, such as *Conus* with its barbed, dart-like radula accompanied by toxic venom, appeared. All these families of predatory gastropods had evolved since the Albian, but reached significant numbers only in the latest Cretaceous. Many heterodont bivalve groups such as the large, extremely diverse, infaunal suspension-feeding superfamily Veneracea and the deposit-feeding Tellinacea, although present during the Cretaceous, underwent rapid development and diversified at the beginning of the Cenozoic to become very important elements in Cenozoic bottom communities. Many of the important groups of reef-building corals appeared in the Late Cretaceous, although they are not particularly common in the British Cenozoic, and diversified during the

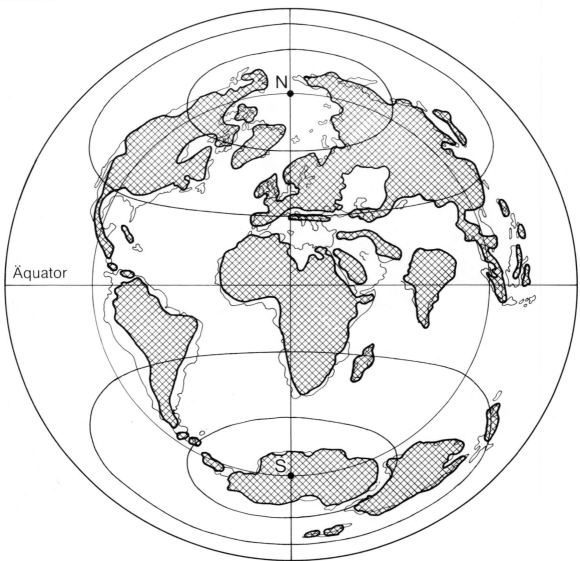

Eocene, whilst many of the older coral groups became extinct during the early Cenozoic.

Fig. q. *The world during the Early Cenozoic. Positions of the continents after Briden et al. 1974.*

PALAEOGEOGRAPHY

The palaeolatitude of England and Germany during most of the Palaeogene was approximately 40° N (Fig. q); the climate throughout this time was certainly warmer than the present and, in general, the faunas and floras indicate subtropical and tropical conditions (Fig. r).

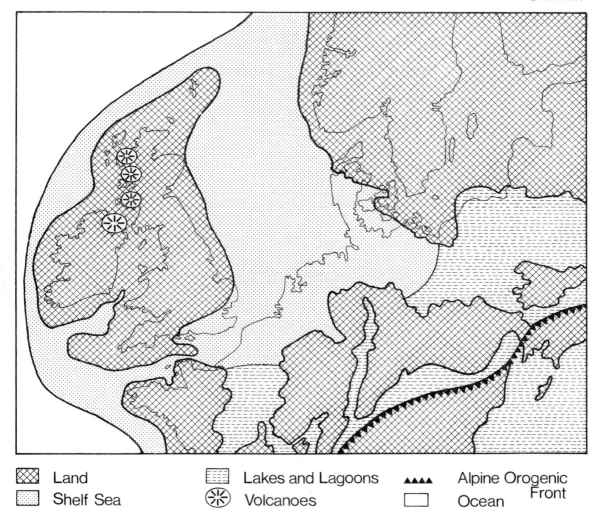

Land

Shelf Sea

Lakes and Lagoons

Volcanoes

Alpine Orogenic Front

Ocean

Fig. 1. *Northern Europe during the Early Cenozoic. (Modified after Ziegler, 1975).*

The history of the north European Cenozoic is effectively the history of the North Sea Basin, where deposition has occurred almost continuously from the Palaeocene to the present.

In the Mesozoic, rifting occurred in the North Sea area associated with the beginning of the crustal separation of the European and the North American-Greenland plates. When plate separation had taken place, and the North Atlantic rifting was under way in Late Palaeocene and Early Eocene times, rifting in the North Sea ceased. It was probably about this time that the Early Cenozoic basalts of Northern Ireland and north-west Scotland erupted.

Neogene:	Pleistocene Pliocene Miocene
Palaeogene:	Oligocene Eocene Palaeocene

Table IX Stratigraphic Divisions of the Cenozoic

Many of these lavas have red weathered tops and are overlain by sediments which suggest terrestrial conditions. Thus it can be assumed that during the time of most of this igneous activity the area between the North Sea and the Atlantic was uplifted. Much of Britain has remained above sea level ever since, though the south of England and East Anglia have been submerged from time to time. The North Sea, however, has continued to subside since the Eocene.

In the Palaeogene (Table IX), the sea covered south-east England, northern France and Belgium, and extended into northern Germany, Denmark and Poland. Most of Britain was above sea level, and the North Sea Basin was closed on the south-west, south, and east by several small land areas (Fig. 4), but had periodic connections with the early Atlantic Ocean along the line of the present western English Channel. To the north, in the vicinity of Norway, there was a probably more permanent connection with the Atlantic, but this is not clear. To the south, the Aquitaine Basin, another shallow marine basin, opened on to the Atlantic, which, further south, joined with the Tethys Ocean. The British Isles then were at the western edge of this sea, and in the Palaeogene beds of south-east England we see alterations of fully marine conditions, estuarine deposits and deltaic fluviatile and marsh sediments, as the shore line of the Palaeogene sea oscillated across south-east England (Curry, 1965, 1967).

To the south of the North Sea Basin lay the gulf of Aquitaine and the Tethys Ocean, and there were intermittent connections between these three areas throughout the Palaeocene.

The Tethyean shallow water faunas are characterized by a high diversity and abundance of nummulitic foraminfera of which only a few species penetrated as far north as the North Sea Basin. The molluscan and echinoid faunas of the Tethys were also much more diverse and many genera such as *Tridacna, Strombus, Lambis, Cypraea,* now typical of the modern tropical Indo-Pacific province, originated at this time. Throughout the Palaeocene of Europe there are beds containing high proportions of Tethyean fossils indicating periodic warming of the sea water.

Hamstead Beds	Marine, estuarine and lacustrine clays
Bembridge Beds and Osborne Beds	Marine and freshwater marls and limestones
Headon Beds	Marine and freshwater sands and clays
Barton Beds	Marine sands and clays
Bracklesham and Bagshot Beds	Marine and fluvial sands
London Clay	Marine clay with a basal sand
Oldhaven and Blackheath Beds	Estuarine sands
Woolwich and Reading Beds	Marine and estuarine clays and sands with freshwater clays
Thanet Beds	Marine sands

Table X   The Palaeogene formations of southern England

At the end of the Cretaceous, the sea had withdrawn eastwards from most of Britain, and the Chalk was gently deformed and tilted towards the south-east. Palaeogene sedimentation in England began with a marine transgression when sands were deposited in the London Basin. Further transgression of the sea covered most of south-east England from the Isle of Wight to East Anglia. Large rivers to the west and north-west deposited deltaic and fluviatile sediments in the western and southern parts of the sea (Reading Beds) whilst further to the east were estuarine mud flats followed by offshore sands (Woolwich Beds) (Table X).

Further deposition of estuarine sediments, again showing progressively reduced salinities in the west, is shown in the Oldhaven and Blackheath beds. The transgression which marked the beginning of the London Clay (which consists of sands below and clay above) spread over the whole of south-east England and extended over northern France into north-west Germany and Denmark. The London Clay sea in the London area was at least 100m deep, but to the west, in the Hampshire area, it was much shallower. The north-eastern part of this sea in the vicinity of Denmark was clearly much deeper, and fine laminated clays were deposited containing few benthic animals but abundant plankton. Volcanoes to the north deposited ash layers in the Denmark area, one of these extending as far south as Harwich in Essex. A connection between the London Clay sea and the Atlantic in the south, in the region of the present western English Channel, may have opened in late London Clay times. In the Hampshire Basin, the London Clay was much more sandy and shallow, and its shoreline facies can be recognized in the Isle of Wight. The margins of the London Clay seas were covered with subtropical plants, and land temperatures seem to have been about 20–25° C, although foraminiferal evidence indicates shallow water temperatures of about 16° C.

Towards the end of London Clay times the water became shallower, and fluviatile and deltaic sediments are present over most of south-east England. The succeeding Bracklesham Beds exemplify well the marginal character of the marine environments close to the shoreline, showing interdigitations of deltaic and fluviatile sediments with normal shallow water nearshore shelf sediments. At the end of Bracklesham times a further marine transgression resulted in south-east England being covered by a shallow shelf sea with a water depth up to about 100m and with clay and sand deposits, the Barton Beds. This sea gradually decreased in depth, apparently through silting up, and turned into the shallow lagoons and freshwater lakes of the lower Headon Beds. A further small marine transgression in middle Headon times was again succeeded by estuarine and lacustrine conditions represented by the Upper Headon, Osborne, Bembridge and Lower Hampstead Beds (Daley, 1972).

PALAEOGENE

108 Palaeocene Marine Sand Community

The depositional environment of this community was an offshore shelf sea with a water depth of probably not more than 50m and a substrate of fine glauconitic sand. This sand community was dominated by infaunal bivalves; various levels within the sediment were occupied by suspension feeding bivalves, while large, but fairly sluggish, *Arctica morrisi* was abundant near the surface of the sand; and the patchily distributed small *Corbula regulbiensis* was buried just beneath it. *Dosiniopsis,* with larger siphons, lived lower down in the substrate, and the sedentary *Panopea intermedia* with large, fused siphons, occurred in the deepest levels. In some sands *Corbula* was the dominant bivalve, often in small aggregations or 'nests'. Gastropods were much less common, but included the detritus-feeding *Aporrhais triangulata,* the semi-infaunal *Euspira* which preyed mainly upon bivalves, and the whelk *Siphonalia,* which may have fed on polychaete worms.

The oldest Cenozoic rocks exposed in Britain are fine glauconitic sands with thin clay bands (Thanet Beds); they are widespread in the London Basin and extend as far as Norfolk, and, although largely unfossiliferous, they do contain communities like the one described here.

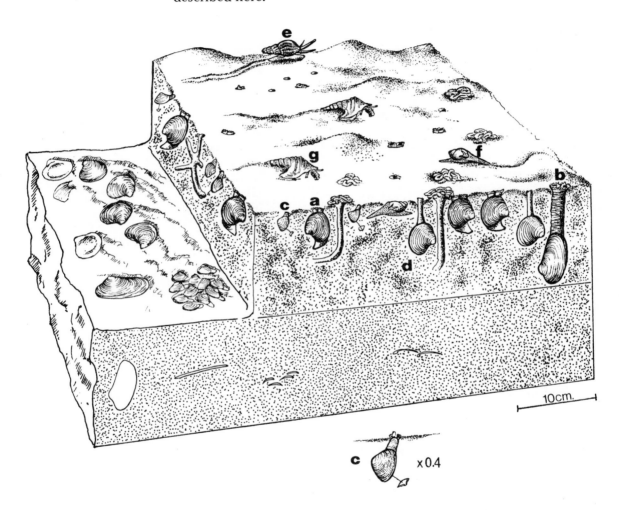

Fig. 108 Palaeocene Marine Sand Community
a *Arctica morrisi* (Mollusca: Bivalvia: Veneroida)
b *Panopea intermedia* (Mollusca: Bivalvia: Myoida)
c *Corbula regulbiensis* (Mollusca: Bivalvia: Myoida)
d *Dosiniopsis bellovacina* (Mollusca: Bivalvia: Veneroida)
e *Siphonalia subnodulosa* (Mollusca: Gastropoda: Neogastropoda)
f *Euspira bassae* (Mollusca: Gastropoda: Mesogastropoda)
g *Aporrhais triangulata* (Mollusca: Gastropoda: Mesogastropoda)

109 Estuarine Sand Community

The community illustrated is typical of early Eocene low salinity, intertidal or shallow sublittoral, estuarine areas with muddy sand substrates. A characteristic feature of this type of environment is the rather low diversity and the high abundance of animals. The dominant element in this community was the shallow-burrowing suspension-feeding bivalve *Corbicula,* which occurred in huge numbers. Also common was the oyster *Ostrea bellovacina.* Gastropods were abundant, and the common forms belong to the superfamily Cerithiacea. They included *Brotia melanoides, Melanopsis antidiluviana* and *Tympanotonos funatus*; these species were characterized by large populations which were extremely variable morphologically. Forms related to these genera abound today in subtropical and tropical estuarine mud flats, where they occur in densities as high as 2—3,000 per square metre, and feed upon organic detritus, diatoms and filamentous algae from the surface layer of sediment. Fish remains, particularly of sharks and rays, can be abundant in this deposit. Comparable communities, though with different genera, are found today in subtropical and tropical estuaries in south-east Asia and West Africa, environments characterized by fluctuating salinities, soft substrates and high productivity.

   The type of community illustrated is recurrent throughout the British Cenozoic deposits, and examples can be found in many European formations. This reconstruction is from the Blackheath Beds which with the Oldhaven Beds, form a sequence of largely sandy sediments overlying the Woolwich and Reading Beds in south-east England.

110 Eocene Marine Mud Community

Although there are some sands at the base, most of the London Clay in the London Basin is formed from a mud sediment deposited in a marine basin open to the north and east, with a water depth of about 150m. The fauna of the Lower London Clay has a fairly low diversity and fossils are often extremely scarce, but some levels show a more abundant fauna; this community is based upon such a horizon at Aveley, Essex (Kirby, 1974).

   The dominant species were the deposit-feeding *Nucula bowerbanki,* the siphonate suspension feeder *Nuculana amygdaloides,* and two species of the lucinoid *Thyasira* which were suspension feeders occupying semi-permanent burrows and connected to the

Fig. 109 Eustarine Sand Community
a *Corbicula cuneiformis* (Mollusca: Bivalvia: Veneroida)
b *Corbicula tellinoides* (Mollusca: Bivalvia: Veneroida)
c *Ostrea bellovacina* (Mollusca: Bivalvia: Pterioida)
d *Brotia melanoides* (Mollusca: Gastropoda: Mesogastropoda)
e *Melanopsis antidiluviana* (Mollusca: Gastropoda: Mesogastropoda)
f *Theodoxus* (Mollusca: Gastropoda: Archaeogastropoda)
g *Tympanotonus funatus* (Mollusca: Gastropoda: Mesogastropoda)
h *Diaphyodus* (Vertebrata: Osteichthyes)

surface by an inhalent tube. The scaphopod *Laevidentalium*, which probably fed mainly upon benthic foraminifera, was also common. Most of the gastropods were semi-infaunal species, which spent much of their time ploughing through the sediment just below the surface. Species of the family Turridae were particularly common; many of them were equipped with highly developed dart-like radula teeth and a toxic venom with which they immobilized their prey. The Turridae feed today mainly upon polychaetes, which are swallowed whole. The gastropod *Euspira* probably fed largely on other molluscs, boring holes in their shells; *Ficus* and *Galeodea* probably ate echinoderms and *Streptolathyrus* probably bivalves. *Solariaxis* has living relatives which live in association with, and feed upon, various soft-bodied coelenterates such as sea anemones. The tubes secreted by the polychaete *Ditrupa* are common throughout the London Clay; although the genus is still extant, very little is known about its living habits, but it was probably a surface-living deposit feeder.

Sunken drifted wood is abundant in the London Clay, and this material, including the palm *Nipa*, is famous as a diverse tropical flora (Davis and Elliott, 1958). The pieces of drifted wood often contain the tubes and valves of the wood-boring bivalve *Teredo*, the ship worm. Pteriid bivalves and the crinoid *Balanocrinus subbasaltiformis* were probably also attached to this wood.

London Clay sediments contain large numbers of otoliths from teleost fishes as well as the teeth of sharks and rays. Benthic invertebrates are the major part of the food for many fish; the behaviour and the number of molluscs, polychaetes and echinoderms would have been controlled largely by predatory fish.

Assemblages similar to this one, containing abundant *Thyasira* and nuculid bivalves, are known today from deep, very quiet, basinal situations on the continental shelf, floored with black mud, such as the deep basins in the northern North Sea and the deeper parts of Scottish sea lochs.

Fig. 110   Eocene Marine Mud Community

a *Odontaspis* (Vertebrata: Chondichthyes)
b *Synodus* (Vertebrata: Osteichthyes)
c *Dirematichthys* (Vertebrata: Osteichthyes)
d *Nucula bowerbanki* (Mollusca: Bivalvia: Palaeotaxodonta)
e *Nuculana amygdaloides* (Mollusca: Bivalvia: Palaeotaxodonta)
f *Pteria papyracea* (Mollusca: Bivalvia: Pterioida)
g *Thyasira angulata* (Mollusca: Bivalvia: Veneroida)
h *Thyasira goodhalli* (Mollusca: Bivalvia: Veneroida)
i *Laevidentalium nitens* (Mollusca: Scaphopoda)
j *Cimomia imperialis* (Mollusca: Cephalopoda: Nautiloidea)
k *Turricula teretrium* (Mollusca: Gastropoda: Neogastropoda)
l *Streptolathyrus cymatodis* (Mollusca: Gastropoda: Neogastropoda)
m *Fisus* (Mollusca: Gastropoda: Mesogastropoda)
n *Galeodea gallica* (Mollusca: Gastropoda: Mesogastropoda)
o *Solariaxis gallica* (Mollusca: Gastropoda: Mesogastropoda)
p *Euspira glaucinoides* (Mollusca: Gastropoda: Mesogastropoda)
q *Ditrupa plana* (Annelida)
r *Teredo* (Mollusca: Bivalvia: Myoida)

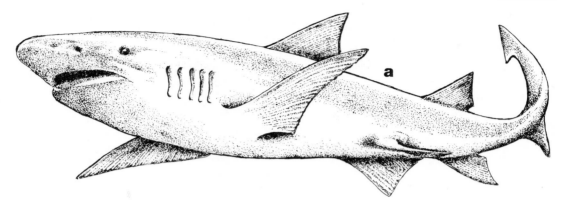

10cm.

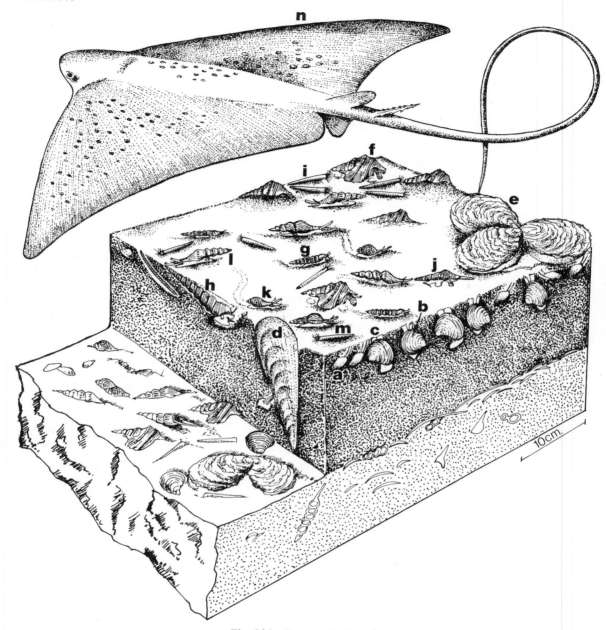

Fig. 111 Eocene Shallow Marine Clay Communities
a *Nuculana amygdaloides* (Mollusca: Bivalvia: Palaeotaxodonta)
b *Pitar sulcatarius* (Mollusca: Bivalvia: Veneroida)
c *Arctica morrisi* (Mollusca: Bivalvia: Veneroida)
d *Pinna affinis* (Mollusca: Bivalvia: Pterioida)
e *Ostrea ovicina* (Mollusca: Bivalvia: Pterioida)
f *Aporrhais sowerbyi* (Mollusca: Gastropoda: Mesogastropoda)
g *Orthochetus elongatus* (Mollusca: Gastropoda: Mesogastropoda)

111 Eocene Shallow Marine Clay Communities

The general character of the Eocene upper London Clay communities is illustrated by this *Aporrhais*-dominated community. It shows a much more diverse and more numerous benthic fauna than the lower London Clay, with a much shallower water aspect. Amongst the epifaunal species there was a wide variety of gastropods, the most abundant of which was *Aporrhais sowerbyi*. This had a thick and heavy shell with a broad apertural lip which extended to become digitate processes. Recent species of *Aporrhais* often live partially buried; the broad lip and process separate the inhalent and exhalent respiratory currents. It probably fed upon diatoms and algal detritus. *Turritella sulcifera,* also common, was an infaunal suspension feeder: detritus particles were collected from the gills and trapped in a mucous string which was then pulled into the mouth. *Tibia* and *Orthochetus* were also vegetarian, while *Crenaturricula* and *Pleurotoma* probably ate polychaetes. The polychaete *Ditrupa* is again abundant.

The bivalves include *Ostrea ovicina* and a large number of infauna, mostly suspension feeders, like the byssate *Pinna*, which, although sedentary, can withdraw deep into its shell and rapidly repair any damage to the shell's exposed posterior. Other common bivalves include the sluggish, shallow-burrowing *Arctica* and the more mobile, siphonate *Nuculana* and *Pitar*.

Teleost fish and rays were abundant, the latter feeding mainly upon infaunal bivalves.

Similar communities to this may be found living today in shallow shelf environments on muddy sands at depths of 10—50m in warm temperate to subtropical areas.

The London Clay of the Hampshire Basin and the higher parts of the succession in the London Basin represent much shallower water conditions than the previous community. There is great variation in the fossil assemblages from bed to bed and this probably represents a mosaic of bottom communities which developed in a shallow shelf sea 20—50m deep. The community illustrated is one of the many diverse assemblages encountered at Shinfield (James et al. 1974).

h *Turritella sulcifera* (Mollusca: Gastropoda: Mesogastropoda)
i *Ditrupa plana* (Annelida)
j *Tibia lucida* (Mollusca: Gastropoda: Mesogastropoda)
k *Ficus londini* (Mollusca: Gastropoda: Mesogastropoda)
l *Pleurotoma* (Mollusca: Gastropoda: Neogastropoda)
m *Streptolathyrus cymatodis* (Mollusca: Gastropoda: Neogastropoda)
n ray (Vertebrata: Chondrichthyes)

Cenozoic

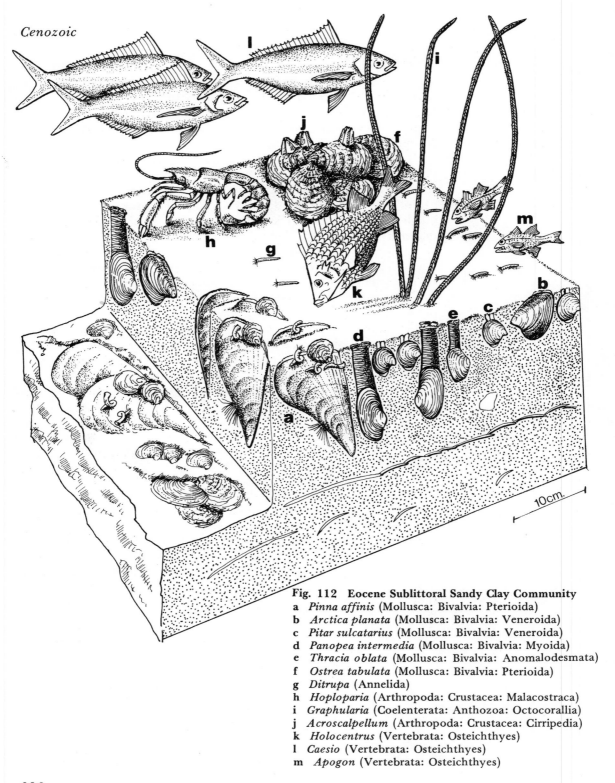

Fig. 112 Eocene Sublittoral Sandy Clay Community
a *Pinna affinis* (Mollusca: Bivalvia: Pterioida)
b *Arctica planata* (Mollusca: Bivalvia: Veneroida)
c *Pitar sulcatarius* (Mollusca: Bivalvia: Veneroida)
d *Panopea intermedia* (Mollusca: Bivalvia: Myoida)
e *Thracia oblata* (Mollusca: Bivalvia: Anomalodesmata)
f *Ostrea tabulata* (Mollusca: Bivalvia: Pterioida)
g *Ditrupa* (Annelida)
h *Hoploparia* (Arthropoda: Crustacea: Malacostraca)
i *Graphularia* (Coelenterata: Anthozoa: Octocorallia)
j *Acroscalpellum* (Arthropoda: Crustacea: Cirripedia)
k *Holocentrus* (Vertebrata: Osteichthyes)
l *Caesio* (Vertebrata: Osteichthyes)
m *Apogon* (Vertebrata: Osteichthyes)

336

112   Eocene Sublittoral Sandy Clay Community

Conspicuous elements here are groups of the infaunal, sedentary bivalve, *Pinna affinis,* which today may be abundant in sheltered, stable substrates, particularly those colonized by marine grasses. The long-siphoned, deep-burrowing bivalves *Panopea* and *Thracia* are also common, with shallow burrowers represented by *Arctica* and *Pitar.* No deposit-feeding bivalves occur in this community.

Epifaunal animals are the oyster, the large barnacle *Acroscalpellum,* the worm *Ditrupa,* the lobster *Hoploparia*, and the erect, stick-like gorgonacean *Graphularia.*

This third London Clay community is found in the Hampshire Basin at Bognor.

113   Eocene Glauconite Sandy Clay Community

The community inhabited a nearshore, shallow water shelf environment with a water depth of between 10–30m. Like most other Cenozoic communities it was dominated numerically by shallow-burrowing, suspension-feeding bivalves; and important elements were the large thick-shelled *Venericardia planicosta, Glycymeris pulvinata,* and *Crassatella,* none of which had siphons. Slightly deeper siphonate burrowers were *Callista laevigata* and *Macrosolen.* Other common infaunal animals were the scaphopod *Dentalium* and the ciliary-feeding gastropods *Mesalia* and *Turritella. Euspira* probably preyed upon shallow-burrowing bivalves. Amongst the epifaunal elements in the community were the ciliary-feeding, limpet-like *Calyptraea,* which probably lived in groups attached to empty shells. *Arca biangula,* a suspension-feeding bivalve, was byssally attached, probably again to empty shells. The large gastropod *Athleta* probably preyed mainly upon bivalves, pulling the valves apart with its large extensible foot. The highly ornamented gastropod *Pterynotus* fed by means of holes drilled through the shells of epifaunal and shallow-burrowing bivalves. *Hemipleurotoma* probably ate infaunal polychaetes. The herbivorous gastropod *Campanile cornucopiae* reached lengths of 50cm, and was one of the largest gastropods ever to live. The solitary ahermatypic coral, *Turbinolia,* probably lived partially embedded in the sediment.

Abundant teeth of sharks and rays, teleost fish otoliths and occasional cuttlebones, indicate a diverse nektonic fauna, largely dependent upon the benthic invertebrates for food.

The Bracklesham Beds, exposed in the Hampshire Basin and the Isle of Wight, consist of two main facies: one of glauconitic sandy clays with a marine fauna, and the other of laminated clays,

Fig. 113 Eocene Glauconite Sandy Clay Community

a *Glycymeris pulvinata* (Mollusca: Bivalvia: Arcoida)
b *Arca biangula* (Mollusca: Bivalvia: Arcoida)
c *Crassatella sowerbyi* (Mollusca: Bivalvia: Veneroida)
d *Venericardia planicosta* (Mollusca: Bivalvia: Veneroida)
e *Callista laevigata* (Mollusca: Bivalvia: Veneroida)
f *Corbula gallica* (Mollusca: Bivalvia: Myoida)
g *Dentalium* (Mollusca: Scaphopoda)
h *Turritella conoides* (Mollusca: Gastropoda: Mesogastropoda)
i *Mesalia* (Mollusca: Gastropoda: Mesogastropoda)
j *Calyptraea aperta* (Mollusca: Gastropoda: Mesogastropoda)
k *Athleta spinosa* (Mollusca: Gastropoda: Neogastropoda)
l *Hemipleurotoma* (Mollusca: Gastropoda: Neogastropoda)
m *Macrosolen hollowaysii* (Mollusca: Bivalvia: Veneroida)
n *Nemocardium* (Mollusca: Bivalvia: Veneroida)
o *Campanile conucopiae* (Mollusca: Gastropoda: Mesogastropoda)
p *Euspira* (Mollusca: Gastropoda: Mesogastropoda)
q *Pterynotus* (Mollusca: Gastropoda: Neogastropoda)
r *Turbinolia* (Coelenterata: Anthozoa: Scleractinia)
s *Sepia* (Mollusca: Cephalopoda: Coleoidea — cuttlefish)
t *Squatina* (Vertebrata: Chondrichthyes)

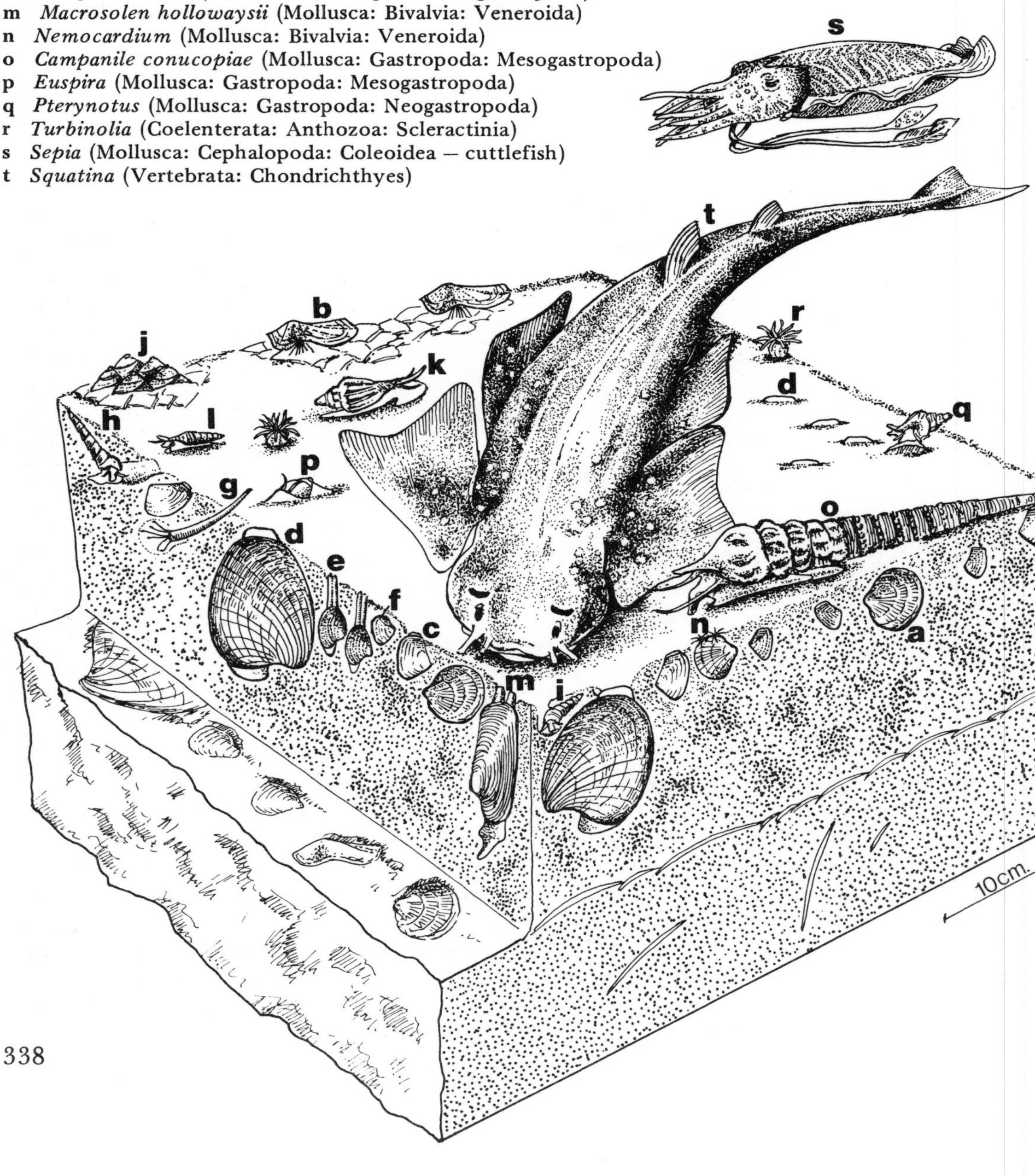

sands and lignite beds of deltaic and marsh origin (Fisher, 1862). Palaeogeographic reconstructions show that at this time the Isle of Wight and the Hampshire areas were at the edge of a shallow shelf sea into which a river system, flowing from the west and north-west, deposited deltaic sediments. During Bracklesham times an embayment formed in the shoreline, with shallow clear warm water, and perhaps sea-grass beds. Later, the embayment silted up, and marshland probably developed all around. The foraminiferal evidence suggests that sea water temperatures were around 18° C and salinities normal or very slightly reduced.

114 Eocene Shelf Silt and Sand Community

Bivalves were very common here, in particular the shallow-burrowing suspension feeders such as the large, thick-shelled *Venericardia planicostata, Crassatella sulcata, Glycymeris deletus* and the small siphonate *Corbula cuspidata.* Deposit feeders were, with the exception of *Nucula,* infrequent. The most abundant epifaunal bivalve was *Chama squamosa*; although young individuals were cemented to pieces of shell debris, the adults may often have been free-living. Although not illustrated, oysters (*Ostrea plicata*), were also common. Other suspension-feeding animals were the gastropod *Turritella* and the ahermatypic solitary coral *Turbinolia.* Gastropods were particularly abundant, and very diverse. Apart from *Turritella,* the vast majority were predators; particularly important were *Athleta* and *Volutocorbis.* Comparable forms living today have a large foot which envelops bivalve prey and pulls the valves apart. *Natica* is a well known predator, feeding upon other molluscs by drilling holes. Other carnivorous gastropods, including *Conus* and *Bathytoma,* were particularly common and diverse; by means of a highly developed dart-like radula they injected a toxic salivary secretion into the (usually polychaete) prey which is then swallowed whole.

The nektonic fauna was diverse, with sharks and rays being particularly common.

Assemblages of organisms in the Barton Beds are extremely diverse, and a large number of species, especially of mollusca, have been recorded (Burton, 1933). The communities are all characteristic of nearshore to offshore shelf habitats developed on silt and sandy substrates; the water depth throughout most of the beds was probably not more than about 50m, and foraminiferal evidence suggests a sea temperature of about 16–18° C; however, the molluscan evidence points to a rather higher temperature. Bed by bed studies have shown quite a large variation in the faunal composition of the communities, but the assemblage illustrated here from the well known *Chama* bed shows the general character of the communities.

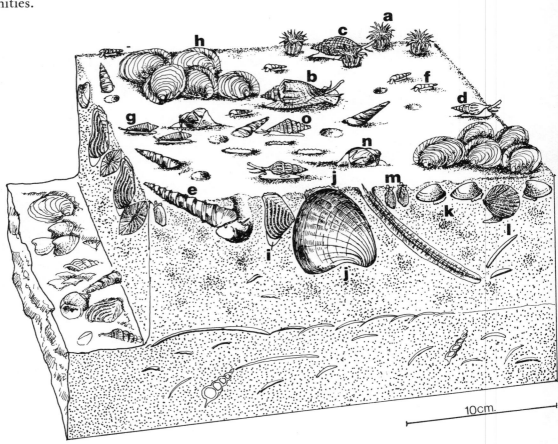

Fig. 114   Eocene Shelf Silt and Sand Community
a *Turbinolia* (Coelenterata: Anthozoa: Scleractinia)
b *Athleta spinosus* (Mollusca: Gastropoda: Neogastropoda)
c *Volutocorbis scabricula* (Mollusca: Gastropoda: Neogastropoda)
d *Pollia labiata* (Mollusca: Gastropoda: Neogastropoda)
e *Turritella imbricataria* (Mollusca: Gastropoda: Mesogastropoda)
f *Bittium semigranosum* (Mollusca: Gastropoda: Mesogastropoda)
g *Conus scrabicula* (Mollusca: Gastropoda: Neogastropoda)
h *Chama squamosa* (Mollusca: Bivalvia: Veneroida)
i *Crassatella sulcata* (Mollusca: Bivalvia: Veneroida)
j *Venericardia planicosta* (Mollusca: Bivalvia: Veneroida)
k *Nucula similis* (Mollusca: Bivalvia: Palaeotaxodonta)
l *Glycymeris deletus* (Mollusca: Bivalvia: Arcoida)
m *Corbula cuspidata* (Mollusca: Bivalvia: Myoida)
n *Natica ambulacrum* (Mollusca: Gastropoda: Mesogastropoda)
o *Bathytoma turbida* (Mollusca: Gastropoda: Neogastropoda)

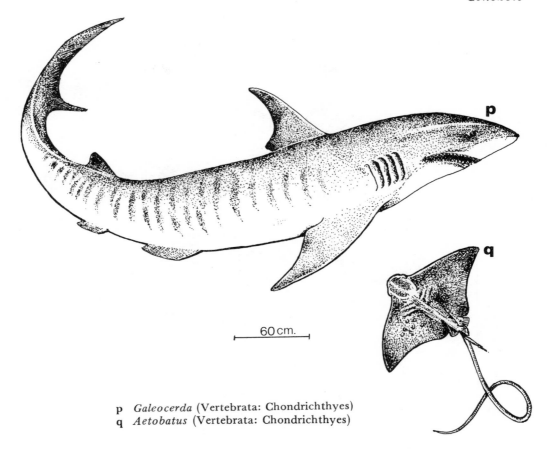

p *Galeocerda* (Vertebrata: Chondrichthyes)
q *Aetobatus* (Vertebrata: Chondrichthyes)

60 cm.

115 Palaeogene Brackish Water Community

The community illustrated was of low diversity, but had large populations of a few species. The substrate surface was dominated by elongate high-spired gastropods like *Batillaria* and *Potamides*. These fed upon the surface layer of the sediment, extracting food

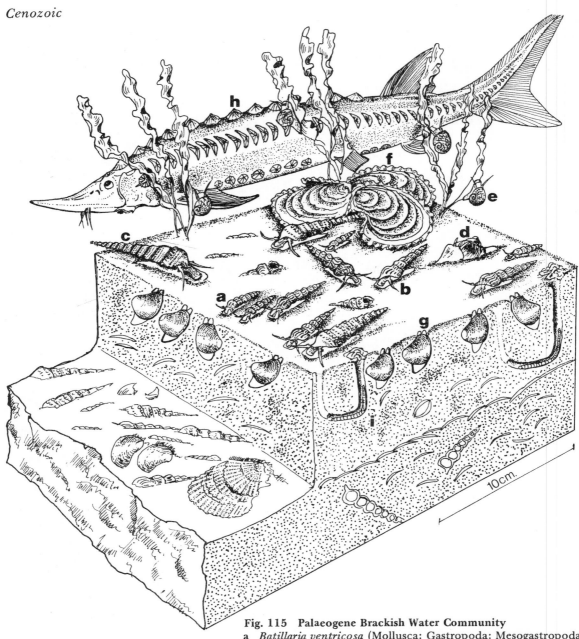

Fig. 115 Palaeogene Brackish Water Community
a *Batillaria ventricosa* (Mollusca: Gastropoda: Mesogastropoda)
b *Batillaria concava* (Mollusca: Gastropoda: Mesogastropoda)
c *Potamides vagus* (Mollusca: Gastropoda: Mesogastropoda)
d *Globularia grossa* (Mollusca: Gastropoda: Mesogastropoda)
e *Theodoxus concava* (Mollusca: Gastropoda: Archaeogastropoda)
f *Ostrea plicata* (Mollusca: Bivalvia: Pterioida)
g *Corbicula obovata* (Mollusca: Bivalvia: Veneroida)
h sturgeon — *Acipenser* (Vertebrata: Osteichthyes)
i polychaete burrows (Annelida)

material, probably diatoms and blue-green algae, from the surface algal mat. Like similar gastropods in modern communities, they exhibited an extremely variable morphology which is thought to be associated with the unpredictability of the estuarine environment; if a traumatic change in physical conditions occurred, at least a part of a very variable population would have a chance of survival. This theory assumes that morphological variation reflects variation in physiological tolerance.

Other gastropods included *Theodoxus,* an algal grazer, and *Globularia,* a predator on infaunal bivalves. Oysters are common at some horizons, and are typical today of quiet estuarine conditions. The shallow-burrowing, suspension-feeding bivalve *Corbicula obovata* dominates the infauna of the community.

Like its more modern relative found in the Caspian Sea, the sturgeon *Acipenser* probably ate mainly polychaete worms, small crustacea and molluscs.

Brackish water communities are common throughout the Headon, Osborne and Hamstead Beds of southern England, where they are interspersed with deltaic, flood plain, lacustrine and freshwater environments.

116 Palaeogene Freshwater Community

In some freshwater lake environments, the sediment is largely composed of the remains of the alga *Chara* and of freshwater snails. The lake bottom was colonized by extensive growths of this branching alga, and the tiny, spirally ornamented, reproductive bodies or nucules are easily recognizable in the sediments. *Chara* is the largest of the freshwater green algae, and forms erect branching growths, often in large, monospecific stands, in calcium-rich waters. The whole plant becomes encrusted with calcium carbonate, which, on the death of the plant, makes a considerable contribution to sediment formation.

The dominant elements in the fauna are freshwater snails, mainly the forms *Galba* and *Planorbina* which fed upon algal films on freshwater plants. The smaller *Viviparus* was a ciliary-feeder. The shallow-burrowing, suspension-feeding bivalve *Corbicula obtusa* is the only abundant component of the infauna.

Vegetation, insects and invertebrates from the surrounding land indicate subtropical to tropical temperatures. Local abundance of land snails indicates emergence from the water in some places.

The Bembridge limestone of the Isle of Wight is a freshwater deposit averaging about 3m in thickness which occurs within a sequence of brackish estuarine, flood plain and lacustrine sediments. The marine communities of the Osborne and Upper Headon Beds are generally similar to the Headon Beds Community illustrated (115), with abundant species characteristic of reduced salinity environments.

Fig. 116 Palaeogene Freshwater Community
a *Galba longiscata* (Mollusca: Gastropoda: Pulmonata)
b *Planorbina discus* (Mollusca: Gastropoda: Pulmonata)
c *Viviparus minutus* (Mollusca: Gastropoda: Mesogastropoda)
d *Corbicula obtusa* (Mollusca: Bivalvia: Veneroida)
e *Chara* (Algae)
f turtle (Vertebrata: Reptilia: Anapsida)
g freshwater fish (Vertebrata: Osteichthyes)

NEOGENE

During the Miocene and most of the Pliocene, the shores of the North Sea Basin lay to the east of Britain, and marine deposition took place in northern Germany and Denmark. The shoreline advanced westwards again in the late Pliocene and early Pleistocene.

From the Miocene onwards, marine faunas consist increasingly of extant species, including, in the north-west European deposits, species found today in the west Atlantic and the Mediterranean. The present day latitudinal sequence of faunas, or provinces, in the west Atlantic was more or less in existence by Pliocene times.

344

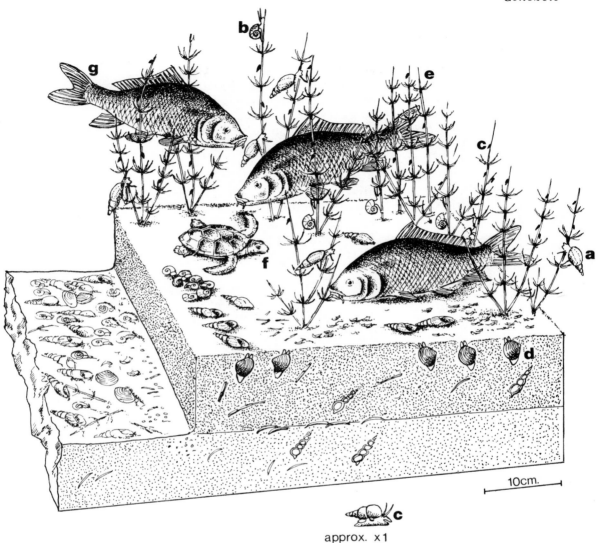

approx. x1

However, the position of the province boundaries fluctuated to the north and south with climatic warming and cooling.

Pliocene and Pleistocene marine sediments in Britain are represented by a series of deposits in East Anglia. These are often small in area, fragmentary, and difficult to relate to a coherent sequence of events or paleogeographic reconstructions. The best known of these deposits are the Coralline Crag of late Pliocene age which occurs in a small area of Suffolk, and the Red Crag which occupies a wider area of East Anglia and Essex.

117 Pliocene Marine Gravel Communities

Bryozoans were particularly abundant in these deposits, and among the species were forms normally found encrusting gorgonian corals. Flat-conical free-living colonies of *Cupuladria canariensis* were common; they were raised clear of the sediment on stilt-like setae. Another bryozoan, *Hippoporidra*, formed botryoidal growths on shells inhabited by hermit crabs (Fig. 117, 7). The molluscs were extremely diverse. They make up the first community we have considered in which many of the species are still living in the north-western Atlantic and the Mediterranean. The epifaunal species included the byssate *Cardita senilis*. The scallop *Chlamys opercularis* was common; it lived on the sediment surface, but was a very active swimmer if disturbed. The brachiopod *Terebratula maxima* was common; it was the largest terebratulid ever known, reaching a length of 11cms.

Infaunal bivalves included the shallow-burrowing gravel-dwelling *Glycymeris*, *Astarte*, *Arctica* and *Venus casina*. Deeper burrowers included *Phacoides borealis* and *Mya truncata*. An even deeper burrower, *Panopea*, was common in places. Gastropods are represented by the ciliary suspension feeder *Turritella communis*, the scavenger *Hinia reticosa*, the bivalve predator *Natica* and the ascidian-feeding *Trivia arctica*.

Similar bryozoa-rich communities to this are found today in shelf environments at depths up to about 50m, on sandy substrates and hardgrounds. The bryozoan *Cupuladria* is restricted today to waters with a mean temperature in excess of 14° C, and thus equivalent to a latitude south of Portugal. The molluscan fauna in general confirms this estimate of palaeotemperature.

Fig. 117   Pliocene Marine Gravel Communities
a *Arctica islandica* (Mollusca: Bivalvia: Veneroida)
b *Mya truncata* (Mollusca: Bivalvia: Myoida)
c *Astarte omali* (Mollusca: Bivalvia: Veneroida)
d *Chlamys opercularis* (Mollusca: Bivalvia: Pterioida)
e *Glycymeris glycymeris* (Mollusca: Bivalvia: Arcoida)
f *Venus casina* (Mollusca: Bivalvia: Veneroida)
g *Lucinoma borealis* (Mollusca: Bivalvia: Veneroida)
h *Cardita senilis* (Mollusca: Bivalvia: Veneroida)
i *Terebratula maxima* (Brachiopoda: Articulata: Terebratulida)
j *Turritella incrassata* (Mollusca: Gastropoda: Mesogastropoda)
k *Emarginula fissurata* (Mollusca: Gastropoda: Archaeogastropoda)
l *Natica millepunctata* (Mollusca: Gastropoda: Mesogastropoda)
m *Hinia reticosa* (Mollusca: Gastropoda: Neogastropoda)
n *Trivia coccinelloides* (Mollusca: Gastropoda: Mesogastropoda)
o *Temnechinus excavatus* (Echinodermata: Echinoidea)
p *Balanus crenatus* (Arthropoda: Crustacea: Cirripedia)
q *Sphenotrochus intermedia* (Coelenterata: Anthozoa: Scleractinia)

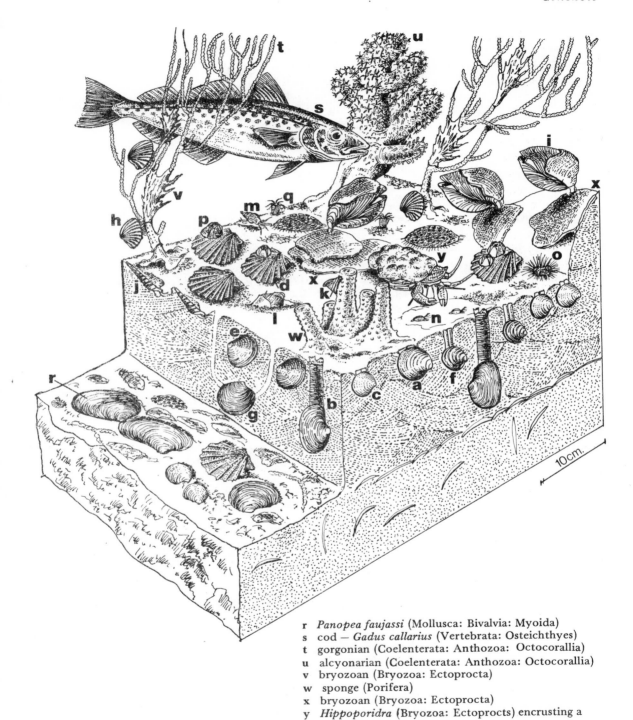

r *Panopea faujassi* (Mollusca: Bivalvia: Myoida)
s cod — *Gadus callarius* (Vertebrata: Osteichthyes)
t gorgonian (Coelenterata: Anthozoa: Octocorallia)
u alcyonarian (Coelenterata: Anthozoa: Octocorallia)
v bryozoan (Bryozoa: Ectoprocta)
w sponge (Porifera)
x bryozoan (Bryozoa: Ectoprocta)
y *Hippoporidra* (Bryozoa: Ectoprocts) encrusting a
   shell inhabited by hermit crab

347

Cenozoic

The fish faunas found here are largely similar to those of the north-west Atlantic of the present day, and include forms such as the cod which feed mainly upon benthic invertebrates such as molluscs and crustacea. These deposits are found only in a restricted area of Suffolk; they consist of cross-bedded gravels and coarse calcarenites, made up largely of bryozoan fragments and broken mollusc shells. Fossils are extremely rich and diverse, but it is obvious that the beds consist of *post mortem* accumulations and represent a variety of actual communities.

118 Pleistocene Sublittoral Sand Community

The community illustrated represents a shallow sublittoral habitat perhaps from 10 to 15m deep, with a sandy substrate and abundant shell debris. Epifaunal animals included clumps of suspension-feeding *Mytilus*, the scallop *Chlamys opercularis* and *Balanus crenatus*. Common gastropods were the scavenging *Nassarius granulatus*, *Nassarius reticosa*, the predatory *Neptunea antiqua* and the sinistrally coiled *Neptunea contraria* which today feed largely on polychaetes and bivalves. The infauna was both abundant and diverse, chiefly suspension-feeding bivalves; particularly abundant were *Glycymeris*, *Macoma*, *Arctica*, *Spisula* and the deeper-burrowing *Phacoides borealis*. Other communities were represented in the Red Crag deposits, and at some horizons an abundance of the cockle *Cerastoderma*, the winkle *Littorina littorea* and the dog whelk *Nucella* indicate intertidal conditions.

A large proportion of the species found in the Red Crag are found in British seas today; some species such as the large whelk *Neptunea despecta* are found in colder Arctic waters, whereas *Neptunea antiqua* is found around British shores. The sinistrally coiled *Neptunea contraria* is found off the Atlantic coast of France, Spain and Portugal and in the Mediterranean. The changing proportions of these three species at various horizons in the Red Crag deposits, with *Neptunea contraria* becoming less abundant higher in the rock sequence as *Neptunea despecta* appears, has been interpreted as indicative of a cooling of seawater temperatures.

The Red Crag sea covered most of East Anglia; the deposits consist mainly of cross-bedded sands with abundant disarticulated bivalve shells. Several divisions of the Red Crag have been recognized, and these follow one another in a progressive sequence from south to north, which probably represents a retreating shoreline. The fossils were obviously transported by the sea from a variety of communities of the intertidal to nearshore shelf.

Fig. 118 Pleistocene Sublittoral Sand Community

a *Glycymeris glycymeris* (Mollusca: Bivalvia: Arcoida)
b *Yoldia oblongoides* (Mollusca: Bivalvia: Palaeotaxodonta)
c *Macoma obliqua* (Mollusca: Bivalvia: Veneroida)
d *Mytilus edulis* (Mollusca: Bivalvia: Mytiloida)
e *Chlamys opercularis* (Mollusca: Bivalvia: Pterioida)
f *Spisula solida* (Mollusca: Bivalvia: Veneroida)
g *Arctica islandica* (Mollusca: Bivalvia: Veneroida)
h *Phacoides borealis* (Mollusca: Bivalvia: Veneroida)
i *Neptunea antiqua* (Mollusca: Gastropoda: Neogastropoda)
j *Neptunea contraria* (Mollusca: Gastropoda: Neogastropoda)
k *Nassarius granulatus* (Mollusca: Gastropoda: Neogastropoda)
l *Nassarius reticosa* (Mollusca: Gastropoda: Neogastropoda)
m *Balanus crenatus* (Arthropoda: Crustacea: Cirripedia)
n cod — *Gadus callarius* (Vertebrata: Osteichthyes)

349

119 Pleistocene Freshwater Sand and Clay Community

This community lived in a deltaic environment where the deposits consisted of sand, gravel, laminated clay and peat layers. Some horizons contain shallow water marine mollusca, and are the result of marine incursions. The peaty sands and organic-rich clays are thought to have been laid down in a slow-moving river or lake rich in plant debris. The lakes were surrounded by mixed oak forests, containing also elms and limes, and inhabited by deer. Temperatures were thought to have been approximately similar to those experienced in southern Britain today, although lower parts of the sedimentary sequence are dominated by pine and birch forest, and were obviously cooler.

The major elements in the community were freshwater mollusca; *Viviparus* and *Valvata* lived upon the sediment surface and the aquatic plants, whereas *Succinea* is found near the lake edges which periodically would have dried out. *Viviparus* is a detritus-feeding gastropod which ploughs along the substrate surface stirring up the sediment; particles are drawn across the gills and collected in a mucous string which is then drawn into the mouth. Both *Viviparus* and *Valvata* incubate eggs within the oviduct which functions as a uterus from which the young snails emerge, hence the name *Viviparus*.

The infaunal components of the community consisted of the sluggish suspension-feeding bivalve *Unio tumidus,* and the small siphonates *Corbicula fluminalis* and *Pisidium amnicus. Unio* and other members of the family Unionidae have an interesting mode of development; the eggs are initially fertilized and incubated within the mantle cavity and gills of the mussel but are soon expelled into the water where they attach themselves by spines or byssal threads to the fins and gills of fish. The fish tissue forms a cyst within which the *Unio* larva develops for up to 36 days; after this period it emerges to begin life as a benthic animal. *Pisidium,* a small bivalve with a single siphon, also incubates its eggs within the mantle cavity, whence they emerge as miniature adults.

The river was also inhabited by giant beavers and by voles; occasionally beaver dams are revealed as the Norfolk cliffs are eroded. This community occurs in the Upper Fresh Water Beds near Cromer, Norfolk.

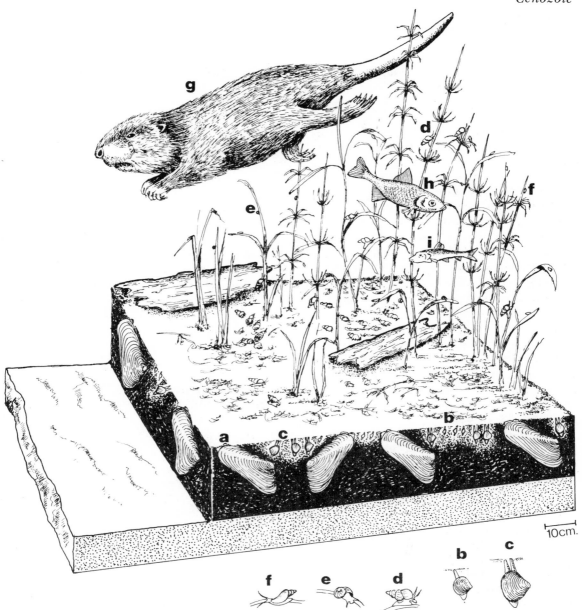

Fig. 119 Pleistocene Freshwater Sand and Clay Community
a *Unio tumidus* (Mollusca: Bivalvia: Unionoida)
b *Pisidium amnicus* (Mollusca: Bivalvia: Veneroida)
c *Corbicula fluminalis* (Mollusca: Bivalvia: Veneroida)
d *Viviparus gibbus* (Mollusca: Gastropoda: Mesogastropoda)
e *Valvata antiqua* (Mollusca: Gastropoda: Mesogastropoda)
f *Succinea putris* (Mollusca: Gastropoda: Pulmonata)
g beaver (Vertebrata: Mammalia)
h roach (Vertebrata: Osteichthyes)
i gudgeon *Gobio gobio* (Vertebrata: Osteichthyes)

351

Present day

Six present communities are illustrated here: four of these from around Britain and two from the tropical Indo-Pacific Province. The tropical communities are more diverse than comparable communities in higher latitudes (Thorson, 1957).

The sands, sandy muds and muds in the shelf seas around the British Isles support a diversity of benthic communities, most of which became established after the beginning of the last major glacial advance about 15,000 years ago.

120   Offshore Muddy Sand Community

The community illustrated is based upon samples taken from the Irish Sea (Jones, 1956), but is typical of offshore muddy sand bottoms all around the British Isles at depths between about 25 and 80m. The community is extremely diverse, but is dominated numerically by infaunal polychaetes and bivalves, although gastropods, crustacea, ophiuroids and echinoids may be common.

Shallow-burrowing, deposit-feeding bivalves such as the siphonate *Abra prismatica* and the palp-feeding *Nucula sulcata* are abundant. Other common bivalves are mainly suspension feeders, and include the shallow-burrowing *Corbula gibba* and *Parvicardium scabrum* as well as deeper burrowers with longer siphons such as *Cultellus* and *Dosinia*. The scaphopod *Dentalium entalis* feeds mainly upon benthic foraminifera living within the sediment.

Polychaetes are extremely abundant, and more than 40 species may occur, many of them deposit feeders. *Pectinaria* is amongst the most common of the larger forms and is interesting, for it forms a tube and has a similar mode of life to the scaphopods. Some of the polychaetes are predators with well developed jaws.

Turritella is a ciliary-feeding gastropod, which lives buried just below the substrate surface and can be very abundant in this community. Other common gastropods are the detritus-feeding *Aporrhais pespelicani,* the predators *Natica* and *Trophon,* which feed upon other molluscs, and several small species such as *Mangelia*

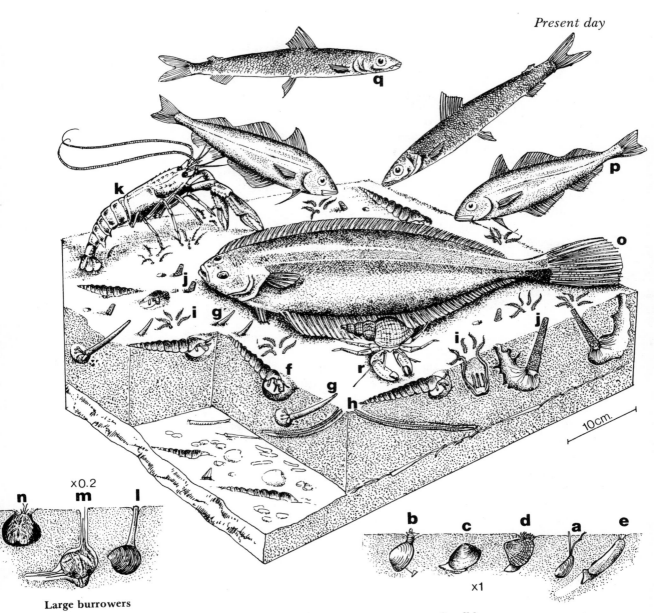

Present day

10cm.

×0.2

Large burrowers

×1

Small burrowers

Fig. 120 Offshore Muddy Sand Community

a   *Abra prismatica* (Mollusca: Bivalvia: Veneroida)
b   *Corbula gibba* (Mollusca: Bivalvia: Myoida)
c   *Nucula sulcata* (Mollusca: Bivalvia: Palaeotaxodonta)
d   *Parvicardium scabrum* (Mollusca: Bivalvia: Veneroida)
e   *Cultellus pellucida* (Mollusca: Bivalvia: Veneroida)
f   *Turritella communis* (Mollusca: Gastropoda: Mesogastropoda)
g   *Dentalium entalis* (Mollusca: Scaphopoda)
h   *Nephthys* (Annelida)
i   *Amphiura* (Echinodermata – Asterozoa)

j   *Pectinaria* (Annelida)
k   *Nephrops norvegica* (Arthropoda: Crustacea: Malacostraca)
l   *Dosinia lupinus* (Mollusca: Bivalvia: Veneroida)
m   *Echinocardium cordatum* (Echinodermata: Echinoidea)
n   *Brissopsis* (Echinodermata: Echinoidea)
o   pole dab (Vertebrata: Osteichthyes)
p   norway pout (Vertebrata: Osteichthyes)
q   *Argentina sphyraena* (Vertebrata: Osteichthyes)
r   *Anapagurus* (Arthropoda: Crustacea: Malacostraca)

353

Present day

which feed upon polychaetes by means of a specialized radula and
a toxic salivary secretion.

The burrowing echinoids *Echinocardium* and *Brissopsis* are
common and the ophiuroid *Amphiura*, a deposit feeder, may be
very abundant, sometimes occurring in such dense aggregations
that the whole sea bottom is a mass of ophiuroid arms. The Dublin
Bay Prawn *Nephrops norvegica* is one of the commoner of the
larger crustacea and inhabits extensive burrow systems (not
shown here). The hermit crab *Anapagurus* is very abundant. Fish
are important predators on the muddy sand benthos, and analysis
of fish diets has shown that the two most common groups in the
community, bivalves and polychaetes, are the most common prey
of the benthic-feeding fish, which are mainly haddock, cod, whiting and pole dabs.

121 Intertidal Mud Community

This community, often known as the *'Macoma balthica'* community, is typical of intertidal mud flats around the shores of
north-west Europe and examples can be seen off the east coast of
Britain, at places such as the outer Thames estuary and the Wash
(Mistakidis, 1951).

The community is developed on inshore muddy sands and
silts stretching out from the intertidal areas to a depth of about
10m; as these facies are frequently found in estuarine areas,
salinities are often fluctuating or reduced. The community is
dominated numerically by deposit-feeding animals which occur in
huge numbers and include the bivalves *Macoma balthica* and
Scrobicularia plana, mobile species with two very long siphons.
The lug worm *Arenicola marina* is a well known deposit-feeding
polychaete, inhabiting 'U' burrows which leave the characteristic surface traces of casts and depressions so conspicuous on intertidal flats.

Mya arenaria, a sedentary suspension feeder with long, thick,
fused siphons is often common in this community; where the
sediment is more sandy, the common cockle *Cerastoderma edule*,
a shallow-burrowing suspension feeder, can be extremely abundant.
Aggregations of the mussel *Mytilus edulis* may be found wherever
small pieces of hard substrate provide it with an anchorage. The
winkle *Littorina littorea* is the most abundant of the larger
gastropods, leaving feeding trails in the mud. The tiny gastropod
Hydrobia ulvae is abundant on the sediment surface and leaves
fine trails of looped furrows. *Hydrobia* occurs in immense numbers; there are usually between 3,000 and 10,000 per square
metre, but densities as high as 60,000 per square metre have been

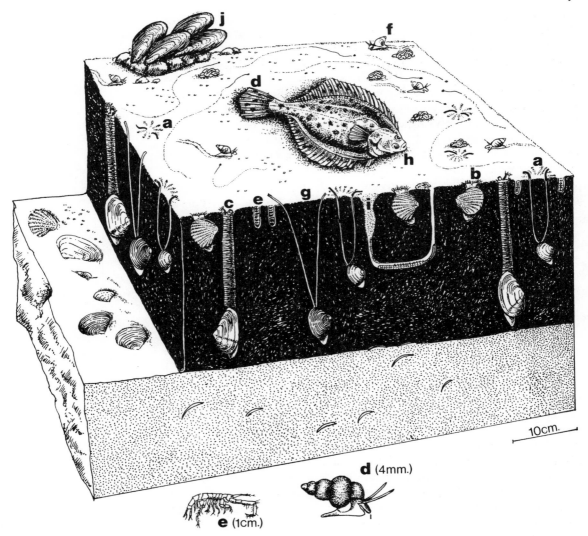

Fig. 121 Intertidal Mud Community

a   *Macoma balthica* (Mollusca: Bivalvia: Veneroida)
b   *Cerastoderma edule* (Mollusca: Bivalvia: Veneroida)
c   *Mya arenaria* (Mollusca: Bivalvia: Myoida)
d   *Hydrobia ulvae* (Mollusca: Gastropoda: Mesogastropoda)
e   *Corophium volutator* (Arthropoda: Crustacea: Malacostraca)
f   *Littorina littorea* (Mollusca: Gastropoda: Mesogastropoda)
g   *Scrobicularia plana* (Mollusca: Bivalvia: Veneroida)
h   flounder — *Pleuronectes flesus* (Vertebrata: Osteichthyes)

recorded. Out of water it is a deposit feeder, consuming mainly the bacterial coatings around the sediment grains and detritus, but in water it has the ability to float on a raft of mucus which holds planktonic material on which the animal can feed. It thus exploits the food resources of both the sediment surface and the inshore plankton. Another deposit feeder is the small amphipod *Corophium volutator* which occurs in very large numbers and inhabits small 'U' burrows up to about 10cm deep.

These intertidal muddy sand communities are important feeding grounds for vertebrates which at high water are mainly flatfish, but at low water include a wide variety of largely migrant wading birds. Cockles and mussels are an important food for the oyster-catcher, and *Hydrobia* and *Corophium* are major items in the diets of the redshank and turnstone.

122   Shell Gravel Community

At the western end of the English Channel, much of the bottom between 20 and 100m consists of coarse shelly sands or shell gravels with many shells; the bottom is unstable and swept by tidal currents.

The community developed on such a bottom is neither very diverse nor abundant (Ford, 1923); it is dominated by suspension-feeding bivalves, particularly *Glycymeris glycymeris* and *Venus fasciata*; other smaller venerids such as *Timoclea ovata* and *Venus* spp. may also be common. *Glycymeris* has notably unspecialized burrowing, feeding and respiratory mechanisms, but is typical of coarse current-swept bottoms in many parts of the world and appears to have been restricted mainly to this type of habitat throughout its history. It lacks siphons, and therefore lives just covered by the sediment, often with the posterior end of the shell exposed; in coarse sediments it burrows rather more deeply but maintains feeding and respiratory currents through the gravel.

Deposit feeders are less common, but include the siphonate bivalves *Arcopagia crassa, Gari tellinella* and *Tellina pusilla.* The cephalochordate *Amphioxus lanceolata* is also a typical animal of this habitat, living buried just beneath the surface of the gravel. Polychaetes, such as *Glycera* and *Lumbriconereis,* and the echinoids *Spatangus purpureus, Echinocardium* and *Echinocyamus* may also be common.

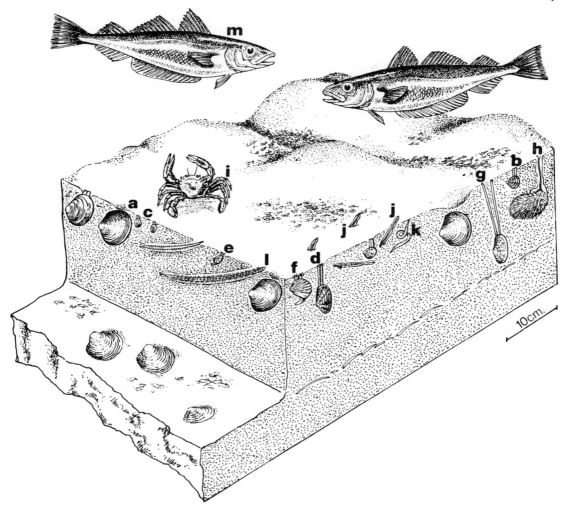

Fig. 122 Shell Gravel Community

a *Glycymeris glycymeris* (Mollusca: Bivalvia: Arcoida)
b *Venus fasciata* (Mollusca: Bivalvia: Veneroida)
c *Timoclea ovata* (Mollusca: Bivalvia: Veneroida)
d *Venerupis rhomboides* (Mollusca: Bivalvia: Veneroida)
e *Parvicardium scabrum* (Mollusca: Bivalvia: Veneroida)
f *Laevicardium crassum* (Mollusca: Bivalvia: Veneroida)
g *Gari tellinella* (Mollusca: Bivalvia: Veneroida)
h *Echinocardium flavescens* (Echinodermata: Echinoidea)
i *Portunus pusillus* (Arthropoda: Crustacea: Malacostraca)
j *Amphioxus* (Chordata: Cephalochordata)
k *Glycera* (Annelida)
l *Lumbriconereis* (Annelida)
m whiting — *Gadus merlangus* (Vertebrata: Osteichthyes)
n *Arcopagia crassa* (Mollusca: Bivalvia: Veneroida)

357

Present day

123 Intertidal Rocky Shore Community

This is the most accessible and most well known of marine communities. The intertidal rocky shore is a steep environmental
gradient between the land and the sea. Plants and animals inhabiting the intertidal zone must be able to tolerate often severe wave
action, alternate emersion and submersion because of tides, and
exposure to the climatic stresses of heat, cold, desiccation, and
freshwater incursions. Despite these deterrents the intertidal zone is
highly productive, and organisms which can tolerate the conditions
are often extremely abundant. Intertidal organisms are often
sharply zoned down the environmental gradient, the position of
their zone depending upon the physiological tolerances of each
organism, and the competitive interactions between the species
(Lewis, 1964).

The community illustrated is typical of the middle parts of the
rocky shore where algae, particularly *Fucus,* are abundant. In
turbulent environments, many of the animals present show obvious
adaptations which help them to avoid being dislodged, or to guard
against damage from wave impact. Thus most of the animals are
thick shelled (such as limpets) or elastic (such as sea-anemones)
and adhere closely to the rock surface, either by cementation
(*Balanus*), byssal attachment (*Mytilus*) or muscular adhesion
(*Patella*). Free-moving animals, such as the gastropods *Littorina*
and *Gibbula*, retreat into cracks and crevices at times of stress.

Most animals on the shore are either suspension feeders living
on plant or animal material (*Mytilus, Balanus, Actinia*) or herbivorous grazers feeding upon algae (*Patella, Littorinia,* and *Gibbula*).
Predators include *Nucella lapillus*, the dog whelk, which feeds on
Mytilus by boring holes into the shell, or on barnacles by penetrating the opercular plates. The shore crab *Carcinus maenas* feeds
mainly upon *Littorina* by cracking the shells. On limestone, sandstone or shale substrates, mechanical rock-boring bivalves, such as
the siphonate *Zirphaea* and *Hiatella*, may be abundant. The rock
is also penetrated by polychaetes and clionid sponges.

The intertidal area is a rich feeding ground for fish at high
water and birds such as oyster-catchers, gulls and migrant waders
at low water.

124 Tropical Marine Grass Beds

Inshore sand areas of the tropics and sub-tropics are frequently
colonized by beds of grass-like marine angiosperms, particularly
forms such as *Thalassia, Cymodocea* and *Thalassodendron*. The

358

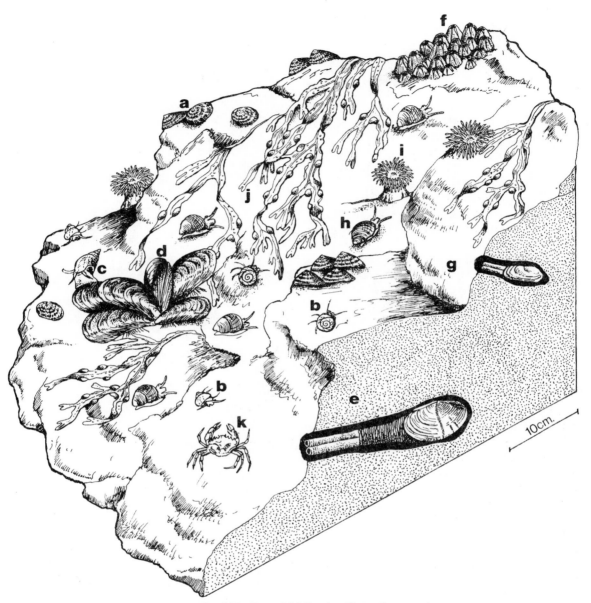

Fig. 123 Intertidal Rocky Shore Community
a *Patella vulgata* (Mollusca: Gastropoda: Archaeogastropoda)
b *Gibbula umbilicalis* (Mollusca: Gastropoda: Archaeogastropoda)
c *Nucella lapillus* (Mollusca: Gastropoda: Neogastropoda — dog whelk)
d *Mytilus edulis* (Mollusca: Bivalvia: Mytiloida)
e *Zirphaea crispata* (Mollusca: Bivalvia: Myoida)
f *Balanus* (Arthropoda: Crustacea: Cirripedia — barnacles)
g *Hiatella arctica* (Mollusca: Bivalvia: Myoida)
h *Littorina littorea* (Mollusca: Gastropoda: Mesogastropoda)
i *Actinia equina* (Coelenterata: Anthozoa — sea anemone)
j *Fucus vesiculosus* (Algae — bladderwrack)
k *Carcinus maenus* (Arthropoda: Crustacea — crab)

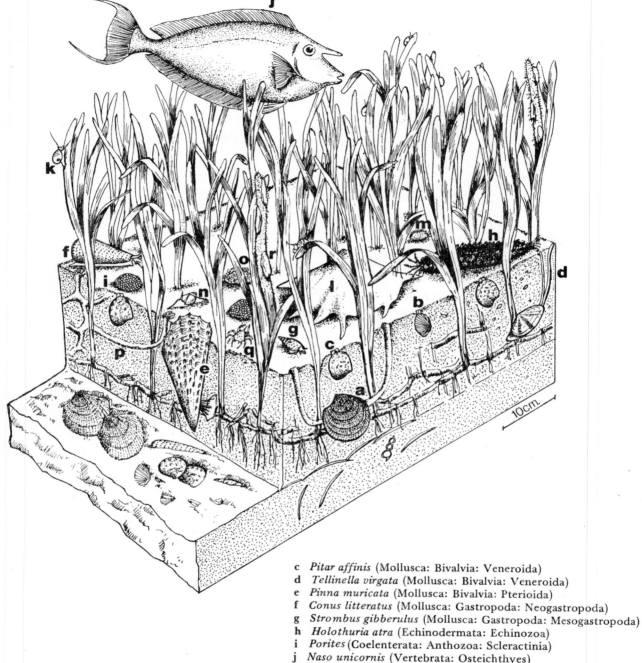

Fig. 124 Tropical Marine Grass Beds

a *Codakia tigerina*
   (Mollusca: Bivalvia: Veneroida)
b *Gafrarium pectinatum*
   (Mollusca: Bivalvia: Veneroida)

c *Pitar affinis* (Mollusca: Bivalvia: Veneroida)
d *Tellinella virgata* (Mollusca: Bivalvia: Veneroida)
e *Pinna muricata* (Mollusca: Bivalvia: Pterioida)
f *Conus litteratus* (Mollusca: Gastropoda: Neogastropoda)
g *Strombus gibberulus* (Mollusca: Gastropoda: Mesogastropoda)
h *Holothuria atra* (Echinodermata: Echinozoa)
i *Porites* (Coelenterata: Anthozoa: Scleractinia)
j *Naso unicornis* (Vertebrata: Osteichthyes)
k *Smaragdia rangiana* (Mollusca: Gastropoda: Archaeogastropoda)
l *Lambis lambis* (Mollusca: Gastropoda: Mesogastropoda)
m *Cypraea annulus* (Mollusca: Gastropoda: Mesogastropoda)
n *Natica marochiensis* (Mollusca: Gastropoda: Mesogastropoda)
o *Tripneustes* (Echinodermata: Echinoidea)
p *Callianassa* burrow (Crustacea)
q *Menaethius monoceros* (Arthropoda: Crustacea: Malacostraca)

communities developed in association with the grass beds are important components of coral reef ecosystems in the Indo-Pacific and Caribbean marine provinces.

The leaves of the grass act as baffles, trapping sediment particles, which are then stabilized by the extensive rhizome and root systems. In this way, sediment can be stabilized in sites of relatively high wave action. Grass bed communities are among the most productive in the marine environment; they support a very diverse biota and organic detritus is abundant.

The community illustrated is based upon a western Indian Ocean grass bed, but the types of animals shown are generally similar to those found throughout the tropics. The stable sediment is an ideal habitat for infaunal animals because there is little risk of major disturbance. Lucinoid bivalves, which, although siphonless, construct mucus-lined inhalent tubes leading to the surface, are particularly characteristic of this habitat, and include such genera as *Codakia* and *Anodontia.* Other infaunal suspension feeders are the byssate sedentary *Pinna muricata* and the siphonate venerids *Pitar* and *Gafrarium.* Deposit feeders may be locally abundant, particularly near the shore, and include *Tellinella staurella* and *Quidnipagus palatum.* Burrowing shrimps (callianassids and alpheids) are abundant, and construct large and complex burrow systems among the roots and rhizomes. Other infaunal animals, not illustrated, include a large variety of burrowing worms.

The epifaunal component of the grass beds community is distinctly stratified, with many foraminifera, amphipods and gastropods such as the small, green, well camouflaged *Smaragdia* living amongst the algae attached to the grass leaves. Other animals, for instance bryozoans, sponges and calcareous hydroids such as *Millepora,* frequently encrust the stems, whilst the substrate surface between the stem and leaf bases is colonized by a wide variety of larger organisms. Deposit-feeding holothurians are abundant; the black *Holothuria* occurs in densities of as many as 3 − 5 per square metre and ingests, by means of its tentacles, huge quantities of sediment, from which it extracts the organic detritus. Large, thick-shelled herbivorous gastropods are common, and include such forms as *Strombus* and *Lambis* with a digitate lip, and predatory gastropods such as *Conus litteratus* which feeds upon infaunal polychaetes. Other herbivores include the echinoid *Tripneustes pileolus* which covers itself with dead shells and coral debris as camouflage.

Corals are not important members of grass bed communities, and tend to be small and restricted to a few genera. *Porites,* a genus particularly tolerant of marginal conditions, often occurs in small unattached ball-like colonies which are rolled by wave action.

The marine grass beds are often intertidal, in which case at high tides they are important feeding grounds for fish such as

Naso and turtles, while at low water egrets and migrant palearctic waders exploit the habitat for small crustacea and gastropods.

125   Tropical Coral Reef Assemblage

Coral reef ecosystems comprise a variety of communities reflecting the range of environmental conditions found across the reef system. The Indo-Pacific community illustrated is found on a coral-dominated hard substrate in a shallow sublittoral area of good circulation but not heavy wave action.

The substrate is a coral-rich limestone which may be a product of late Pleistocene rather than recent reef growth. The epifauna is dominated by scleractinian corals, particularly species of the genus *Acropora,* which is very important in forming the framework of the reef. *Acropora* is particularly characteristic of sites of good circulation and low turbidity. Corals, although containing symbiotic algae, feed mainly upon zooplankton which they catch by means of stinging cells or mucus strings. Also containing symbiotic algae, but suspension-feeding, is the large, thick-shelled bivalve *Tridacna,* firmly anchored by an enormous byssus. These giant clams are believed to have evolved from forms resembling cockles.

Many other epifaunal animals live beneath the corals, including echinoids, crustacea, gastropods, ophiuroids and a host of encrusting sponges, tunicates and bryozoans. Such habitats are characterized by richness in species but by small populations. Because of the antagonistic reactions few animals are able to live on the surfaces of the living corals. The calcareous rock surface is penetrated by a wide variety of boring and nestling animals including the sponge *Cliona,* the boring bivalves *Lithophaga* and *Gastrochaena,* polychaetes and sipunculid worms and the echinoid *Echinometra.*

The huge and diverse fish populations of the reefs exert important influence on the habitat, behaviour and distribution of the reef invertebrates. Some fish graze upon the living coral, some on calcareous algae, while many others feed on particular types of invertebrates. The habitats and diet of reef animals are highly specialized and this is thought to be the reason why so many species can be accommodated in reef communities. This kind of specialization can only take place where environmental conditions and food supply are stable and predictable.

Fig. 125 Tropical Coral Reef Assemblage

a   *Tridacna gigas* (Mollusca: Bivalvia: Veneroida)
b   *Heterocentrotus mamillathus* (Echinodermata: Echinoidea)
c   *Echinometra* (Echinodermata: Echinoidea)
d   *Stichopus chloronotus* (Echinodermata: Echinoidea)
e   *Ophiocoma* (Echinodermata: Asterozoa)
f   *Conus textile* (Mollusca: Gastropoda: Mesogastropoda)
g   *Aspidosiphon* (Sipunculida)
h   *Lithophaga* (Mollusca: Bivalvia: Mytiloida)

i   eunicid and nereid worms (Annelida)
j   *Rocellaria lamellosa* (Mollusca: Bivalvia: Myoida)
k   *Botula silicula* (Mollusca: Bivalvia: Mytiloida)
l   *Lithotrya* (Arthropoda: Crustacea: Cirripedia)
m   *Phascolosoma* (Sipunculida)
n   *Balistes* (Vertebrata: Osteichthyes — trigger fish)
o   *Pomacanthus* (Vertebrata: Osteichthyes)
p   *Dascyllus* (Vertebrata: Osteichthyes)
q   *Holocentrus spinifer* (Vertebrata: Osteichthyes)
r   *Atergatis floridus* (Arthropoda: Crustacea: Malacostraca)

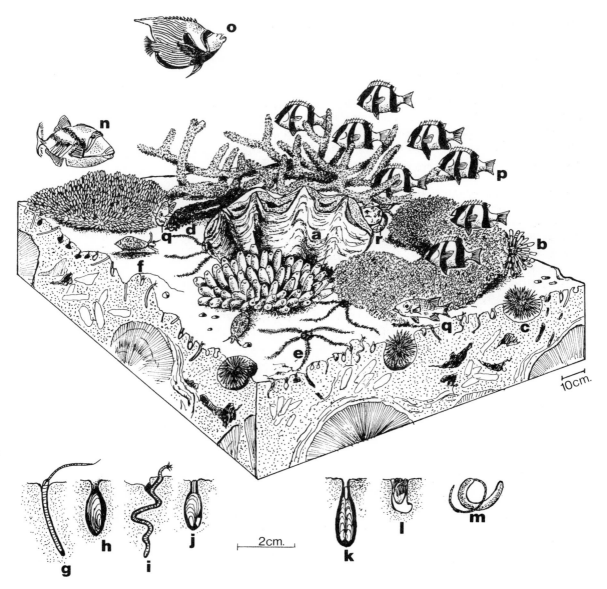

PRESERVATION POTENTIAL OF MODERN COMMUNITIES

Modern communities contain a great variety of organisms which differ considerably in their preservation potential. Animals with well calcified rigid skeletons such as molluscs will obviously have the best chance of fossilization; others such as echinoderms have well calcified skeletons, but are made up of very small pieces and are thus prone to disarticulation and dispersal. Many crustacea have articulated and poorly calcified skeletons which cannot be easily preserved.

Some of the most abundant organisms in modern communities such as polychaetes, turbellaria, nematodes, sipunculid worms and many coelenterates have muscular, hydrostatic skeletons and any hard parts are only microscopic, for instance jaws, skin spicules and setae. These soft-bodied organisms can be preserved only under special circumstances and usually no trace is left of these abundant animals. Although they leave no important hard parts the activities of these animals are often preserved as trace-fossils, burrows, grazing trails, and rock and shell borings. Nevertheless there is considerable loss of information during fossilization and the fact that polychaetes and other worms must have been the most abundant and diverse group of animals is rarely evident in fossil communities.

Glossary

Many geological and zoological words not explained fully in the text are described briefly. Names of animals and plants are listed in the index at the end of the book, and selected taxa are described in the Classification of Organisms.

Abyssal Relating to the ocean floor (at depths greater than about 2,000 m).

Adductor muscle In bivalves and brachiopods: a muscle used for closing the valves.

Ambulacrum In echinoids: area of plates which are perforated by pores for the tube feet.

Antenna In arthropods and worms: a sensory appendage.

Appendage In arthropods and worms: an organ for locomotion, feeding, feeling or breathing; usually a pair on each segment.

Aptychus In ammonites: one of a pair of plates which closed the aperture.

Aragonite An orthorhombic mineral form of calcium carbonate; less stable at normal temperatures and pressures than calcite.

Arenaceous Applying to a sediment consisting of sand-size particles.

Argillaceous Applying to a sediment consisting of clay-size particles. An argillaceous sandstone is a muddy sandstone.

Articulate Jointed; in brachiopods: possessing teeth on the hinge where the valves open.

Assemblage A group of organisms which occur together.

Association A recurring assemblage, which may be a community or part of a community.

Backreef The area behind a reef (away from the open sea).

Basin An area where sediments accumulate, usually associated with contemporaneous subsidence.

Benthos Animals and plants living on the sea floor. (Adjective: benthic)

Bioherm An accumulation of organic remains on the site where the organisms lived; distinguished from other sediments by the fact that the majority of fossils have not been transported by currents, and thus bedding is often poorly developed.

Biota The fauna and flora.

Bioturbated Disturbed by burrowing animals; the bedding planes of many bioturbated sediments are convoluted or obscured.

Bituminous Containing bitumen (hydrocarbons) or other carbonaceous matter.

Body fossil The fossil remains of an organism (as opposed to a trace, trail or imprint).

Boreal The northern parts of the Earth.

Botryoidal In the form of a bunch of grapes.

Brachial valve In brachiopods: the valve containing the support for the lophophore (the brachial skeleton).

Byssus In bivalves: horny threads for attachment to the sea floor or other anchorage.

Calcarenite A limestone composed of sand-sized particles, usually shell fragments.

Calcareous Made of, or rich in, calcium carbonate.

Calcilutite A limestone consisting of clay-size particles; often thought to be of algal origin. (Synonyms: micrite, calcite mudstone)

Calcisphere A microscopic hollow calcareous sphere; these spheres are important rock formers of uncertain biological affinities, though probably algal.

Calcite A hexagonal mineral form of calcium carbonate; usually more stable than aragonite.

Carapace In arthropods: the dorsal covering.

Carbonate A compound containing the CO_3 ion; calcium carbonate ($CaCO_3$) is the most abundant carbonate in marine environments. Carbonate sediments usually consist of the calcareous skeletons of marine organisms.

Carbonate mudstone Synonymous with calcilutite and micrite.

Glossary

Carnivore An animal which feeds on other animals.

Cartilage Elastic tissue (gristle) present in most vertebrates.

Cast The replacement of a fossil by some other material.

Cephalon In trilobites: the head-shield.

Chamosite A dark mineral, related to chlorite, which is common in many sedimentary iron ores.

Chert Cryptocrystalline silica; a silicous rock often composed of radiolaria.

Chitin A tough organic compound. It can combine with other substances to form a large variety of materials, e.g. the chitino-phosphatic shells of some inarticulate brachiopods.

Class A major division of a phylum.

Clastic Relating to sediments composed of rock fragments (e.g. conglomerate, sand, silt) or clay minerals.

Clay A fine grained sediment composed mostly of clay minerals (hydrous aliminium silicates) and other material less than 4 microns in diameter.

Community A group of organisms living in the same habitat (at the same time, and in the same area). In palaeontology, the word is normally used for the part of the community which is preserved (as body fossils or as trace fossils).

Concretion A mass of more resistant material precipitated in a rock, usually a concentration (round a nucleus, which may often be a fossil) of some compound already present in the rock (e.g. calcareous concretions in a clay; flints in chalk).

Conglomerate A sedimentary rock composed of rounded fragments that are normally greater than 2 mm in diameter, set in a fine grained matrix.

Coquina A deposit of drifted shells.

Cracker A term for a large calcareous concretion in a sandstone, applied in particular to concretions in some Mesozoic sandstones in the south of England.

Cross-bedding Inclined bedding planes within a thicker bedded unit, which reflect deposition on a sloping surface.

Cusp In vertebrates: a projection on a tooth.

Deltaic sequence Sediments laid down in the vicinity of a delta, which characteristically show coarsening upward changes in grain size over several metres or tens of metres (as the delta front advances).

Dental plates In brachiopods: a pair of plates near the umbo which extend from the teeth to the floor of the pedicle valve.

Deposit feeder An alternative term for detritus feeder.

Detritus feeder An animal which feeds on organic matter coating grains of sediment or mixed with the sediment.

Dextral On the right (as opposed to *sinistral*).

Diagenesis Alterations to a sediment after initial deposition.

Diversity The number of species present in an assemblage.

Dorsal Strictly speaking, relating to the top of an animal, but in some invertebrates the dorsal direction is conventional (e.g. in bivalves, the hinge line is always taken to be on the dorsal side).

Ecogroup Successive communities in a similar habitat.

Ecology The study of the relations between organisms and their environments.

Epeiric sea A sea on the continental shelf or covering part of a continent.

Epifauna Animals living on the surface of the sea floor.

Epiplankton Organisms living attached to larger plankton or to floating objects (e.g. driftwood).

Epizoan An organism living attached to another organism.

Euryhaline Relating to animals which can tolerate a broad range of salinity.

Eustatic Relating to world-wide changes in sea level.

Facies A term applied to rocks or fossils (e.g. sandy facies, shelly facies) to indicate an association of lithological and/or palaeontological characteristics.

Family A group of genera with common ancestry.

Fauna The animals present in an area at a particular time.

Faunal province A large area of the world, containing numerous communities, which is isolated from other faunal provinces by some (complete or partial) barrier to migration.

Filament An elongate or thread-like arrangement of cells of organs.

Filter feeder An animal which extracts organic matter suspended in sea water by setting up a current through a feeding organ, which filters out the food required.

Flagellum A thread-like structure used to set up water currents in some invertebrates.

Flint Black chert, common as nodules in the Chalk of southern England.

Fluviatile Relating to a river.

Forereef The area on the seaward side of a reef.

366

Fossil The skeleton, impression or trace of an organism preserved in a rock. Trace fossils (burrows, trails, etc.) are classified separately from body fossils (shells, bones, etc.).

Functional morphology The interpretation of the function of an organism by reference to its shape.

Genal spine In trilobites: a spine extending backwards from the side of the cephalon.

Genus A group of closely related species (plural: genera).

Glabella In trilobites: the axial region of the cephalon.

Glauconite A green iron silicate, common in many shallow marine sands (e.g. the Upper Greensand of southern England).

Gondwanaland The large Palaeozoic continent consisting of South America, Africa, India, Australia and Antarctica.

Graben An elongate depression bounded by faults.

Guard In belemnites: the solid calcite counterweight.

Habitat The space in which a community lives.

Hardground A cemented surface exposed on the sea floor, particularly common in some shallow water carbonate sequences.

Herbivore An animal which feeds mainly on plant material.

Hermatypic In corals: a form which depends on the presence of microscopic algae in its tissues.

Heteromorph In ammonites: a form in which the coiling is not a regular plane spiral.

Hinge line In bivalves and brachiopods: a line (passing through the teeth and sockets) about which the two valves articulate.

Iapetus Ocean An ocean which extended from Norway to Connecticut during Lower Palaeozoic time, and which did not finally close until the Devonian.

Inarticulate In brachiopods: a primitive class with no teeth.

Infauna Animals which live within the sediment on the sea floor.

Interarea In brachiopods: that part of the pedicle valve between the umbo and the hinge line.

Invertebrate An animal without a bony or cartilaginous internal skeleton.

Ironstone A sedimentary rock rich in iron minerals.

Island arc A chain of islands (predominantly volcanic) which occurs above a region where ocean floor material is descending below the Earth's crust.

Kingdom One of the three largest divisions of organisms, viz. protists, plants and animals.

Lag deposit A residual accumulation of coarser particles left after the finer material has been removed by currents.

Lamina The thinnest recognizable unit in a bedded sediment.

Larva The earliest growth stage of many invertebrates; the larvae of many bottom dwelling invertebrates are pelagic.

Ligament In bivalves: an elastic substance which opens the valves when the adductor muscles are relaxed.

Limestone A rock composed of calcium carbonate, usually the skeletons of marine organisms.

Limonite Hydrous iron oxide, usually the result of weathering of other iron compounds.

Lithification The chemical and physical processes which convert an unconsolidated sediment into a solid rock.

Lithology The physical character of a rock.

Lophophore An organ with tentacles which sets up currents for feeding and respiration.

Lumachelle An accumulation of shells in a sediment.

Magnetic stripes Areas of oceanic floor which have alternatively reversed magnetic polarity, and which are symmetrical about a mid-ocean spreading centre.

Mantle Fleshy tissue which lines the shell of some invertebrates, and from which additions to the shell are secreted. (Also: that part of the Earth which lies below the crust and outside the core)

Massif An area which subsides less than the surrounding areas, and where sediments (of a particular time interval) are either thin or absent.

Mesentery In corals: a radial fleshy structure situated between adjacent septa.

Micrite A limestone composed of clay-size particles (calcilutite).

Mould The cavity or impression left after removal of a fossil by solution.

Muscle scars Marks on the interior of a shell where muscles were attached.

Nekton Swimming animals.

Niche The habitat and ecological setting of an organism.

Nodule A concretion.

Notochord In chordates: a cylindrical sheath forming a flexible support for the back.

Oncolite A concentrically layered calcareous structure formed by blue-green algae.

Oolite A rock composed of ooliths. The small spheres resemble a fish roe (hence the name).

Oolith A small sphere built up of concentric

layers (usually calcium carbonate) due to successive episodes of inorganic precipitation around a nucleus (often a shell fragment).

Operculum A lid covering the aperture of some shelled invertebrates.

Order A group of families with certain characters in common, which suggest that they are related.

Organism A plant, an animal, or a protist.

Ornament Ribs, tubercles, or other irregularities on the surface of a shell. In spite of the name, ornament probably always has some function.

Orogeny The formation of mountains, usually accompanied by deformation of the rocks in the region concerned.

Pangea The Permian continent formed from the fusion of all the large continental masses on the Earth.

Pedicle A stalk, developed in various invertebrate epifauna for support.

Pedicle opening In brachiopods: the opening in the larger (pedicle) valve through which the pedicle emerges.

Pelagic Relating to organisms which are free-swimming or floating.

Period The time during which the rocks of a stratigraphical system were formed, e.g. the Jurassic Period was the time when rocks of the Jurassic System were formed.

Phanerozoic The eon comprising the Palaeozoic, the Mesozoic and the Cenozoic eras.

Photosynthesis The formation of carbohydrates from water and carbon dioxide in tissues exposed to light; the method normally employed in plants.

Phragmacone In cephalopods: the chambered shell; in belemnites, this is the conical shell which fits into the front end of the guard.

Phylum A major subdivision of a kingdom (plural: phyla).

Phytoplankton Plankton which employ photosynthesis.

Plankton Floating organisms; much of the plankton consists of minute protists and invertebrate larvae.

Plant A multi-cellular organism normally subsisting by photosynthesis.

Population The members of a species in a defined area.

Process An organ, part of an organ, or part of the skeleton which projects outwards from the principal surface.

Protein A group of complex carbohydrates; some proteins form soft fleshy matter, while others form hard substances.

Protista The kingdom which includes most single-celled organisms.

Radula In gastropods and some other molluscs: an organ bearing rasp-like teeth used in scraping, tearing or boring for food.

Realm A large area of the world with characteristic faunas.

Reef Either a bioherm (q.v.) or a ridge of hard rock.

Regression The retreat of the sea from a land area.

Rib A raised ridge on a shell.

Rostral spine A spine projecting forwards from the head.

Sabkha A flat coastal area subject to occasional inundation.

Salt plug A structure resulting from the upward movement of a salt mass.

Sand A sediment containing grains between 0.06 and 2 mm in diameter. Most sands contain quartz grains, but the grains can be of any mineral.

Scavenger An animal which feeds on the dead remains of other animals.

Sediment Solid material that has been transported and deposited, or material that has been precipitated from water. Most sediments have been transported by water, air or ice, but some result from mass flow.

Septum A plate or partition within the skeleton. In corals: a vertical radial plate. In cephalopods: a plate separating the internal chambers. (Plural: septa).

Series A major division of a stratigraphical system.

Shale An indurated fissile rock consisting mostly of clay grade minerals.

Siderite A rhombohedral form of iron carbonate.

Silt A sediment containing grains between 4 and 60 microns in diameter.

Sinistrally coiled Coiled in a left-handed manner. In gastropods: those forms which, when the spire is upwards and the aperture opening towards the observer, have the aperture on the left (most gastropods have the aperture on the right, i.e. they are coiled *dextrally*).

Siphon A tubular organ for conveying water. In bivalves it consists of two tubes (like a double-barrelled shotgun), one for the water going in, and one for the water going out.

Siphuncle In cephalopods: a tube extending to the apex of the shell, which passes through openings in the septa.

Slump A mass of sediment which has slid down a slope.

Species A group of organisms which can interbreed and produce fertile offspring. In fossils, members of the same species can only be deduced from morphological gradations between variants.

Spondylium In brachiopods: internal plates in the pedicle valve of pentamerids and some other groups.

Spreite Sedimentary laminations between the arms of a 'U'-shaped burrow.

Stage A biostratigraphical subdivision of a system, commonly based on a succession of biostratigraphical zones.

Stenohaline Relating to animals which can only tolerate a narrow range of salinity; most stenohaline animals are confined to the open sea.

Stolon A tube.

Substrate The sediment or rock surface on the sea floor.

Sulcate In brachiopods: with a folded anterior margin. In ammonites: with a ventral groove.

Suspension feeder An animal which feeds on organic matter suspended in sea water. Most suspension feeders are filter feeders.

Suture A line. In cephalopods: the line where a septum joins the outer shell. In trilobites: the facial suture is the line on the dorsal headshield (cephalon) along which the carapace split prior to the animal's moulting.

Swell An area which, over a period of time, subsides less than the surrounding areas, and is thus covered by thinner sedimentary sequences.

Symbiotic Relating to different organisms which live in close association with each other.

System The rocks formed during a geological period.

Tabula A more or less horizontal plate extending across the interior of a coral. (plural: tabulae).

Taxon A classification unit in biology, e.g. species, genus, family, order. (plural: taxa).

Tectonic Relating to geological structures.

Telson In arthropods: a posterior plate or spine.

Terrigenous Derived from the land.

Tethys An ocean which, during the Mesozoic, lay to the south of Europe and much of Asia, and to the north of Africa and India.

Theca A cup-like structure in many animals.

Thixotropic clay A clay which becomes weaker when disturbed and more coherent when left undisturbed.

Thorax In arthropods: the part of the body between the head and the tail, usually with well marked segments.

Tillite A lithified boulder clay, deposited by ice.

Tooth In bivalves and brachiopods: a projection on the hinge line which fits into a socket in the opposite valve, and thus serves to prevent slippage between the valves.

Trace fossil A sedimentary structure resulting from the activity of an animal (e.g. burrow, trail, bore).

Transgression The spread of the sea over a land area.

Trophic Relating to the food chain in animals.

Turbidite A sediment deposited by a turbidity current; the lower part of each bed usually grades up into a mud.

Turbidity current A turbid current with a high density due to suspended rock fragments.

Umbilicus That part of a planispiral shell within the outermost whorl.

Umbo In shells which grow by marginal accretion: the point where shell growth starts. (plural: umbones).

Valve A single part of an invertebrate skeleton; usually external.

Venter The ventral part of an animal. In cephalopods: the outer part of each whorl.

Ventral Strictly speaking relating to the underside of an animal, but in some invertebrates the ventral side is conventional (e.g. in brachiopods, the pedicle valve is always considered to be ventral).

Viviparous Giving birth to living young (as opposed to laying eggs).

Zone In geography and ecology: a region with some distinctive features. In stratigraphy: the rocks deposited during an interval of time which can be recognized by characteristic fossils.

References

Allen, J. A. (1958) On the basic form and adaptations to habitat in the Lucinacea (Eulamellibranchia). *Phil. Trans. R. Soc. B.*, **241**, 421–84.

Allen, P. (1976) Wealden of the Weald: a new model. *Proc. Geol. Assoc.* **86** (for 1975), 389–436.

Allen, P. Keith, M.L., Tan, F. C. and Dines, P. (1973) Isotopic ratios and Wealden environments. *Palaeontology*, **16**, 607–621.

Anderson, F. W. (1963) Ostracod faunas in the Weald Clay. In *Geology of the country around Maidstone*, Worssam, B. C. (Ed.), *Mem. Geol. Survey, U.K.*, 16–19.

Anderson, F. W. (1967) Ostracoda from the Weald Clay of England. *Bull. Geol. Survey, Gt. Britain*, **27**, 237–269.

Arkell, W. J. (1933) *The Jurassic System in Great Britain.* Clarendon Press, Oxford, 681p.

Arkell, W J. (1956) *Jurassic geology of the world.* Oliver and Boyd, Edinburgh and London, 806p.

Barnes, C. R. and Fåhraeus, L. E. (1975). Provinces, communities, and the proposed nektobenthic habit of Ordovician conodontophorids. *Lethaia*, **8**, 133–149.

Barthel, K. W. (1969) Die obertithonische, regressive Flachwasser Phase der Neuberger Folge in Bayern. *Abh. Bayer, Akad. Wissensch. Math-naturw. Kl., N.F.*, **142**, 1–174.

Bates, D. E. B. (1969) Some aspects of the Arenig faunas of Wales. In *The Precambrian and Lower Palaeozoic rocks of Wales,* Wood, A. (Ed.), Univ. of Wales Press, Cardiff, p. 155–159.

Bathurst, R. G. C. (1959) The cavernous structure of some Mississippian stromatactis reefs in Lancashire, England. *J. Geol.*, **67**, 506–521.

Batten, D. J. (1975) Wealden palaeoecology from the distribution of plant fossils. *Proc. Geol. Assoc.*, **85**, 433–458.

Bergström, J. (1973) Organization, life and systematics of trilobites. *Fossils and Strata*, 2, 1–69.

Bergström, S. M. and Cooper, R. A. (1973). *Didymograptus bifidus* and the trans-Atlantic correlation of the Lower Ordovician. *Lethaia*, **6**, 313–339.

Bernoulli, D. and Jenkyns, H. C. (1974) Alpine, Mediterranean, and Central Atlantic Mesozoic facies in relation to the early evolution of the Tethys: In *Modern and Ancient geosynclinal sedimentation*, Dott, R. H. and Shaver, R. H., (Eds.). *Soc. Econ. Paleont. Miner. Spec. Publ.,* 19, 129–160.

Berry, W. B. N. and Boucot, A. J. (1967) Continental stability – a Silurian point of view. *Journ. geophys. Research,* **72**, 2254–2256.

Berry, W. B. N. and Boucot, A. J. (1970) Correlation of the North American Silurian rocks. *Geol. Soc. Amer. Special Paper,* **102**, 289p.

Berry, W. B. N. and Boucot, A. J (1972) Silurian graptolite depth zonation. *24th International Geol. Congress, Montreal*, 7, 59–65.

Black, W. W. (1954) Diagnostic Characters of the Lower Carboniferous Knoll-Reefs in the North of England. *Trans. Leeds Geological Association*, 6, 262–97.

Boucot, A. J. (1953) Life and death assemblages among fossils. *Am. Journ. Sci.*, **251**, 25–40.

Bowen, Z. P., Rhodes, D. C. and McAlester, A. L. (1974) Marine benthic communities in the Upper Devonian of New York. *Lethaia*, 7, 93–120.

Bretsky, P. W. (1970) Late Ordovician benthic marine communities in north-central New York. *New York State Mus. Sci. Serv. Bull.*, 44, 1–34.

Briden, J. C., Drewry, G. E. and Smith, A. G., (1974) Phanerozoic equal-area world maps. *J. Geol.*, **82**, 555–574.

Bromley, R. G. and Surlyk, F. (1973) Borings produced by brachiopod pedicles, fossil and Recent. *Lethaia*, 6, 349—365.

Brunton, C. H. C. (1966) Silicified Productoids from the Viséan of County Fermanagh. *Bull. Br. Mus. nat. Hist. (Geol.)*, 12, 173—243.

Burton, E. St. J. (1933) Faunal horizons of the Barton Beds in Hampshire. *Proc. Geol. Assoc.*, 44, 131—167.

Calver, M. A. (1968) The distribution of Westphalian marine faunas in northern England and adjoining areas. *Proc. Yorks. geol. Soc.*, 37, 1—72.

Calvert, S. E. (1964) Factors affecting distribution of laminated diatomaceous sediments in Gulf of California. In Marine Geology of the Gulf of California — a symposium, T.H. Van Andel and G.C. Shor Jr. (Eds.), *Am. Assoc. Petrol. Geol. Mem.*, 3, 311—30.

Casey, R. (1961) The stratigraphical palaeontology of the Lower Greensand. *Palaeontology*, 3, 487—621.

Casey, R. (1973) The ammonite successions of the Jurassic-Cretaceous boundary in eastern England. *Geol. J., Special Issue*, 5, 193—266.

Casey, R. (1971) Facies, faunas and tectonics in late Jurassic-early Cretaceous Britain. In *Faunal provinces in space and time*, Middlemiss, F. A., Rawson, P. F. and Newall, G. (Eds.). *Geol. J. Special Issue*, 4, 153—168.

Casey, R. and Rawson, P. F. (Eds.)(1973) *The Boreal Lower Cretaceous. Geol. J. Special Issue* 5.

Chaloner, W. G. and Lacy, W. S. (1973). The distribution of Late Palaeozoic floras. In *Continents and organisms through time*, Hughes, N. F. (Ed.), *Special Paper in Palaeontology*, 12, 271—290.

Chave, K. E. (1964) Skeletal durability and preservation. In *Approaches to paleoecology*. Imbrie, J. and Newell, N. D. (Eds.). Wiley, New York.

Clarkson, E. N. K. (1966) Schizochroal eyes and vision of some Silurian acastid trilobites. *Palaeontology*, 9, 1—29.

Clarkson, E. N. K. (1966a) Schizochroal eyes and vision in some phacopid trilobites. *Palaeontology*, 9, 464—487.

Cloud, P. E. (1968) Pre-Metazoan evolution and the origins of the Metazoa. In *Evolution and Environment*, E. T. Drake (Ed.), Yale Univ. Press, New Haven, 1—72.

Cloud, P. E. (1976) Beginnings of biospheric evolution and their biogeochemical consequences. *Paleobiology*, 2, 351—387.

Cocks, L. R. M. (1970) Silurian brachiopods of the superfamily Plectambonitacea. *Bull. Br. Mus. nat. Hist. (Geol.)*, 19, 141—203.

Cocks, L. R. M., Holland, C. H., Rickards, R. B. and Strachan, I.,(1971) A correlation of Silurian rocks in the British Isles. *Q. Jl. geol. Soc. London*, 127, 103—136.

Cowie, J. W. (1974) The Cambrian of Spitzbergen and Scotland. In *Cambrian of the British Isles, Norden and Spitzbergen*, Holland, C.H. (Ed.). Wiley, London, New York, Sydney and Toronto, 123—155.

Craig, G. Y. and Hallam, A. (1963) Size frequency and growth-ring analysis of *Mytilus edulis* and *Cardium edule*, and their palaeoecological significance. *Palaeontology*, 6, 731—50.

Craig, G. Y. and Oertel, G. (1966) Deterministic models of living and fossil populations of animals. *Q. Jl. geol. Soc. London*, 122, 315—55.

Crimes, T. P. (1970) A facies analysis of the Cambrian of Wales. *Palaeogr. Palaeoclimatol. Palaeoecol.*, 7, 113—170.

Curry, D. (1965) The Palaeogene Beds of southern England. *Proc. Geol. Assoc.*, 76, 151—173.

Curry, D. (1967) Problems of correlation in the Anglo-Paris-Belgian Basin. *Proc. Geol. Assoc.*, 77, 437—467.

Daley, B. (1972) Macroinvertebrate assemblages from the Bembridge Marls (Oligocene) of the Isle of Wight, England and their environmental significance. *Palaeogeog. Palaeoclimatol. Palaeoecol.*, 11, 11—32.

Davis, A. G. and Elliott, G. F. (1958) The palaeogeography of the London Clay sea. *Proc. Geol. Assoc.*, 68, 255—277.

Dean, W. T., Donovan, D. T. and Howarth, M. K., (1961). The Liassic ammonite zones and subzones of the north-west European Province. *Bull. Br. Mus. nat. Hist. (Geol.)*, 4, 438—505.

Dixon, E. E. L. and Vaughan, A. (1911) The Carboniferous succession in Gower (Glamorganshire). *Q. Jl. geol. Soc. Lond.*, 67, 477—571.

Fischer, A. (1964) The Lofer cyclothems of the Alpine Triassic, In *Symposium on cyclic sedimentation*, Merriam, D. W. (Ed.). *Kansas Geol. Surv. Bull.*, 169, 107—149.

References

Fisher, O. (1862) On the Bracklesham Beds of the Isle of Wight basin. *Q. Jl. geol. Soc. Lond.*, 18, 65—94.

Ford, E. (1923) Animal communities of the level sea-bottom in the waters adjacent to Plymouth. *Jl. mar. Biol. Assoc. U.K.*, 13, 164—224.

Ford, T. D. (1965) The palaeoecology of the Goniatite Bed at Cowlow Nick, Castleton, Derbyshire. *Palaeontology*, 8, 186—91.

Fortey, R. A. (1975) Early Ordovician trilobite communities. *Fossils and Strata*, 4, 330—352.

Garrett, P. (1970) Phanerozoic stromatolites: noncompetitive ecologic restriction by grazing and burrowing animals. *Science*, 169, 171—173.

George, T. N. et al. (1976) A correlation of Dinantian rocks in the British Isles. *Special Report No. 7., Geol. Soc. Lond.*

Ginsburg, R. N. (Ed.), (1975) *Tidal Deposits. A casebook of Recent examples and fossil counterparts.* Springer-Verlag, New York, Heidelberg, Berlin, 428p.

Glaessner, M. F. (1971) Geographic distribution and time range of the Ediacara Precambrian fauna. *Bull. Geol. Soc. Amer.*, 82, 509—514.

Glaessner, M. F. and Wade, M. (1966) The late Precambrian fossils from Ediacara, South Australia. *Palaeontology*, 9, 599—628.

Goldring, R. (1964) Trace-fossils and the sedimentary surface in shallow marine environments. In *Developments in Sedimentology 1: Deltaic and Shallow Marine Deposits*, Van Straaten, L. M. J. U. (Ed.), Elsevier, Amsterdam, 136—143.

Hakansson, E., Bromley, R. G. and Perch-Nielsen, K. (1974) Maastrichtian Chalk of northwest Europe: a pelagic shelf sediment. *Spec. Pub. Int. Assoc. Sediment*, 1, 211—233.

Hallam, A. (1963) Observations on the palaeoecology and ammonite sequence of the Frodingham Ironstone (Lower Jurassic). *Palaeontology*, 6, 554—574.

Hallam, A. (1965) Observations on Marine Lower Jurassic stratigraphy of North America, with special reference to United States. *Bull. Am. Assoc. Petrol. Geol.*, 49, 1485—1501.

Hallam, A. (1966) Depositional environment of British Liassic ironstones in the context of their facies relationships. *Nature*, 209, 1306—1309.

Hallam, A. (1969) A pyritized limestone hardground in the Lower Jurassic of Dorset (England). *Sedimentology*, 12, 231—240.

Hallam. A. (1972) Diversity and density characteristics of Pliensbachian-Toarcian molluscan and brachiopod faunas of the North Atlantic. *Lethaia*, 5, 389—412.

Hallam, A. (Ed.) (1973) *Atlas of palaeobiogeography.* Elsevier, Amsterdam, London and New York. 531p.

Hallam, A. (1975) *Jurassic Environments.* Cambridge University Press, 269p.

Hallam, A. and Sellwood, B. W. (1976) Middle Mesozoic sedimentation in relation to tectonics in the British area. *J. Geol.*, 84, 301—321.

Halstead, L. B. (1975) *The evolution and ecology of the dinosaurs.* Peter Lowe, London, 115p.

Hamilton, D. (1961) Algal growths in the Rhaetic Cotham Marble of Southern England. *Palaeontology*, 4, 324—333.

Hancock, J. M. (1967) Some Cretaceous-Tertiary marine faunal changes. In *The Fossil record*, Harland, W. B. et al. (Eds.), Geol. Soc. London, 91—104.

Hancock, N. J., Hurst, J. M. and Fürsich, F. T., (1974). The depths inhabited by Silurian brachiopod communities. *Jl. geol. Soc. London*, 130, 151—156.

Harland, W. B., Holland, C. H., House, M. R., Hughes, N. F., Reynolds, A. B., Rudwick, M. J. S., Satterthwaite, G. E., Tarlo, L. B. and Willey, E. C. (Eds.) (1967) *The Fossil record.* Geol. Soc. London, 828p.

Hartman, W. D. and Goreau, T. F. (1970). Jamaican coralline sponges: their morphology, ecology and fossil representatives. *Zool. Soc. London Symposium*, 25, 205—243.

Hauff, B. (1953) *Das Holzmadenbuch.* Rau, Öhringen, 181p.

Hedgpeth, J. W. (Ed.) (1957) *Treatise on marine ecology and paleoecology, Vol. 1, Ecology, Mem. Geol. Soc. Amer.*, 67, 1296p.

Hodson, F. and Ramsbottom, W. H. C. (1973) The distribution of Carboniferous goniatite faunas in relation to suggested continental reconstructions for the period. In *Continents and organisms through time*, Hughes, N. F. (Ed.), *Special Paper in Palaeontology*, 12, 321—329.

Hoffman, P. (1974) Shallow and deep water stromatolites in Lower Proterozoic platform-to-basin facies change, Great Slave Lake, Canada. *Bull. Am. Assoc. Petrol. Geol.*, 58, 856—867.

Holdsworth, B. K. (1966) A preliminary study of the palaeontology and palaeoenvironment of some Namurian limestone 'bullions'. *Mercian Geol.,* 1, 315—37.

House, M. R. (1967) *Continental drift and the Devonian System.* University of Hull, Hull, 24p.

House, M. R. (1975) Facies and time in Devonian tropical areas. *Proc. Yorks. geol. Soc.,* 40, 233—288.

Hughes, N. F. (Ed.) (1973) *Organisms and continents through time. Special paper in Palaeontology,* 12, 329p.

Hutchinson, G. E. (1961) The biologist poses some problems. In *Oceanography,* M. Sears (Ed.). Amer. Assoc. Adv. Sci., Washington, D.C., 85—94.

Imlay, R. W. (1957) Paleoecology of Jurassic seas in the western interior of the United States. *Mem. Geol. Soc. Amer.,* 67, 469—504.

Imlay, R. W. (1965) Jurassic marine faunal differentiation in North America. *J. Paleont.,* 39, 1023—1038.

Jackson, J. B. C., Goreau, T. F. and Hartman, W. D. (1971) Recent brachiopod-coralline sponge communities and their paleocologic significance. *Science,* 173, 623—625.

James, J., Ward, D.J., and Cooper, J. (1974). A temporary exposure of fossiliferous London Clay (Eocene) at Shinfield, Berkshire. *Proc. Geol. Assoc.,* 85, 49—64.

Jefferies, R. P. S. and Minton, P. (1965) The mode of life of two Jurassic species of 'Posidonia'. *Palaeontology,* 8, 156—185.

Jenkyns, H. C. (1971) Speculations on the genesis of crinoidal limestones in the Tethyan Jurassic. *Geol. Rundschau,* 60, 471—488.

Jones, N. S. (1956) The fauna and biomass of a muddy sand deposit off Port Erin, Isle of Man. *Jl. animal Ecol.,* 25, 217—252.

Jukes-Browne, A. J. and Hill, W. (1900) *The Cretaceous rocks of Britain,* vol. 1, *Gault and Upper Greensand. Mem. Geol. Survey U.K.*

Kauffman, E. G. (1969) Form, function and evolution. In *Treatise on Invertebrate Paleontology,* Part N. R.C. Moore (Ed.), Mollusca 6, Bivalvia, 129—205.

Kennedy, W. J. and Garrison, R. E. (1975) Morphology and genesis of hardgrounds and nodular chalks in the Upper Cretaceous of southern England. *Sedimentology,* 22, 311—386.

Kevan, P. G., Chaloner, W. G. and Savile, D. B. O. (1975) Interrelationships of early terrestrial arthropods and plants. *Palaeontology,* 18, 391—417.

Kilenyi, T. I. and Allen, N. W. (1968) Marine brackish bands and their microfauna from the lower part of the Weald Clay of Surrey and Sussex. *Palaeontology,* 11, 141—162.

King, L. C. (1958) Basic palaeography of Gondwanaland during the Late Palaeozoic and Mesozoic eras. *Q. Jl. geol. Soc. London,* 104, 47—77.

Kirby, R. I. (1974) Report of Project Meeting and Field Meeting to Aveley, Essex. *Tertiary Times,* 2, No. 2, 53—67.

Kummel, B. and Teichert, C. (Eds.) (1970) *Stratigraphic boundary problems: Permian and Triassic of West Pakistan.* University of Kansas Press, Lawrence, 474p.

Lambert, R. St. J. (1971) The pre-Pleistocene time-scale — a review. In *The Phanerozoic time-scale — a supplement.* Harland, W. B. and Francis, E. H. (Eds.). *Geol. Soc. London Special Pubn.,* 5, Pt. 1, 9—31.

Lapworth, C. (1878) The Moffat Series. *Q. Jl. geol. Soc. London,* 34, 240—346.

Lapworth, C. (1879) On the tripartite classification of the Lower Palaeozoic rocks. *Geol. Mag.* (2), 6, 1—15.

Leeder, M. (1975) Lower Border Group (Tournaisian) limestones from the Northumberland Basin. *Scot. J. Geol.* 11, 151—67.

Lewis, J. R. (1964) *The ecology of rocky shores.* English Univ. Press, London.

McKerrow, W. S. and Cocks, L. R. M. (1976) Progressive faunal migration across the Iapetus Ocean. *Nature,* 263, 304—306.

McKerrow, W. S., Johnson, R. T. and Jakobson, M. E. (1969) Palaeoecological studies in the Great Oolite at Kirtlington, Oxfordshire. *Palaeontology,* 12, 56—83.

McKerrow, W. S. and Ziegler, A. M. (1971) The Lower Silurian paleogeography of New Brunswick and adjacent areas. *J. Geol.,* 79, 635—646.

McKerrow W. S. and Ziegler, A. M. (1972) Palaeozoic Oceans. *Nature Phys. Sci.,* 240, 92—94.

Mistakidis, M. M. (1951) Quantitative studies on the bottom fauna of Essex oyster grounds. *Gt. Brit. Fishery Invest.* ser. II, 17, 1—47.

Moorbath, S. (1975) The geological significance of Early Precambrian rocks. *Proc. Geol. Assoc.,* 86, 259—279.

References

Muir, M. D. and Grant, P. R. (1976) Micropalaeontological evidence from the Onverwacht Group, South Africa. In *The Early History of the Earth*, B. F. Windley (Ed.). Wiley, London, New York, Sydney, Toronto, 595–604.

Nestler, H. (1965) Die Rekonstruktion der Lebensraumes der Rügener Schreibekreide-Fauna (Unter-Maastricht) mit Hilfe der Paläoökologie und Paläobiologie. *Geologie*, 14,147p.

Neuman, R. B. (1972) Brachiopods of Early Ordovician volcanic islands. *24th International Geol. Congress, Montreal*, 7, 297–302.

Palmer, T. J. and Fürsich, F. T. (1974) The ecology of a Middle Jurassic hardground and crevice fauna. *Palaeontology*, 17, 507–524.

Pattison, J., Smith, D. B. and Warrington, G., (1973) A review of late Permian biostratigraphy in the British Isles. In Logan, A. V. and Mills, L. V. (Eds.), The Permian and Triassic systems and their mutual boundary. *Mem. Can. Soc. Petrol. Geol.*, 2, 220–260.

Purser, B. H. (1969) Syn-sedimentary marine lithification of Middle Jurassic limestones in the Paris Basin. *Sedimentology*, 12, 205–230.

Ramsbottom, W. H. C. (1973) Transgressions and regressions in the Dinantian: a new synthesis of British Dinantian stratigraphy. *Proc. Yorks. geol. Soc.*, 39, 567–607.

Ramsbottom, W. H. C. (1974) Dinantian. In *Geology and Mineral Resources of Yorkshire*. Rayner, D. H. and Hemingway, J. E. (Eds.). Yorkshire Geological Soc.

Ramsbottom, W. H. C., Rhys, G. H. and Smith, E. G. (1962) Boreholes in the Carboniferous rocks of the Ashover district, Derbyshire. *Bull. geol. Surv. Gt. Brit.*, 19, 75–168

Raup, D.M. and Stanley, S.M. (1971) *Principles of paleontology*. 388p. Freeman, San Francisco.

Romer, A. S. (1966) *Vertebrate paleontology*. Univ. Chicago Press, Chicago and London, 468p.

Rose, G. N. and Kent, P. E. (1955) A *Lingula*-bed in the Keuper of Nottinghamshire. *Geol. Mag.*, 92, 476–480.

Ross, C. A. (1967) Development of Fusulinid (Foraminifers) faunal realms. *J. Paleont.*, 41, 1341–54.

Rudwick, M. J. S. (1964) The function of zigzag deflexions in the commissures of fossil brachiopods. *Palaeontology*, 7, 135–171.

Rudwick, M. J. S. (1970) *Living and fossil brachiopods*. Hutchinson, London, 199p.

Rushton, A. W. A. (1974) The Cambrian of Wales and England. In *Lower Palaeozoic rocks of the world*, Holland, C. H. (Ed.), 2, 43–121, pl. 1–3. Wiley, London, New York, Sydney, Toronto.

Ryther, J. H. (1963) Geographic variations in productivity. In *The Sea*, Hill, M. N., (Ed.), 2, 374–80. Interscience, New York.

Schäfer, W. (1972) *Ecology and palaeoecology of marine environments*. Oliver and Boyd, Edinburgh, 586p.

Schopf, T. J. M. (1974) Permo-Triassic extinctions: relation to sea-floor spreading. *J. Geol.*, 82, 129–143.

Schwarzacher, W. (1961) Petrology and Structure of some Lower Carboniferous reefs in northwestern Ireland. *Bull. Am. Assoc. Petrol. Geol.*, 45, 1481–1503.

Scoffin, T. P. (1971) The conditions of growth of the Wenlock reefs of Shropshire (England). *Sedimentology*, 17, 173–219.

Seilacher, A. (1967) Bathymetry of trace fossils. *Marine Geol.*, 5, 413–428.

Seilacher, A., Drozdzewski, G. and Haude, R., (1968) Form and function of the stem in a pseudoplanktonic crinoid *(Seirocrinus)*. *Palaeontology*, 11, 275–282.

Sellwood, B. W. (1970) The relation of trace-fossils to small-scale sedimentary cycles in the British Lias. In *Trace Fossils*, Crimes, T. P. and Harper, J. C. (Eds.) *Spec. Edition Geol. J.*, 489–504.

Sellwood, B. W. (1972) Regional environmental changes across a Lower Jurassic stage-boundary in Britain. *Palaeontology*, 15, 125–57.

Sellwood, B. W. and Hallam, A. (1975) Bathonian volcanicity and North Sea rifting. *Nature*, 252, 27–28.

Sellwood, B. W. and Jenkyns, H. C. (1975) Basins and swells and the evolution of an epeiric sea (Pleinsbachian to Bajocian of Great Britain). *Jl. geol. Soc. Lond.*, 131, 373–388.

Sheehan, P. M. (1973) The relation of Late Ordovician glaciation to the Ordovician-Silurian changeover in North American brachiopod faunas. *Lethaia*, 6, 147–154.

Smart, J. G. O., Bisson, G., and Worssam, B. C., (1966) *Geology of the country around Canterbury and Folkestone*. *Mem. Geol. Surv. G.B.*, 337p.

Smith, A. G. and Briden, J. C. (1977) *Mesozoic and Cenozoic palaeocontinental maps*. Cambridge Univ. Press. 64p.

Smith, D. B. (1970) The palaeogeography of the English Zechstein. In *Third Symposium on Salt* v. 1: Rau, J. L. and Dellwig, L. F. (Eds.). *Northern Ohio Geol. Soc.*, 20–23.

Smith, D. B. (1972) Permian and Trias. In *Geology of Durham County*, Hickling, G. (Ed.). *Northumberland Natur. Hist. Soc., Trans.*, 66–91.

Stanley, K. O., Jordan, W. M. and Dott, R. H., Jr. (1971) New hypothesis of early Jurassic paleogeography and sediment dispersal for Western United States. *Bull. Am. Assoc. Petrol. Geol.*, 55, 10–19.

Stanley, S. M. (1970) Relation of shell-form to life habits of the Bivalvia (Mollusca). *Mem. Geol. Soc. Am.*, 125, 296p.

Stanley, S. M. (1976) Fossil data and the Precambrian-Cambrian transition. *Am. Journ. Sci.*, 276, 56–76.

Steele-Petrović H. M. (1975) An explanation for the tolerance of brachiopods and relative intolerance of filter-feeding bivalves for soft muddy bottoms. *J. Paleont.*, 49, 552–556.

Stehli, F. G. (1971) Tethyan and boreal Permian faunas and their significance. In *Paleozoic perspectives; a paleontological tribute to G. Arthur Cooper*, Dutro, J. T. (Ed.). *Smithsonian Contr. Paleobiol.*, 3, 337–45.

Steinich, G. (1967) Sedimentstrukturen der Rugener Schreibekreide. *Geologie*, 16, 570–583.

Suess, E. (1885–1901) Das Antlitz der Erde. [*The face of the Earth*. Translated by H. B. C. Sollas. Clarendon Press, Oxford, 1904].

Surlyk, F. (1972) Morphological adaptations and population structures of the Danish Chalk Brachiopods (Maastrichtian) Upper Cretaceous. *Biol. Skr.*, 19, 57p.

Surlyk, F. and Birkelund, T. (1977) An integrated stratigraphical study of fossil assemblages from the Maastrichtian White Chalk of N.W. Europe. In *Concepts and Methods of Biostratigraphy*, Kauffman, E.G. and Hazel, J.E. (Eds.), Dowden, Hutchinson & Ross, Stroudsburg, Pa.,

Sutton, J. (1968) Developments of the continental framework of the Atlantic. *Proc. Geol. Assoc.*, 79, 275–303.

Tarlo, L. B. H. (1967) Vertebrata. In *The Fossil record*, Harland, W. B. et al. (Eds.). Geol. Soc. London, 628.

Taylor, K. and Rushton, A. W. A. (1971) The pre-Westphalian geology of the Warwickshire

Coalfield. *Bull. geol. Surv. G.B.*, 35, 1–150, pls. 1–12.

Thorson, G.(1957) Bottom Communities. *Geol. Soc. Amer. Mem.*, 67, 461–534.

Townson, W. G. (1975) Lithostratigraphy and deposition of the type Portlandian. *Jl. geol. Soc. Lond.*, 131, 619–638.

Trechmann, C. T. (1913) On a mass of anhydrite in the Magnesian Limestone at Hartlepool, and on the Permian of south-eastern Durham. *Q. Jl. geol. Soc. Lond.*, 69, 184–218

Tucker, M. E. (1974) Sedimentology of Palaeozoic pelagic limestones: The Devonian Griotte (Southern France) and Cephalopodenkalk (Germany). In *Pelagic sediments: on land and under the sea*, Hsü, K.·J. and Jenkyns, H. C. (Eds.), *Spec. Pubn.*, 1, *Internat. Assoc. Sedimentol*, 72–92.

Valentine, J. W. (1973) *Evolutionary paleoecology of the marine biosphere*. Prentice-Hall, Englewood Cliffs, New Jersey, 511p.

Walker, A. D. (1969) The reptile fauna of the 'Lower Keuper' Sandstone. *Geol. Mag.*, 106, 470–476.

Walker, K. R. (1972) Community ecology of the Middle Ordovician Black River Group of New York State. *Bull. Geol. Soc. Amer.*, 83, 2499–2524.

Walter, M. R. (1972) Stromatolites and the biostratigraphy of the Australian Precambrian and Cambrian. *Special Paper in Palaeontology*, 11, 190p, 33 plates.

Warrington, G. (1971) The stratigraphy and palaeontology of the 'Keuper' Series of the central Midlands of England. *Q. Jl. geol. Soc. Lond.*, 126, 183–224.

West, I. M. (1960) On the occurrence of celestine in the Caps and Broken Beds at Durlston Head, Dorset. *Proc. Geol. Assoc.*, 71, 391–401.

West, I. M. (1964) Evaporite diagenesis in the Lower Purbeck Beds of Dorset. *Proc. Yorks. geol. Soc.*, 34, 315–30.

West, I. M. (1975) Evaporites and associated sediments of the basal Purbeck Formation (Upper Jurassic) of Dorset. *Proc. Geol. Assoc.*, 86, 205–226.

Whittaker, R. H. (1969) New concepts of kingdoms of organisms. *Science*, 63, 150–160.

Whittington, H. B. and Hughes, C. P. (1973) Ordovician trilobite distribution and geography. In *Organisms and Continents through time*,

References

Hughes, N. F. (Ed.), *Special Paper* in *Palaeontology*, **12**, 235—240.

Wilson, R. B. (1974) A study of the Dinantian marine faunas of southeast Scotland. *Bull. geol. Surv. Gt. Br.*, **46**, 35—65.

Wilson, R. C. L. (1968) Upper Oxfordian palaeogeography of Southern England. *Palaeogeog. Palaeoclimatol. Palaeoecol.* **4**, 5—28.

Wilson, R. C. L. (1975) Atlantic opening and Mesozoic continental margin basins of Iberia. *Earth and Plan. Sci. Letters*, **25**, 33—43.

Wolfenden, E. B. (1958) Paleoecology of the Carboniferous reef complex and shelf limestones in northwest Derbyshire, England. *Bull. Geol. Soc. America*, **69**, 871—98.

Yancey, T. E. (1975) Permian marine biotic provinces in North America. *J. Paleont.*, **49**, 758—66.

Ziegler, A. M. (1965) Silurian marine communities and their environmental significance. *Nature*, **207**, 270—272.

Ziegler, A. M. (1966) The Silurian brachiopod *Eocoelia hemisphaerica* (J. de C. Sowerby) and related species. *Palaeontology*, **9**, 523—543.

Ziegler, A. M. (1970) Geosynclinal development of the British Isles during the Silurian Period. *J. Geol.*, **78**, 445—479.

Ziegler, A. M. Cocks, L. R. M. and Bambach, R. K. (1968) The composition and structure of Lower Silurian marine communities. *Lethaia*, **1**, 1—27.

Ziegler, A. M., Cocks, L. R. M. and McKerrow, W. S. (1968a) The Llandovery transgression of the Welsh Borderland. *Palaeontology*, **11**, 736—782.

Ziegler, A. M., Rickards, R. B. and McKerrow, W. S. (1974) Correlation of the Silurian rocks of the British Isles. *Geol. Soc. Amer. Special Paper*, **154**, 154p.

Ziegler, P. A. (1975) North Sea Basin history in the tectonic framework of north-western Europe. In *Petroleum and Continental Shelf of north-west Europe*, Woodland, A. W. (Ed.). Applied Science Publishers, Barking, Essex, 131—148.

Ziegler, P. A. (1975a) The geological evolution of the North Sea area and the tectonic framework of North Western Europe. *Norges Geol. Unters.*, **316**, 1—27.

Index

Index

379

Index

Index

Acknowledgments

All the illustrations are by Mrs E. Winson; the authors are very grateful for the artistic skill, biological knowledge and patience she has shown in producing the drawing from our crude instructions, Mr P. Deussen drew the maps and other text-figures. We also owe thanks to Leslie and Tony Birks-Hay of Alphabet and Image who first suggested an illustrated book on fossils, and who guided us through many problems as the project developed. Specialist advice was generously given by the following: Dr N. Fannin (Devonian sediments), Dr R. A. Fortey (trilobites), Dr L.B. Halstead (vertebrates), Dr J. M. Hurst (Ordovician and Silurian brachiopods), Dr M. R. Leeder (Carboniferous), Dr N. J. Morris (molluscs), Mr D. Mundy (Carboniferous reefs), Dr T. J. Palmer (Mesozoic communities), Mr J. Pattison (Permian communities), Dr R. B. Rickards (graptolites), Dr A. W. A. Rushton (Cambrian and Ordovician faunas), Mr A. E. Timms (Carboniferous reef-slope), and Dr R. Watkins (Silurian communities).